9th International Conference on Engineering and Technology (ICET-2021)

IOP Conference Series: Materials Science and Engineering
Volume 1163

Hat Yai, Thailand
27 May 2021

ISBN: 978-1-7138-4626-0
ISSN: 1757-8981

Printed from e-media with permission by:

Curran Associates, Inc.
57 Morehouse Lane
Red Hook, NY 12571

Some format issues inherent in the e-media version may also appear in this print version.

Additional copies of this publication are available from:

Curran Associates, Inc.
57 Morehouse Lane
Red Hook, NY 12571 USA
Phone: 845-758-0400
Fax: 845-758-2633
Email: curran@proceedings.com
Web: www.proceedings.com

TABLE OF CONTENTS

IOP Conf. Series: Materials Science and Engineering 1163 (2021) 011001 doi:10.1088/1757-899X/1163/1/011001

Preface

About the Conference

The Faculty of Engineering, Prince of Songkla University (PSU) in collaboration with PSU Energy Systems Research Institute (PERIN) are very pleased to invite your researchers and students to join the 9th International Conference on Engineering and Technology (ICET-2021). The date for the ICET-2021 is on May 27, 2021 through an online platform.

The main objective of this conference is to provide a stage for industrial players, academicians, researchers, engineers and scientists from various countries to share their ideas and latest research in various fields of emerging engineering and technology. This conference focuses on key areas in energy conversion, renewable energy, robotic & automation, mechanics & mechatronics, materials & civil engineering, and environmental sustainability.

Co-Organizers

This conference is co-organized byour partners as follows:

- University of Novi Sad (UNS), Serbia
- Shanghai Jiao Tong University (SJTU), China
- UniversitiSains Malaysia (USM), Malaysia
- UniversitiTunHussien Onn Malaysia (UTHM), Malaysia
- Universiti Kuala Lumpur (UniKL), Malaysia
- SSBT's College of Engineering & Technology (SSCOET), India
- Kanazawa University, Japan
- Musashino University, Japan

IOP Conf. Series: Materials Science and Engineering 1163 (2021) 011001 doi:10.1088/1757-899X/1163/1/011001

Conference Topics

T1 Energy Conversion and Renewable Energy

T2 Mechanics, Robotic and Automation

T3 Electrical and Mechatronics

T4 Materials and Civil Engineering

T5 Logistic, Industrial and Manufacturing

T6 Computerization and Information Technology

T7 Chemical and Environmental Sustainability

T8 Biomedical Science and Engineering

Organizing Committee:

Chairperson

Assoc. Prof. Dr. TanitChalermyanont, PSU, Thailand

Managerial Chairperson

Prof. Dr. Sumate Chaiprapat, PSU, Thailand

Co-Chairperson

1. Asst. Prof. Dr. JuntakanTaweekun, PSU, Thailand
2. Assoc. Prof. Dr. WiriyaThongruang, PSU, Thailand
3. Assoc. Prof. Dr. CharongpunMusikavong, PSU, Thailand
4. Asst. Prof. Dr. KlangduenPochana, PSU, Thailand
5. Asst. Prof. Dr. PrapasMuangjunburee, PSU, Thailand

Co-organizing committee members

1. Prof. Dr. TeerapolSrichana, PSU, Thailand
2. Prof. Dr. DejanJaksic, UNS, Serbia
3. Prof. Dr. Yanjun Dai, SJTU, China
4. Assoc Prof Dr. Jamaluddin Abdullah, USM, Malaysia
5. Assoc. Prof. Dr. Shahruddin bin Mahza, UTHM, Malaysia
6. Assoc. Prof. Dr. ShahrilNizam Mohamed Soid, UniKL, Malaysia
7. Dr. Kishor S. Wani, SSCOET, India

IOP Conf. Series: Materials Science and Engineering 1163 (2021) 011001 doi:10.1088/1757-899X/1163/1/011001

International Advisory Committee:

1. Prof. Dr. SurasakSangkhathat, PSU, Thailand
2. Assoc. Prof. Dr. ThitiwornChusong, PSU, Thailand
3. Asst. Prof. Dr. SurapongChatpun, PSU, Thailand
4. Dr. RaphatphornNavakanitworakul, PSU, Thailand
5. Dr. ThanmasinIngviya, PSU, Thailand
6. Dr.TonghathaiPhairatana, PSU, Thailand
7. Dr. PasaratKhongkow, PSU, Thailand
8. Dr. WorradornPhairuang, PSU, Thailand
9. Asst. Prof. PiratKhunkitti, KKU, Thailand
10. Prof. Dr. RadeDoroslovacki, UNS, Serbia
11. Prof. DrAleksandarKupusinac, UNS, Serbia
12. Prof. Dr Boris Dumnic, UNS, Serbia
13. Prof. Dr. SrdjanKolakovic, UNS, Serbia
14. Prof. Dr. DarkoStefanovic, UNS, Serbia
15. Prof. Dr. DejanUbavin, UNS, Serbia
16. Prof. Dr. BranislavBorovac, UNS, Serbia
17. Assoc. Prof. Dr. Goran Sladic, UNS, Serbia
18. Assoc. Prof. Dr. Mas FawziMohd Ali, UTHM, Malaysia
19. Assoc. Prof. Dr. NorzelawatiAsmuin, UTHM, Malaysia
20. Dr. MohdFaizalMohideen Batcha, UTHM, Malaysia
21. Dr. AkmalNizam Mohammed, UTHM, Malaysia
22. Dr. Mohammad Kamil Abdullah, UTHM, Malaysia
23. Dr. DjamalHisseinDidane, UTHM, Malaysia
24. Dr. ZamriNoranai, UTHM, Malaysia
25. Prof. Dr. S. A. Khan, IIUM, Malaysia
26. Prof. Ir. Dr. Faizal Mustapha, UPM, Malaysia
27. Assoc. Prof. Ir. Dr. KamarulArifin Ahmad, UPM, Malaysia
28. Prof. Ir. Dr. MohdZulkifly Abdullah, USM, Malaysia
29. Assoc. Prof. Ir. Dr. Parvathy Rajendran, USM, Malaysia
30. Dr. Mohamad Aizat Abas, USM, Malaysia
31. Dr. MohdSharizal Abdul Aziz, USM, Malaysia

32. Dr. Abdullah Aziz Saad, USM, Malaysia
33. Dr. MohdAzmi Ismail, USM, Malaysia
34. Dr. Yu KokHwa, USM, Malaysia
35. Dr. Muhammad Khalil Abdullah, USM, Malaysia
36. Assoc. Prof. Dr. Abdul Rahim Othman, UTP, Malaysia
37. Assoc. Prof. Dr. ShahrulKamaruddin, UTP, Malaysia
38. Dr. Sanjay P. Shekhawat, SSCOET, India
39. Dr. Paresh J. Shah, SSCOET, India
40. Dr. Girish K. Patnaik, SSCOET, India
41. Dr. Mujahid F. Husain, SSCOET, India
42. Dr. Prashant V. Thakre, SSCOET, India
43. Prof. Dr. Ishak Abdul Azid, UniKL, Malaysia
44. Assoc. Prof. DrShahrilNizam Mohamed Soid, UniKL, Malaysia
45. Dr. Muhammad Najib Abdul Hamid, UniKL, Malaysia
46. Dr. MohdNorzaimiChe Ani, UniKL, Malaysia
47. Dr. ShahrizanYusoff, UniKL, Malaysia

The summary of important statistics of the papers in the conference is as follows:

- A total number of submissions received: 52 papers (abstract)
- A total number of submissions sent for review: 42 papers
- A total number of submissions accepted: 36 papers
- Acceptance rate (number of submissions accepted / number of submissions received X 100): 69.23 %
- Average number of reviews per paper: 2-3 reviews
- Total number of reviewers involved: 155 reviewers

IOP Conf. Series: Materials Science and Engineering 1163 (2021) 011001 doi:10.1088/1757-899X/1163/1/011001

A Message from the President of the
Prince of Songkla University
Asst. Prof. Dr. NiwatKeawpradub

The 9th International Conference on Engineering and Technology (ICET-2021) is the achievement of an enduring relationship between Faculty of Engineering, Prince of Songkla University and Faculty of Technical Sciences from the University of Novi Sad. To enhance our collaboration then, a first international conference was organized in the year 2002. Nineteen years has passed, this year we haveorganized ICET conference with our expanding partners from India, Japan,Malaysia, and China that have enjoyed a friendly and productive collaboration in the past decade. We are honored to be a part of this one. With so many researchers and participants from various fields of engineering and sciences, I am confident that this conference can provide a perfect environment for the gathering of ideas and experiences which could essentially lead to future research collaborations and the development of new technologies and innovations. Not only will these academic exchanges support the critical national and international agenda on Global Warming and Economic Crisis, but they will also help promote the development of applicable technology that serve from regional communities to small and large industries alike.

At this time, I would like to once again congratulate the Faculty of Engineering and the Energy Systems Research Institute of the Prince of Songkla University with 8 partnering institutions from Serbia, China, Malaysia, India and Japanfor working together in putting forth the ICET-2021 in this special atmosphere during the COVID pandemic.Your valuable efforts have made the ICET-2021 possible.I am also delighted to learn that the organizing committee successfully realigned the format of our conference to virtual and online conference as this online format will enable us to collaborate after the conference seamlessly.On behalf of the Prince of Songkla University, I would like to wish all participants an enjoyable experience at this event and we look forward to being a part of such a commendable activity that will lead to the development of a better tomorrow for everyone. Thank you.

IOP Conf. Series: Materials Science and Engineering 1163 (2021) 011001 doi:10.1088/1757-899X/1163/1/011001

A Message from the Dean of the Faculty of Engineering,
Prince of Songkla University
Assoc. Prof. Dr. TanitChalermyanont

It is a great honor for the Faculty of Engineering at Prince of Songkla University to have the opportunity to host the The 9thInternational Conference on Engineering and Technology (ICET-2021) which is held invirtual format at the Faculty of Engineering, Prince of Songkla University, Thailand on May 27, 2021on the theme "Toward Disruptive Future Technology". This event also coincides with the celebration of the 55th year anniversary for our Faculty of Engineering at Prince of Songkla University which will come in September, 2022.We apologize for the change of venue from Krabi to online due to COVID-19. Otherwise, it would have been the wonderful getaway and learning experience in a scenic Krabi province.

It has been a difficult time for us all as the world is threatened by the pandemic COVID-19. Most activities have been delayed or canceled, and people are suffering the disease and its impacts, especially the national economic shrinkage and families having a hard time to make a living. As such, we pledge to donate a portion of our conference registration to charitable organizations to help fight COVID-19. Despite that, our main initiative of this international conference remains resilient as a forum for researchers from the field of sciences and technology from around the world to come together, share their valuable findings and take part in stimulating discussions. We anticipate that all of our participants, whether they are from Europe, Asia, or othercontinents learn from one another and perhaps have a chance to develop fruitful collaborations that will enhance the development of the society as a whole.

Moving toward a new decade, we now face the even more challenging issues than in the past. The conference organizers have insightfully offered topics, such as environment preservation, sustainable energy alternatives, waste management,IoT, and biomedical science and engineering as a forum for sharing ideas and developing solutions to these critical problems. I truly hope this conference will be the starting point of finding these global solutions together. Furthermore, I believe that with the sincere andcontinuing efforts of delegates from all participating universities, we will be moving closer towards the goal of research excellence for global impact, which is the utmost aim of all educational institutions around the world.

IOP Conf. Series: Materials Science and Engineering 1163 (2021) 011001 doi:10.1088/1757-899X/1163/1/011001

Last but not least, on behalf of the Faculty of Engineering, Prince of Songkla University, I would like to take this opportunity to express our sincere thanks for all the co-organizers of ICET-2021 and to congratulate all on the success of bringing forth this marvelous conference. I would like to express my appreciation to members of the advisory committee, and organizing committee, as well as everyone from the Faculty of Engineering at PSU and our correspondences from all partners who have contributed their valuable time and efforts in this conference. I wish all of you a successful and stimulating meeting or wherever you are on- line attending this conference during this COVID-19 pandemic. Stay safe with good health, Thank you

****

IOP Conf. Series: Materials Science and Engineering 1163 (2021) 011001 doi:10.1088/1757-899X/1163/1/011001

A Message from the Managerial Chairman of ICET-2021, PSU
Prof. Dr. Sumate Chaiprapat,
Assoc. Dean for Research and International Relations

In light of the coronavirus disease 2019 (COVID-19) outbreak, therewas a great uncertainty of all mass gathering activities. Our 9th International Conference on Engineering and Technology was originally planned to take place in August 2020 and it took us considerable efforts to move it to May 2021.Nevertheless, it is a great honor for ICET-2021 committee to organizeit as an online virtual conference this year. We still firmly believed it will provide ample opportunities for us to showcase therecent research findings and share your experience with other participants, and networking with universities, organizations, and business sector of Thailand in a new way. This year we set forth the conference theme "Toward Disruptive Future Technology" which reflects the atmosphere in the fast changing human society as a whole. We are honored to have the reputable keynote speakers both from academia and business sector to share their experiences and knowledge.I am grateful for the support from the Energy Systems Research Institute in jointly coordinating this conference with the Faculty of Engineering here at the Prince of Songkla University (PSU). Theconference is a result of the long collaborative relations of the Faculty of Engineering, PSU withFaculty of Technical Sciences from the University of Novi Sad under MOU, and delightfully this year we have expanded collaboration to our partners in co-organizing this ICET-2021 together. Their names are as follows:

- Shanghai Jiao Tong University (SJTU), China
- UniversitiSains Malaysia (USM), Malaysia
- UniversitiTunHussien Onn Malaysia (UTHM), Malaysia
- Universiti Kuala Lumpur (UniKL), Malaysia
- SSBT's College of Engineering & Technology (SSCOET), India
- Kanazawa University, Japan
- Musashino University, Japan

ICET-2021 offers a forum for researchers from the fields of engineering, sciences and technology from around the world to come together, share their research findings and take part in stimulating discussions. We hope that our participants learn from one another and develop collaborations that

IOP Conf. Series: Materials Science and Engineering 1163 (2021) 011001 doi:10.1088/1757-899X/1163/1/011001

lead to an enduring fruitful relationship.I am glad to report that over 100 participants have attended the conference with 56 papers. With the collaboration with the Songklanakarin Journal of Science and Technology (SCOPUS indexed), some papers are selected to publish in the supplementary issue of the journal.

On behalf of conference organizing committee, I would like to express my appreciation to members of the advisory committee, and organizing committee, as well as everyone from the Faculty of Engineering at PSU and our correspondences from all partners who have contributed their valuable time and efforts in putting this conference together. I wish all of you a successful on-line attending this conference during this COVID-19 pandemic.We wish all participants a good health wherever you may be. Stay safe with good health, and see you again in ICET-2023. Thank you.

****

IOP Conf. Series: Materials Science and Engineering 1163 (2021) 011001 doi:10.1088/1757-899X/1163/1/011001

Keynote Speaker

Waste to Energy in Thailand

Mr. WorawitLerdbussarakam
Senior Vice President
TPI POLENE Power Public Company Limited

Abstract

Currently, cities worldwide are increasing generate the municipal solid waste (MSW). Thailand generated about 28 million tons of solid waste per year. Waste-generation rates will more than double over the next twenty years. The global impacts of solid waste are growing fast. Solid waste is a large source of methane, a powerful GHG that has particularly strong short-term impacts. Locally, uncollected solid waste contributes to flooding, air pollution and publichealth impacts.

Thailand's waste has been sorted out from the beginning and waste has been recycled increased from the previous year and more waste being disposed of in the right way. The proportion of waste recycled is still less than it should be, so Waste to Energy can be effective solution and higher portion. There are many ways to convert waste into energy. In addition, to eliminating waste from overflowing the city, it also reduces foreign energy imports, land filled gas, Incinerator Plant or RDF sorting Plant

Today the direction of MSW waste to energy in Thailand is supported by Government policy by Increase of the power generation from waste in Power Development Plan (PDP) and Subsidize the electricity energy price in higher FIT .

TPI POLENE Power Public Company Limited (TPIPP) is the largest waste to Energy plant in Thailand. TPIPP have MSW power capacity 180 MW. can produce the RDF (Refuse Derived Fuel) from MSW to alternative fuel as coal replacement in Power Plant and Cement Plant, more than 3,000 ton per day. The incinerator plants 'capacity of MSW incinerator is 3,750 ton per day. TPIPP's advantage can manage as ZERO WASTE Plant, the waste from the power plant used raw materials for cement plant. The MSW and Presorted MSW deliver from many nearby provinces to TPIPP' Plant in Saraburi. TPIPP is the company which several technologies and advanced operation experience for Waste to Energy.

IOP Conf. Series: Materials Science and Engineering 1163 (2021) 011001 doi:10.1088/1757-899X/1163/1/011001

Keynote Speaker

AI Transformation

Prof. Dr. VirachSornlertlamvanich
Asia AI Institute (AAII), Faculty of Data Science, Musashino University
Faculty of Engineer, Thammasat University
Chair of RUN Digital Cluster
virach@gmail.com, http://virach.com/

Abstract

What is AI? What are we expecting from AI? What is our AI challenge in the Thailand context? Currently technology to realize AI become unveiled in many domains of implementation. The news about the success in Self-driving cars, natural language question-answering systems, or even the personal assist gadgets are publicized almost every day. Some people are aware of losing their jobs to AI. Some are questioning whether the today AI is truly imitate the human intelligence. However, it is quite certain that many domains of application are benefitting from the machine capability improvement based on the current AI technology. It is understood that digitization and the digital form of the data are the source of the intelligent pattern for machine to make decision and response to the human activities i. e. road traffics are visualized from the result of GIS information and naturally routed by the traffic prediction for the best route, automatic answering machine is replaced by natural language speech conversation composed from the trained knowledge based on the speaker emotion and intention.

Digitized Thailand project (2009-2013) was a first move of challenges toward the tremendous growth of data. Storage for collected data was still high and data collection technology was also limited. We started from archiving cultural and language data which we aimed to create a national data platform for learning and inheriting the wisdom or human understanding. Costume design, performing arts, weaving pattern, archaeological site point cloud, music, ethnic languages and such were studied and made up as a guideline for knowledge base design. The project has grounded technology alliance and data standard as well as some concrete implementations in the Ministry of Culture project of cultural knowledge archiving for instance.

IOP Conf. Series: Materials Science and Engineering 1163 (2021) 011001 doi:10.1088/1757-899X/1163/1/011001

Today, the advancement of AI is providing a great opportunity for ones to explore especially in its open source philosophy in providing the datasets and experiment software libraries. With the widespread of Python programming as a de facto standard, the communities can share the datasets and implementation methodologies. This situation also encourages the AI implementation in various domains. We put the practical process into a guideline for AI transformation for the upcoming AI introducing initiatives. The talk will end up with AI challenges which are left for getting closer to human intelligence in many aspects.

****

IOP Conf. Series: Materials Science and Engineering 1163 (2021) 011002 doi:10.1088/1757-899X/1163/1/011002

Peer review declaration

All conference organisers/editors are required to declare details about their peer review. Therefore, please provide the following information:

- **Type of peer review: Single-blind / Double-blind / Triple-blind / Open / Other**

Single-anonymous: authors' identities are known to the reviewers, reviewers' identities are hidden from authors

Double-anonymous: author and reviewer identities are hidden to each other

Triple-blind: author and reviewer identities are hidden to each other, and from the Editor(s)

Open: author and reviewer identities are known to each other

- **Describe criteria used by Reviewers when accepting/declining papers. Was there the opportunity to resubmit articles after revisions?**
  : The manuscript's content is within the scope and aims of the conference, and addresses advances in its discipline.
  The manuscript is clearly written and with good English proficiency.
  The methodology and experimental design are clearly declared with sufficient information.
  Results and findings are presented and interpreted with clarity supported by structured figures and tables.

- **Conference submission management system:**

  : https://icet2021.eng.psu.ac.th/submission/

- **Number of submissions received:**
  : A total number of submissions received: 52 papers (abstract)

- **Number of submissions sent for review:**

  : A total number of submissions sent for review: 42 papers

- **Number of submissions accepted:**

  :A total number of submissions accepted: 36 papers

- **Acceptance Rate (Number of Submissions Accepted / Number of Submissions Received X 100):**

IOP Conf. Series: Materials Science and Engineering 1163 (2021) 011002 doi:10.1088/1757-899X/1163/1/011002

:Acceptance rate (number of submissions accepted / number of submissions received X 100): 69.23 %

- **Average number of reviews per paper:**

: Average number of reviews per paper: 2-3 reviews

- **Total number of reviewers involved:**

: Total number of reviewers involved: 155 reviewers

- **Any additional info on review process (ie plagiarism check system):**
  :The plagiarism software "Turnitin" was used to check all manuscript before reviewing under 30 percent similarity index.

- **Contact person for queries:**
  :Asst. Prof. Jantakan Taweekul
  E-mail : juntakan.t@psu.ac.th

Please submit this form along with the rest of your files on the submission date written in your publishing agreement.

The information you provide will be published as part of your proceedings.

Assessment of wind energy potential of Hat Yai (Songkhla), Thailand

I Kamdar[1], **J Taweekun**[2,3]
[1] Energy Technology Program, Faculty of Engineering, Prince of Songkla University, Hat Yai, Songkhla 90112, Thailand.
[2] Department of Mechanical and Mechatronics Engineering, Faculty of Engineering, Prince of Songkla University, Hat Yai, Songkhla 90112, Thailand.
Abstract. Due to environmental problems associated with fossil fuels and rising need for energy, wind power renewable energy will play a vital role in transformation of future energy structure in Thailand. Exploitation of wind energy for sustainable development, wind resource assessment plays an important role in wind power utilization. This case study presents wind resource assessment using the wind atlas analysis and application program (WAsP), in order to assess the potential of wind energy in Hat Yai, Thailand. Ten-minutes interval wind data observed by the Thai Meteorological Department recorded during the period 2017-2019, have been used to harness adequate wind power density and wind energy potential. The annual mean wind speed is 3.5 m/s at 10 m height above the ground. The prevailing direction southwest, south and northeast of the wind rose are very pronounced. The WAsP analysis estimates about 2,731.28 MWh of the total net annual energy production for the specific sites using the Enercon E-18 (80 kW) wind turbine of 1.92 MW capacity. The findings of this study indicate the possibility of the small-scale wind farm project for electricity generation in future.

Keywords: Sustainability, Wind resource assessment, Wind turbine, WAsP, Hat Yai.

1. Introduction

Fossil fuels, including coal, oil and natural gas, have been the primary energy source for many years. However, concerns regarding climate change and CO_2 emissions have encouraged the search for alternative energy systems that can reduce the greenhouse gases emissions. Energy goals such as the 2020 Climate and Energy Package across the countries have targeted energy projects that are independent of fossil fuels. To fulfill these goals and regarding climate change, wind energy has been found as a promising and clean technology with large potential [1].

Wind turbines are becoming familiar in the world for electricity generation. The electrical energy produced by wind turbines are without liberating CO_2, radioactive pollutants, smog, or acid rain. The global cumulative installed capacity of wind energy reached 651 GW in 2019 and it is expected that a new capacity of over 355 GW will be added during 2020 2024 [2]. The capability of these projects was in MW range and most of them were onshore and offshore wind turbines. One of the significant issues for micro-wind farms is wind mapping. Hence, accurate assessment of the wind pattern of the selected places is essential for the wind farm output [3].

IOP Conf. Series: Materials Science and Engineering 1163 (2021) 012001 doi:10.1088/1757-899X/1163/1/012001

The energy scenario in terms of increasing demand annually in Thailand is not much different in comparison to other countries. The electricity generation of Thailand in 2012 was 32.6 GW which reached to 42.163 GW by 2017, with over 67% being produced from natural gas. It is expected by Thailand's Ministry of Energy power development plan 2010 (PDP 2010) that the total installed capacity of electricity generation will reach 70.868 GW in Thailand by 2030. Nevertheless, the aim of the Alternative Energy Development Plan (AEDP) of Thailand under PDP (2015-2036) is to replace fossil fuels up to 30% by 2036. This objective clearly defines the road map of renewable energy in Thailand [4].

In this research, the raw wind data from Thai Meteorological Department (TMD), along with the terrain data from Shuttle Radar Topography Mission (SRTM) and the surface roughness length information from Global Wind Atlas (GWA) warehouse map is used to conduct this study using the Wind Atlas and Analysis Application Program (WAsP). The objective of this study is to do feasible measures for wind resource assessment.

2. Methodology

2.1. Overview of study area and meteorological data description

Hat Yai is a city in the southern Thailand with an area of about 852.7 km^2 and hosts many tourists each year. It is the largest city of Songkhla province and has a population of about 156,802. The topography of Hat Yai is generally a vast plain. The city is adjacent to Songkhla Lake in the north while Sankalakiri mountain range passes across border of the city in southwest. Figure 1 shows a map of Hat Yai.

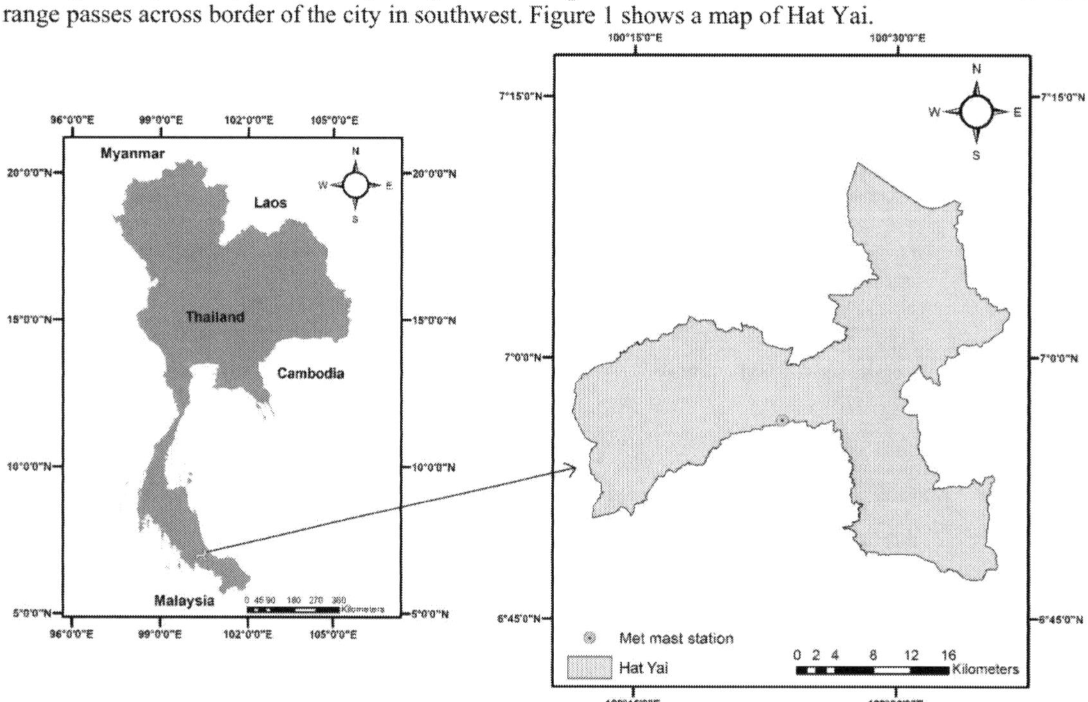

Figure 1. Study area. (adapted source: [5]).

This study analyzes the meteorological data in selected Hat Yai area situated in the southern part of Thailand. Wind data measurement for Hat Yai met mast has been taken at a standard height of 10 m above the ground for the wind resource assessment. The altitude of measurement tower is 57 m above sea level with

IOP Conf. Series: Materials Science and Engineering 1163 (2021) 012001 doi:10.1088/1757-899X/1163/1/012001

geographical coordinates (6° 56' 24.0''N, 100° 23' 31.9''E, Figure 2). The quality of wind data is found above 95% in the covered 3-years but there have been some missing data during some periods. At least one-year observation data is recommended for wind resource assessment to avoid monthly and seasonal bias [6]. In consideration of monthly and seasonal bias, the data used in this study was measured each 10-min over a period of 3-years from January 2017 until the end of 2019.

Figure 2. Overview of the satellite imagery of Hat Yai.

2.2. Elevation and roughness maps

The map projection is the Universal Transverse Mercator (UTM) coordinate system, Zone 47 and the datum is WGS-1984. The elevation and roughness maps of Hat Yai were prepared using the WAsP Map Editor tool in WAsP program (Figures 3 and 4). In order to show elevation and roughness changes of the area, a map extension of about 50 km x 40 km in WAsP Map editor was used for both maps. Most of the study area is fairly smooth except mountain ridges in the southwest and northeast with the highest elevation of about 920 m above sea level. There is total 7558 elevation contours in Figure 3. The southwest mountain ridge axis is almost perpendicular to the wind direction coming from the northeast which make these ideal sites for wind farm projects [7].

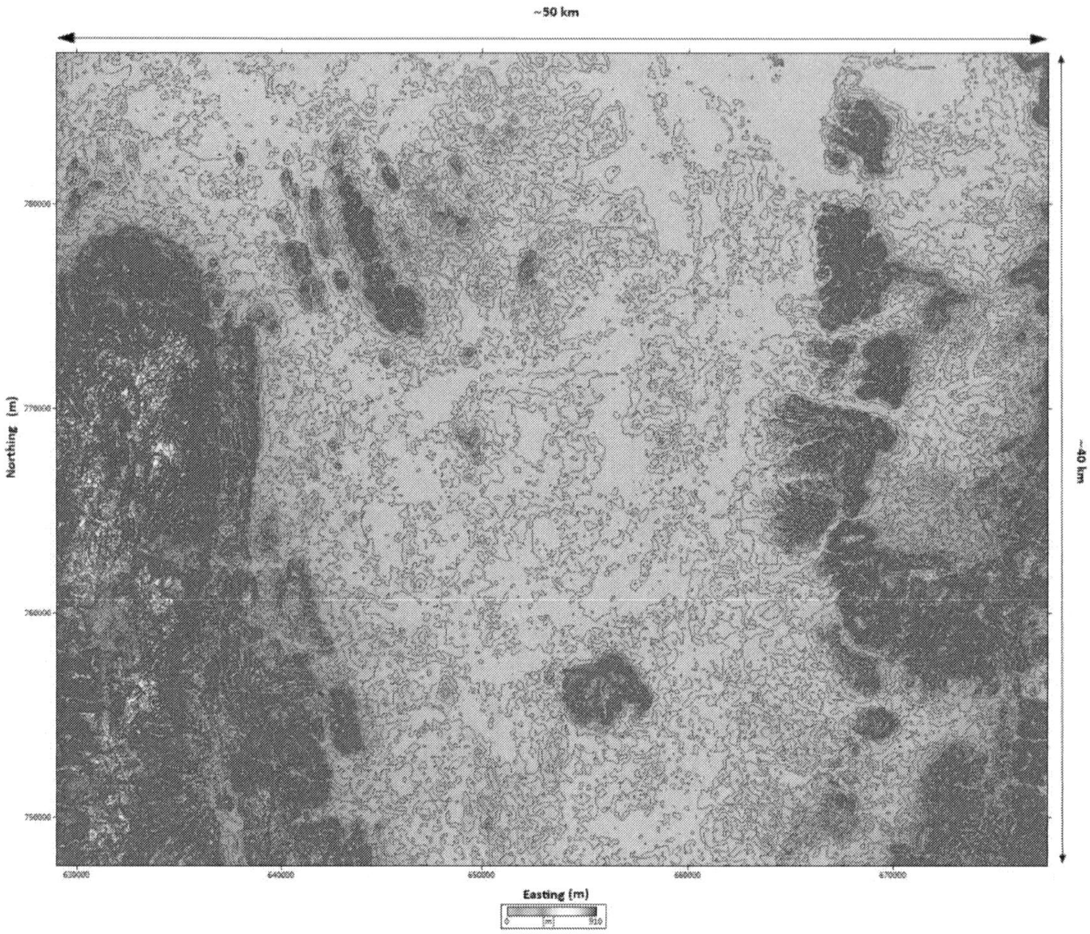

Figure 3. Digital map of Hat Yai depicting elevation.

Global Wind Atlas (GWA) warehouse map server and Google Earth maps were used for different landcover categories to classy the roughness lengths. The study area is classified into 6 different roughness lengths which are given in table 1.

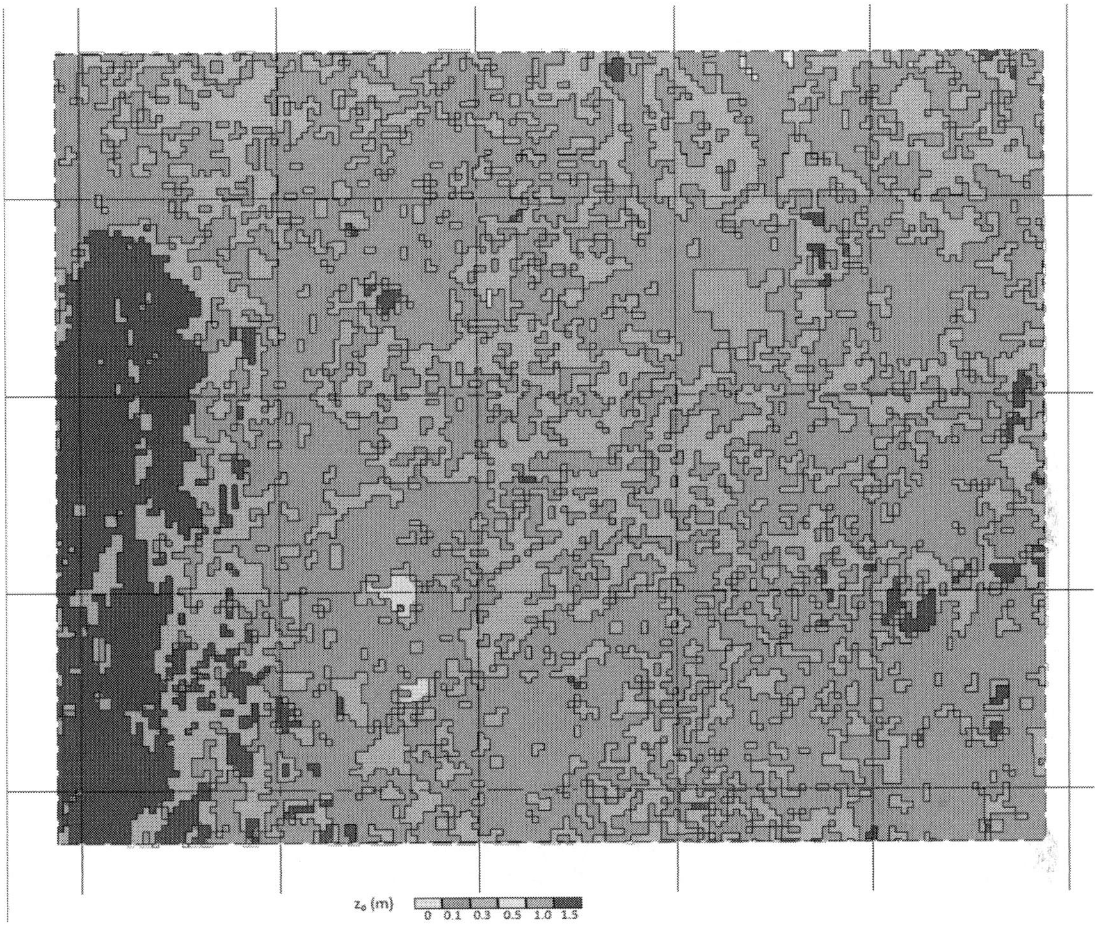

Figure 4. Digital map of Hat Yai depicting roughness.

Table 1. Values of surface roughness for different landcover.

S/N	Land Cover Class Name	Roughness Length (m)
1	Water surface	0.0000
2	Croplands, Shrubland	0.1000
3	Mosaic natural vegetation / cropland	0.3000
4	Flooded forest	0.5000
5	Urban areas	1.0000
6	Forests	1.5000

IOP Conf. Series: Materials Science and Engineering 1163 (2021) 012001 doi:10.1088/1757-899X/1163/1/012001

3. Results and discussion

3.1. Wind climate analysis

Table 2 shows the omni-directional and statistical inspection of wind characteristics of the processed data. The fitted parameters are used with respect to the observed wind climate and are representative of a Weibull distribution that fits to all-sector wind speed histogram. The fitted distribution is not further used for wind resource calculations. In WAsP program, the emergent distribution is used to calculate the wind resources [8]. The weighted sum of the Weibull distribution from all the direction sectors is represented by emergent distribution. Ultimately, the combined parameters are also representative of a Weibull distribution. This distribution is used to match the power density and mean wind speed with the weighted sum of the sector-wise power densities and mean wind speeds, respectively [8]. The discrepancy between the source and emergent mean power densities is 0% while the difference between the source and emergent mean wind speeds is 3.2%.

Table 2. Wind statistics of study area in all-sectors in the wind resource assessment.

Wind statistics	Weibull-A (m/s)	Weibull-k	Mean speed (m/s)	Power density (W/m^2)
Source data	-	-	1.87	17
Fitted	2.1	1.22	1.93	17
Emergent	-	-	1.93	17
Combined	2.1	1.22	1.93	17

Figure 5 shows the wind rose and the omni-directional Weibull distribution. The most prevailing wind direction in Hat Yai is from southwest, south and northeast. Table 3 shows the sector-wise Weibull parameters, mean power density and mean wind speed. From the table, 60

W/m^2 W/m^2, respectively. The second windiest direction is 30
W/m^2 The least
frequent wind directions are 120 , 150 , 270 , 300 and 330

Table 3. Sectors-wise wind statistics of the processed data.

f	
A (m/s)	
k	
U (m/s)	
P (W/m^2)	
[a]ΔU	

[a] represents difference in observed and calculated wind speed using Weibull distribution (%).

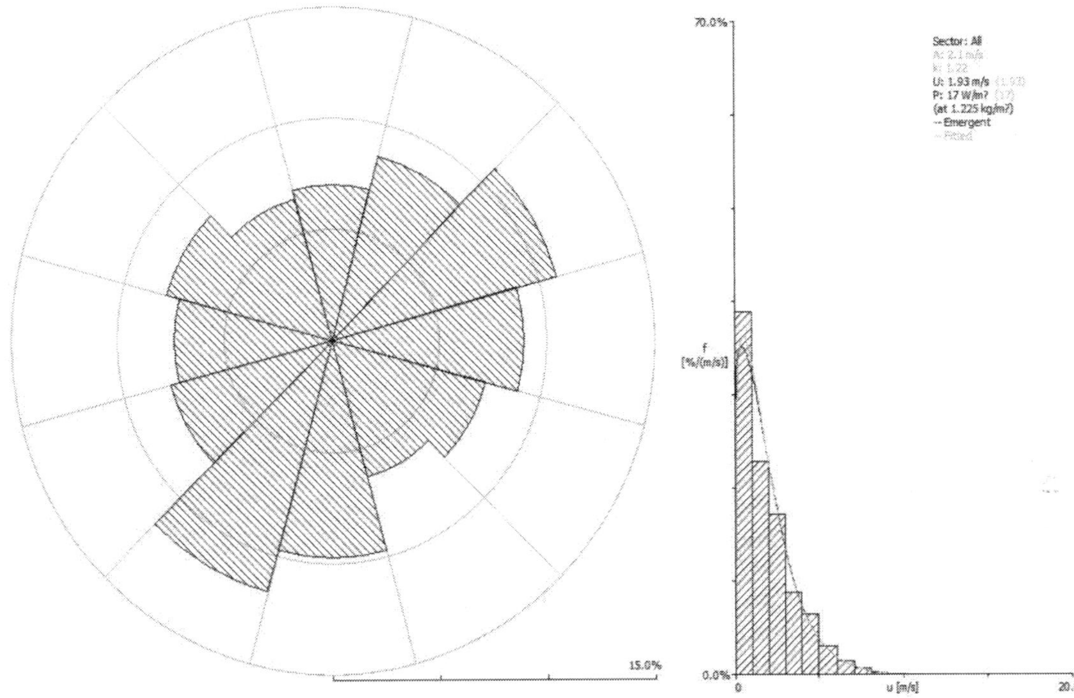

Figure 5. Wind rose and Weibull distribution of the study area.

3.2. Wind resource assessment by WAsP

WAsP analysis was carried out for the rapid population growth and increasing infrastructure in Hat Yai. Three locations for prospective wind farm were scrutinized by analyzing 3-years wind data for future investment in wind energy. Figures 6 and 7 show the extrapolative average wind speeds and power densities in resource maps of Hat Yai. Enercon E-18 wind turbine with a rating of 80 kW have been marked at 24 sites for power analysis. These sites were carefully chosen in terms of mean average wind speed, wind power density and access to roads and electrical transmission lines. The distance among all wind turbine sites have been kept at 10 m rotor diameter and a hub height of 30.1 m. The specifications of the Enercon E-18 wind turbine are given in table 4 while power curve is shown in Figure 8. The wind map indicates good potential of wind energy for the southwest mountain ridge in Hat Yai. The central part of the city shows a poor potential for wind farm development due to urban infrastructure.

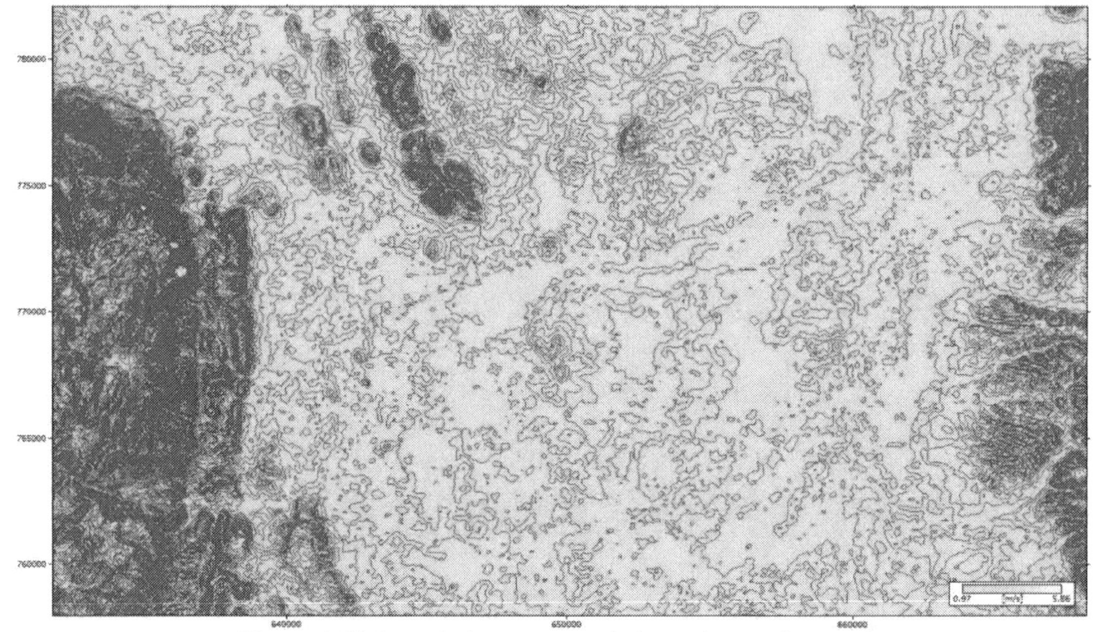

Figure 6. High resolution average wind speed map of Hat Yai.

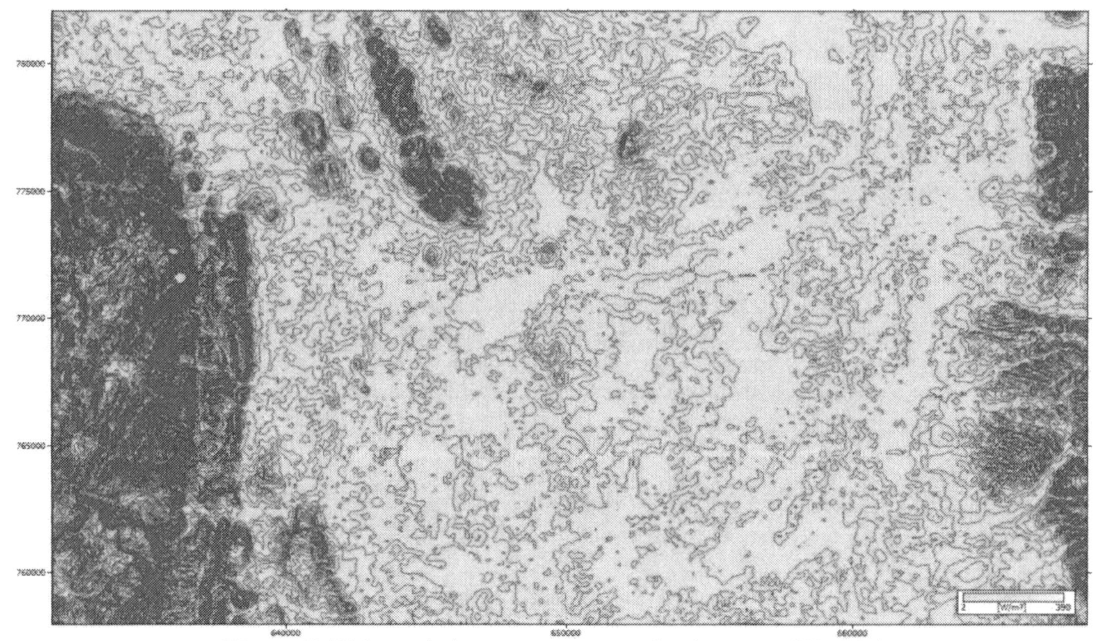

Figure 7. High resolution average power density map of Hat Yai.

IOP Conf. Series: Materials Science and Engineering 1163 (2021) 012001 doi:10.1088/1757-899X/1163/1/012001

Table 4. Enercon E-18 wind turbine specifications.

Rotor diameter	Hub height	Cut-in speed	Cut-out speed	Survival wind speed	Rated power	Rated wind speed
18 m	30.1 m	2.5 m/s	25.0 m/s	67.0 m/s	80 kW	12.0 m/s

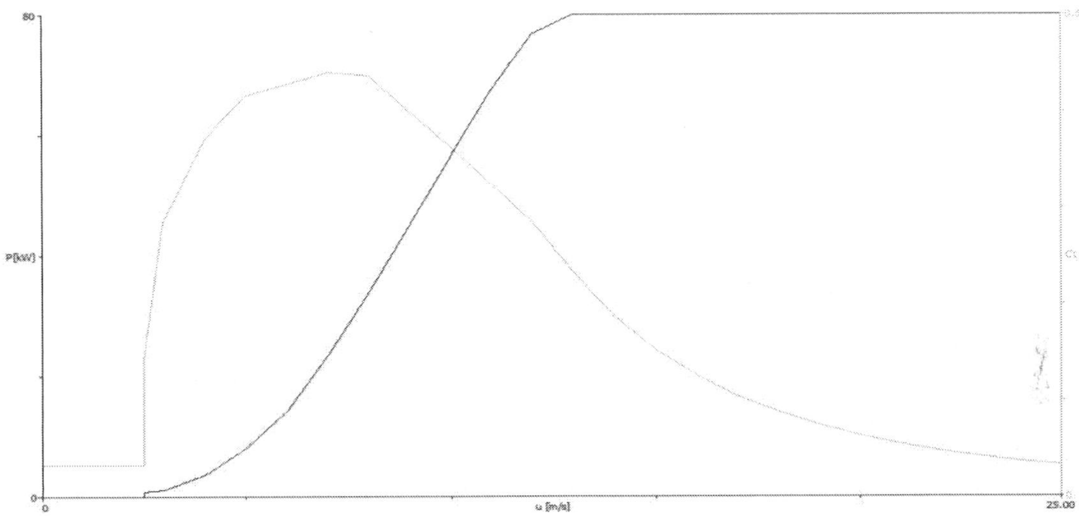

Figure 8. Power curve of Enercon E-18 wind turbine.

The net annual energy production (AEP) for the prospective turbine cluster 1' wind farm including 8 sites are given in table 5. The mean wind speed and mean power density of the prospective turbine cluster 1' wind farm are 5.1 m/s and 285 W/m^2, respectively. Turbine site 006 shows the maximum capacity factor of 24.9% in this prospective wind farm. The average net AEP for the prospective turbine cluster 1' wind farm is around 145 MWh. Figure 9 displays the resource map of the prospective turbine cluster 1' wind farm.

Table 5. Statistical analysis of prospective turbine cluster 1' wind farm.

[illegible]	[illegible]	[illegible]	[illegible]	[illegible]	[illegible]
[illegible]	[illegible]	[illegible]	[illegible]	[illegible]	[illegible]
[illegible]	[illegible]	[illegible]	[illegible]	[illegible]	[illegible]
[illegible]	[illegible]	[illegible]	[illegible]	[illegible]	[illegible]
[illegible]	[illegible]	[illegible]	[illegible]	[illegible]	[illegible]
[illegible]	[illegible]	[illegible]	[illegible]	[illegible]	[illegible]
[illegible]	[illegible]	[illegible]	[illegible]	[illegible]	[illegible]
[illegible]	[illegible]	[illegible]	[illegible]	[illegible]	[illegible]
[illegible]	[illegible]	[illegible]	[illegible]	[illegible]	[illegible]
[illegible]	[illegible]	[illegible]	[illegible]	[illegible]	[illegible]

[b] represents Ruggedness Index.

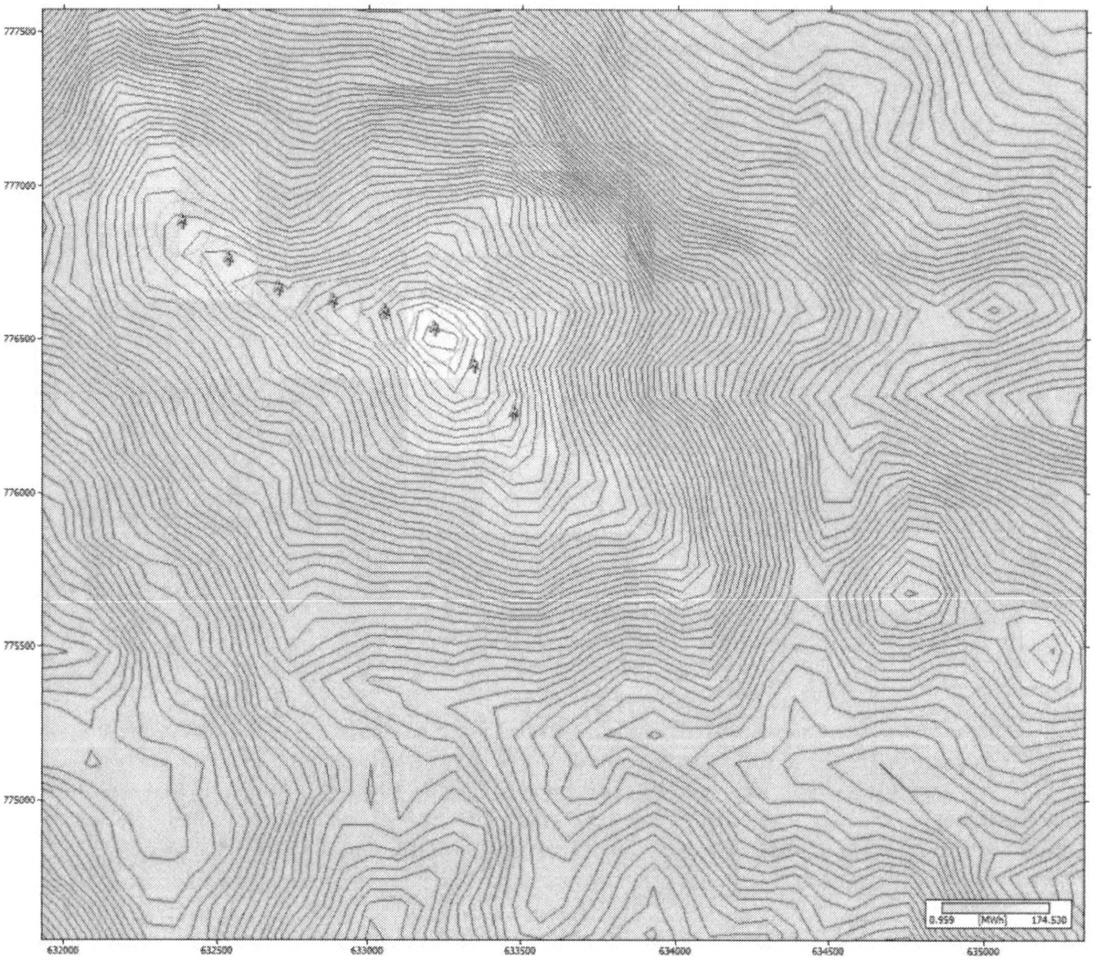

Figure 9. High resolution map of prospective turbine cluster 1' wind farm.

The net annual energy production (AEP) for the prospective turbine cluster 2' wind farm including 8 sites are given in table 6. The mean wind speed and mean power density of the prospective turbine cluster 2' wind farm are 4.3 m/s and 178 W/m^2, respectively. Turbine site 011 shows the maximum capacity factor of 18.1% in this prospective wind farm. The average net AEP for the prospective turbine cluster 2' wind farm is around 104 MWh. Figure 10 depicts the resource map of the prospective turbine cluster 2' wind farm.

Table 6. Statistical analysis of prospective turbine cluster 2' wind farm.

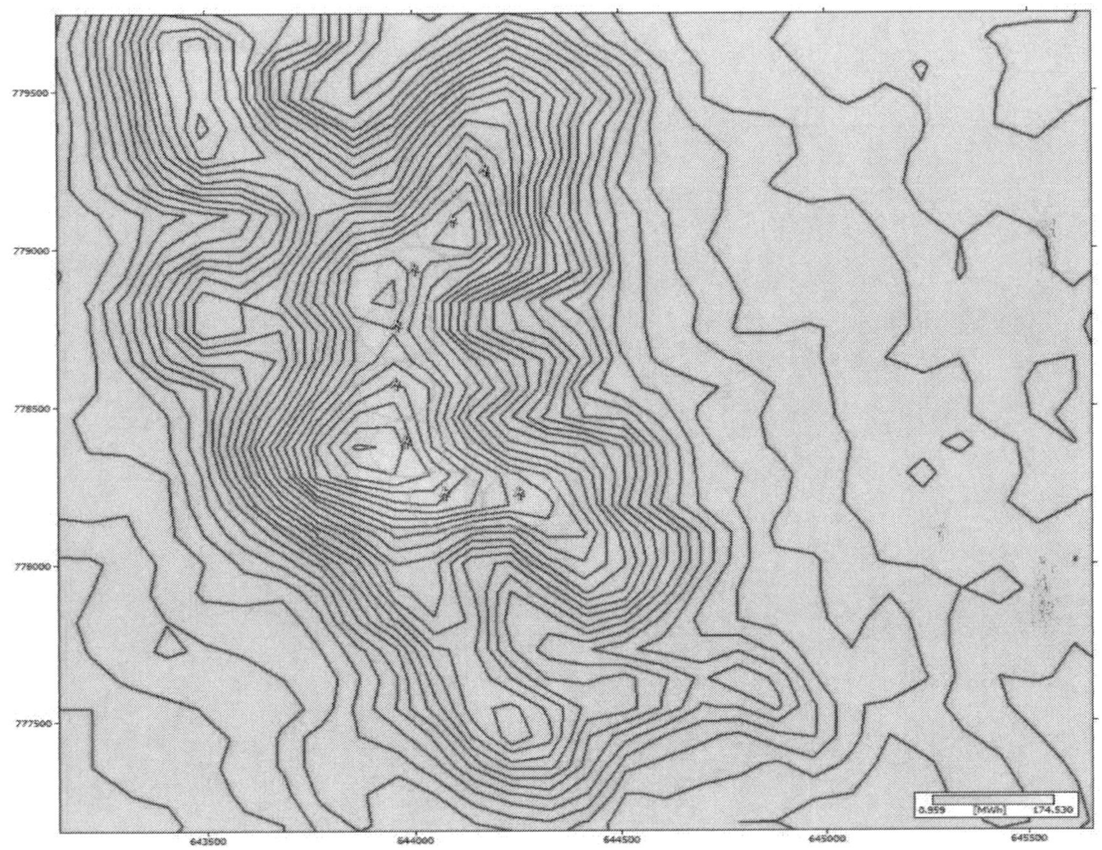

Figure 10. High resolution map of prospective turbine cluster 2' wind farm.

The net annual energy production (AEP) for the prospective turbine cluster 3' wind farm including 8 sites are given in table 7. The mean wind speed and mean power density of the prospective turbine cluster 1' wind farm are 3.9 m/s and 154 W/m^2, respectively. Turbine site 022 shows the maximum capacity factor of 14.3% in this prospective wind farm. The average net AEP for the prospective turbine cluster 3' wind farm is around 91 MWh. Figure 11 shows the resource map of the prospective turbine cluster 3' wind farm.

IOP Conf. Series: Materials Science and Engineering 1163 (2021) 012001 doi:10.1088/1757-899X/1163/1/012001

Table 7. Statistical analysis of prospective turbine cluster 3' wind farm.

[illegible]	[illegible]	[illegible]	[illegible]	[illegible]	[illegible]
[illegible]	[illegible]	[illegible]	[illegible]	[illegible]	[illegible]
[illegible]	[illegible]	[illegible]	[illegible]	[illegible]	[illegible]
[illegible]	[illegible]	[illegible]	[illegible]	[illegible]	[illegible]
[illegible]	[illegible]	[illegible]	[illegible]	[illegible]	[illegible]
[illegible]	[illegible]	[illegible]	[illegible]	[illegible]	[illegible]
[illegible]	[illegible]	[illegible]	[illegible]	[illegible]	[illegible]
[illegible]	[illegible]	[illegible]	[illegible]	[illegible]	[illegible]
[illegible]	[illegible]	[illegible]	[illegible]	[illegible]	[illegible]

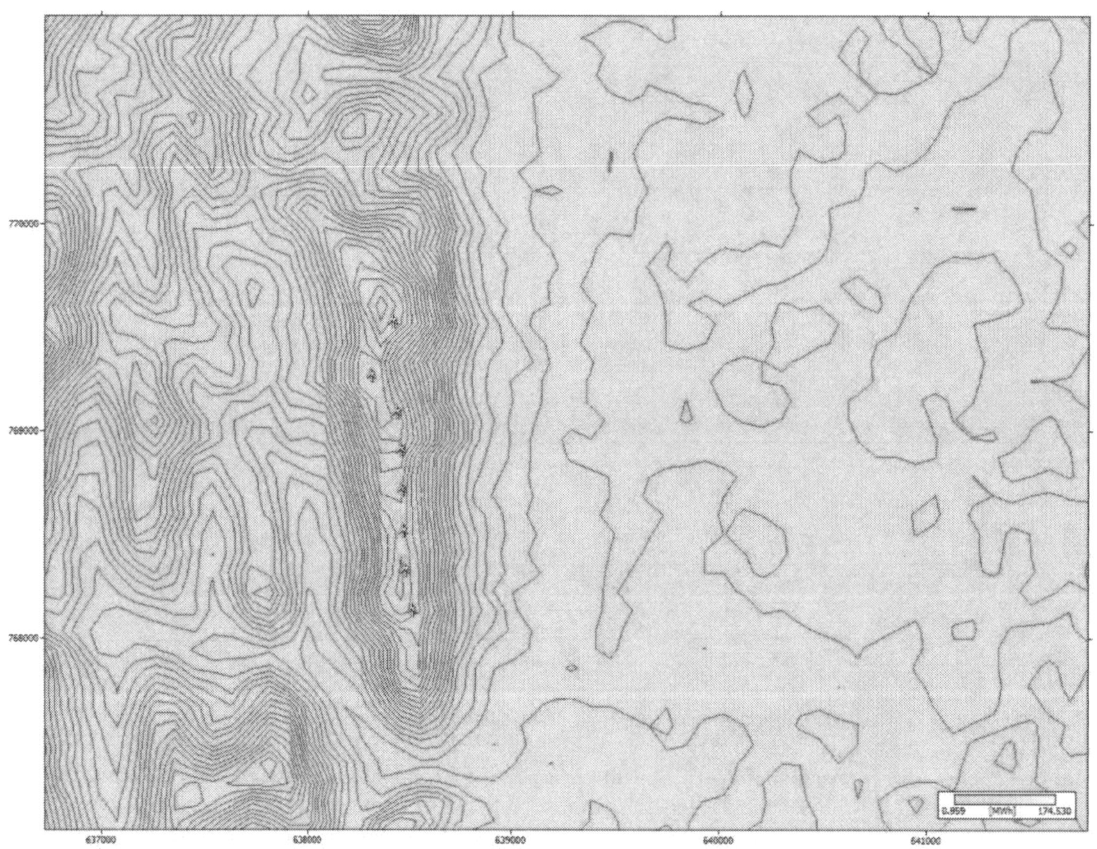

Figure 11. High resolution map of prospective turbine cluster 3' wind farm.

4. Conclusions

The wind resource assessment in Hat Yai has been carried out in the three best locations for the future wind farms. The overall wind speed in the hilly and nearby surrounding areas is about 3.0 m/s to 5.8 m/s according to the data analysis. This range of wind speed regime falls in the low-medium class. The prevailing wind direction in the area is southwest, south and northeast. Due to availability of small and flat lands, southwest wind direction towards Sankalakiri mountain range is best for small-scale wind farm projects in future. In the past, some of the wind turbines installed in Thailand were unsuitable according to the weather conditions. Those wind turbines imported were constructed for windier conditions. They were steel structured that only rotated in storms. Subsequent technological developments have resulted in the creation of low-speed and decentralized wind turbine which can be installed anywhere from homes to offices. Hence, further investigation must be carried for the development of an integrated renewable energy system.

The wind resource assessment in this study provides a scientific basis for development of wind farm projects in future to reduce dependence of Thailand's urban cities such as Hat Yai on fossil fuel-based generation. The approach used in this research can also be reproduced for wind resource assessment in other regions of the world.

Acknowledgements

This work was supported by the grants of Interdisciplinary Graduate School (IGS) and Energy Conservation and Promotion Fund Office (Project number: RE-2-0037), Prince of Songkla University, Hat Yai, Songkhla 90112, Thailand. The authors acknowledge the help of the Thai Meteorological Department (TMD) for sharing valuable wind data.

References

[1] P. Hevia-Koch, J. Ladenburg, Where should wind energy be located? A review of preferences and visualisation approaches for wind turbine locations, Energy Research & Social Science (2019). doi:10.1016/j.erss.2019.02.010.

[2] Global Wind Report, 2019. Accessed: November 30, 2020. Available: https://gwec.net/global-wind-report-2019/

[3] Z. Tasneem, A. Al Noman, S.K. Das, D.K. Saha, M.R. Islam, M.F. Ali, M.F. R Badal, M.H. Ahamed, S.I. Moyeen, F. Alam, An analytical review on the evaluation of wind resource and wind turbine for urban application: Prospect and challenges, Developments in the Built Environment (2020). doi:10.1016/j.dibe.2020.100033.

[4] S. Ali, J. Taweekun, K.K. Techato, J. Waewsak, S. Gyawali, S. Ali, S., Taweekun, J., Techato, K., Waewsak, J., Gyawali, GIS based site suitability assessment for wind and solar farms in Songkhla, Thailand, Renewable Energy. 132 (2019) 1360–1372. doi:https://doi.org/10.1016/j.renene.2018.09.035.

[5] I. Kamdar, S. Ali, A. Bennui, K. Techato, W. Jutidamrongphan, Municipal solid waste landfill siting using an integrated GIS-AHP approach: A case study from Songkhla, Thailand, Resources, Conservation & Recycling 149 (2019) 220–235. doi:10.1016/j.resconrec.2019.05.027.

[6] D.H. Ko, S.T. Jeong, Y.C. Kim, Assessment of wind energy for small-scale wind power in Chuuk State, Micronesia, Renewable & Sustainable Energy Reviews (2015).

doi:10.1016/j.rser.2015.07.160.

[7] H.L. Wegley, J. V. Ramsdell, M.M. Orgill, R.L. Drake, A siting handbook for small wind energy coversion systems., (1980).

[8] N.G. Mortensen, Wind resource assessment using the WAsP software (2018). DTU Wind Energy.

AN IMPROVEMENT OF PRODUCTIVITY BY REAL TIME MACHINE MONITORING SYSTEM: A case study of printing industry

Sawat Pararach[1], Apiwat Muttamara[2], Phisan Kaewprapha[3]

[1,2]Industrial Engineering, Thammasat University, Thailand
[3]Electrical and Computer Engineering, Thammasat University, Thailand

E-mail: kphisan@engr.tu.ac.th

Abstract. Digital technologies and automation integration is one of the promising directions to increase productivity in the manufacturing. A consulting project to promoting the adoption of digital technology under the policy of the Department of Industrial Promotion, and Thailand 4.0 to help small and medium entrepreneurs (SMEs) to assess and gain readiness to the use of cyber-physical including digital technology. The real-time data gathering system for monitoring and displaying information is presented. For example, IoT based sensors is applied in digital counting of machine-made goods. This data can be used to analyze the root cause of a manufacturing process problem, for example, time lost due to discontinuity of work or setting up. Best practices are observed at a printing factory. The implementation shows the effective data acquisition even the running speed of a machine is high. The owner and manager gain real-time access to the dashboard through their mobile phone or personal computer. Therefore, they can manage their jobs instantly and remotely. Moreover, the stored data can be downloaded in Excel format. The data can be used in process waste analysis. The numerical results of using this system, overall equipment productivity and efficiency (OEE) were increased by 10% to 15%, approximately.

1. Introduction

The advancement in the internet of things (IoTs) technology has potential applications in various industry sectors. The technology enables manufacturers to gain access to precious and on-time production data via sensors and connected sensing nodes. This lays down the infrastructure as the starting point for the transition to Industry 4.0. In Thailand. The Department of Industrial Promotion (DIP) is one of the taskforces to accelerate the adoptions of technologies, especially for the small to medium enterprises (SMEs). Due to high competition and uncertainty in the current country's and world's economy amid the outbreak of the COVID-19, manufactures scramble to be self-reliant and adaptive seeking to improve efficiency in every aspect or the business. To improve productivity and competitiveness in the market, manufacturers must be making transition to for better adaptive, real-time and automated process and become more efficient and data-centric taking action and making decision as much as possible. However, moving toward industry 4.0 is not just a simple step, it requires dedication of resources and effort such as budget and human. In current situation where budget cuts and expense tightening is inevitable, investing in the 4.0 technology seems to be out of reach. SMEs are, therefore, behind and the investments are sometimes prohibitively beyond the return of investment point to consider whole factories adoption of such technology.

SMEs currently fall into between industry 2.0 – 3.0, aka., 2.5 where machine are operated separately and controlled manually by humans. The transition to from current industry evolutions requires innovative technologies and mass adoption in the eco-system [1]. Companies have been trying to improve the whole operation to digital based operations. From business to manufacturing, for example, utilizing software to handle business flow and resources (known as ERP system). Sub systems Material Resource Planning (MRP) and Manufacturing Execution System (MES) are also included. This system control flow of data through software. Gaining access to this software is also another challenge for companies. Production control and management of SMEs rely on paper based and individual employee skills. Without software and digital system, data gathering, which is a crucial step toward industrial 4.0 is still in paper based and pose delayed, unreliable of data and sometime those data will be not useful after a while and collecting those data for analysis or even to collecting them require labor intensive work where company wastes their resources on.

Thus, to accelerate industrial 4.0 adoption, this paper presents a case study how IoT sensing system is designed and implemented and integrating with printing machine to gather data and monitor production in real-time as well as integrating the data into company's own production management system and knowledge transfer for further adoption of IoTs and digital technologies. One of the contribution of this work is how the promoting of industry 4.0 is done via advising and showcase with actual implementation. Challenges such as technical and management issues are also addressed.

This paper is organized as follows: The second section shows related work while section 3 provides company's background, problem statement, challenge in apply machine monitoring system. Section 4 gives detail design implementation and data integration and usage. Section 5 provides the process improvement and the application of data analysis. Section 6 provides improvement and productivity results and discussion and further improvement. Finally, section 7 concludes this work.

2. Related Work

Promoting and assisting small to medium businesses to entering or at least considering of transforming to industry 4.0 is a challenging task. Companies executives usually have visions regarding how the ideal manufacturing should be or how the problems and issues should be handled or resolved. The obstacle, however, lies on the management and technological difficulties in addition to the investment and expenses priorities. Thu realization effort is where SMEs seeks some assisting from government agencies and organizations. Acquiring manufacturing standard such as ISOs are often subsidised or trained/advices by experts and industrial specialists. The same approach applies to digital and automation technologies, energy, lean process, etc. Focusing on the general aspects of transforming to industry 4.0, one of the approaches is to form an institution advocating for promoting and transferring technologies. In [2], LeanLab is a platform that facilitate and provide evaluation and planning as well as testbed for industry to gain insight into transition to industry 4.0. The process consisting of 5 steps which include preparation, analysis, opportunity evaluation and implementation step. There are several areas where transition must be done, for example asset utilization, labor, quality, supply chain management, etc. One of the asset utilizations is machine utilization, which can be done in real-time via remote monitoring and control system [3]. This area is where machine monitoring system and IoTs can be applied and integrated to gain data and utilized in a timely manner.

IoTs-based machine monitoring is a mature technology, however, implementing, deploying and transitioning, as well as raising awareness of technologies are still new to SMEs . In [4], provides insight into how implementing IoTs in industrial environment is challenging from both managements down to day to day working operation due to limitations and constraints such as equipment communication and proprietaries software, mixed of PLCs sub system. [4] demonstrate the application and adoption home-applicant based IoTs technologies to industrial environments. Home-applicants based IoTs are finished product with eco-system that satisfies user's experience. The study aims to follow the approach by design easy to setup, easy to operate while maintain wide range of application. The main focus is to design their own IoTs edge devices and communication to be used with standard cloud based IoT platform. In our work, however, we focus on integrating low-cost, low learning curve, ready-to-use, standard and open source platform that's available with ease of transferring knowledge to the company with aim to educate as well as assisting them to rely on their available technical resources such as technician and retain with new IoT skill on the actual working environments. We aslso provide analysis an provide planning for their next steps investment. This allow the company to evaluate and know the gap between current situation and target environment. They may then estimate resource to archive that.

3. Company background and production process

We work with a printing company where machineries are outdated. The printing business maintain its customers pool via long-term relationship such as banks, utilities companies. Their products include receipts paper, invoices. The printing processes include preparing raw paper, ink then the printing plate preparation. As shown in Fig. 1, continuous paper is then fed to the machine, then folded. The machine counts number of pages per box then cut and pack the paper into the provided boxes.

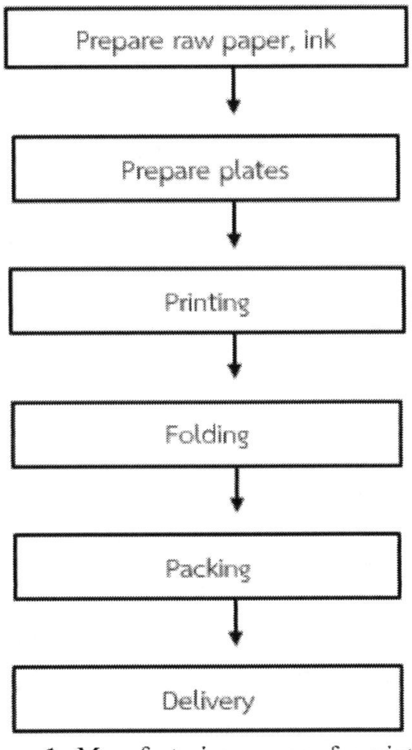

Figure 1. Manufacturing process for printing.

In term of digital technology level in the current company. The company itself is consider active in developing and attempting to adapt and integrate digital and IoTs technology to its process. They have developed their own production control system software based on SQL database and Visual basic software platforms. It is interesting to observed that they also employ long-time IT employee who is a technician with basic software knowledge and has since became main developer who develop the in-house system that work seamlessly for the worker and provide reports to executives. Hence the company has successfully moved from paper based data collecting into computer based data gathering and printing process management. Customer invoice and orders are transformed into work order that will be planed into the system. Then workers at each printing station, each shift, get the information in advance and plan for the work, order the raw material, recording in the warehouse/inventory system as well as opening the job. After the actual work is done, date time, duration, break down, wastes, defects and finish goods are recorded into the system. In this current system, the company and production that is with the software is consider ahead of other SMEs where flow of data is still on papers. This consider to be the advantageous and it is the key success to transferring to Industry 4.0.

Even though the machineries and equipment are not the main focus in transforming to the Industry 4.0 due to the business itself is old and hasn't changed for decades, also the fact that new printing business model are unknown in the future, the company is still trying to stay competitive and to stay as efficient as possible. Hence their main focus is on the data gathering and data usage for business management, analysis, cost-control and forecasts. The system serves well with current day-to-day operation, however, it has some major drawbacks, worker at stations may not record data at the moment it happens, data may come as summary at the end of shifts which is sometime not very useful to be used as process improvement. The company needs real-time monitoring system where workers and business owner share the same data. More importantly the data that will be obtained are more accurate than hand-recorded, which is prone to errors. Thus the vision of this company is clear, the management prefer real-time machine monitoring system where they can take a quick capture of the current working state of the production line. The new IoTs based monitoring should show how the order has been produced and fulfilled. Fig. 2 show the printing stations.

Figure 2. The printing station and folding station

4. System Architecture and Process Improvement

From the requirements gathered from managing team. It is clear that machine monitoring via IoTs technology will suit the needs from both managing and working force. Executive benefit on real-time and anywhere access to data via cloud-based monitoring system. Workers know there working efficiency and use the data to improve their working procedures. The process engineer can calculate the overall equipment effectiveness (OEE) to understand and to improve for better resource utilization which will finally reflect the cost reduction and productivity improvement.

Even though the company has its own IT developer however, the IT does not have basic knowledge regarding IoTs platform and modern technologies. Meanwhile, they also have technician who operate and maintain the machine. These people are expert in PLC system and they can invent control mechanism based on PLC and digital I/O. The printing station has its own counter where counting paper piece and paper bulk are recorded on screen, however, this information cannot be linked to the database and production management platform.

We then form a team which consists of IT and technician to work together to come up with IoT-enable architecture. The first step is to transfer knowledge regarding IoTs and how modern IoT devices and sensor can be used. We then design a connection between PLC's counting and a micro controller device which will be used as a communication device that collect data from the machine. The data consist of 1) counting piece of paper signal 2) bulk complete signal. The two information will allow users to know how many boxes of finish goods are produced. The real-time information is then mapped to the open work order job. Figure 3 shows the architecture below.

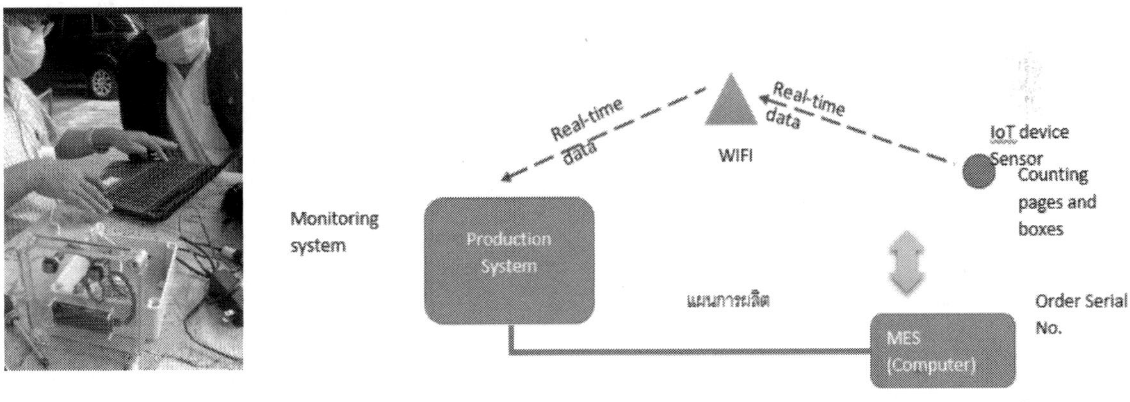

Figure 3. The development of IoT edge communication device.

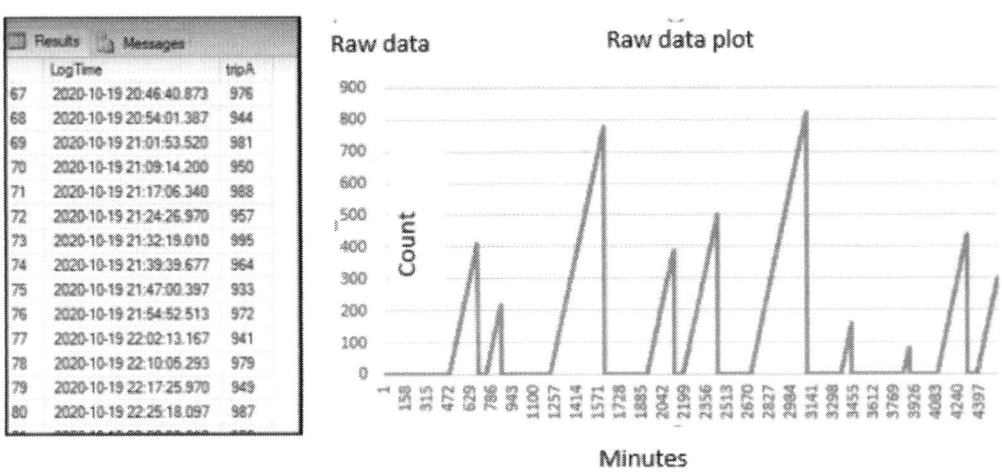

	LogTime	tripA
67	2020-10-19 20:46:40.873	976
68	2020-10-19 20:54:01.387	944
69	2020-10-19 21:01:53.520	981
70	2020-10-19 21:09:14.200	950
71	2020-10-19 21:17:06.340	988
72	2020-10-19 21:24:26.970	957
73	2020-10-19 21:32:19.010	995
74	2020-10-19 21:39:39.677	964
75	2020-10-19 21:47:00.397	933
76	2020-10-19 21:54:52.513	972
77	2020-10-19 22:02:13.167	941
78	2020-10-19 22:10:05.293	979
79	2020-10-19 22:17:25.970	949
80	2020-10-19 22:25:18.097	987

Figure 4. Raw data obtained from the system.

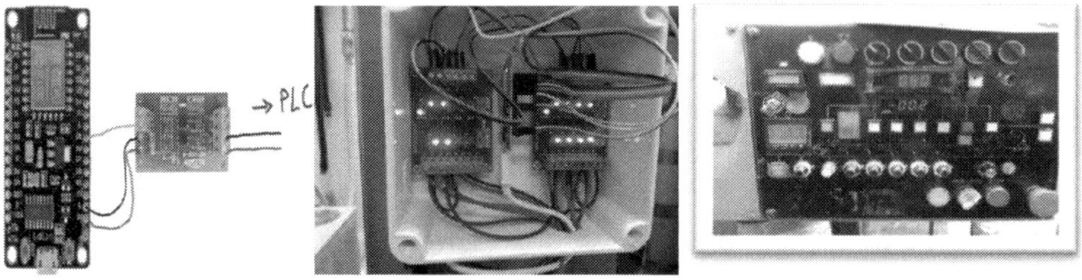

Figure 5. 1) The nodemcu (esp8266) microcontroller 2) opto-couple 3) PLC controller

The installation is done at printing station K11. The PLCs counting signal is tapped by the micro controller which will transmit data though WIFI via low overhead UDP protocol to the SQL server that listen and insert counting data into its database. The increment counting method is used (e.g. car odometer), a few packages loss during transmission won't affect the recorded data and the counting as number of paper is accumulated and transmitted the next interval time which is set to be every 10 second. The speed of the printing can be as fast as 300-500 pages per minute. The cloud server then processes and format the data into human readable showing information as graph as shown in figure 4. The microcontroller used as edge IoT device is a well-known nodemcu (esp8266), a system-on-chip unit, with WIFI available is connected to PLC pulse signal through optical isolator circuit which transmit information through light signal that separates 5-Volt MCU apart from 24-Volt PLC. The overall cost of making 1 unit is relatively low. The reliability of the unit is excellent in the current working environment where electromagnetic noise is quite low.

After this first implementation, the company has their know-how and is able to create, program install and integrate the system to other stations.

5. Process Improvement

After obtaining raw counting data from K11 machine, the data is then processed and displayed on dashboard as a graph in figure 4. Moreover, the data is also mapped back to the work order created during that time. The manager can now know the status of the current order. Sale persons can also get the real-time information of the their current in-progress work. The planning department can also benefit from the actual data so they can plan in case of break-down and get warning when machine stop for a certain period of time longer than a set threshold through line notification. The report is shown in figure 6. This is where managing team can monitor work-in-progress instantly. Any unusual event can be detected and notified right away as depicted in figure 7.

From figure 6, the data is obtained automatically via IoT device, it can be seen that for continuous working process, each box of printing paper requires 7-8 minutes to finish 1 batch. However, in the case of machine failures, interruptions or breaks can be seen as a record. Any setup time can be seen where the counting in 1 batch is low. For example, the current batch is 1000 pages per batch, if the user setup and spend a trial print for 100 pages, the worker will reset to start the actual batch. This setup information is also recorded separately as a row in the report.

Hence, this allows us to compute the overall equipment effectiveness (OEE) which is a standard tool for measuring productivity and capacity of a machine and how effective the machine performs.

$$\text{OEE} = \text{Availability} \times \text{Performance} \times \text{Quality} \quad (1)$$

where

Availability means active time to total time when the machine is running and when the machine is idle or broken (planned/unplanned stop).

Performance is the speed or time the machine can produced compared to the maximum design speed. Older machine may have to run at lower than designed speed, otherwise it might affect the quality.

Quality measures good products against overall output produced by the machine.

The 3 factors are weight as 100% or 1.0. The ultimate score is also 100% or 1.0 when the machine run perfectly non-stop.

According to the formula we are able to compute the OEE of sample from figure 6.

OEE = Availability (0.82) x Performance (0.87) x Quality (0.99) = 0.71 (2)

The availability is obtained from overall runtime which is equal to 187 minutes. The active time is 153 minutes. Performance is when the machine run at the fastest speed at current period (although, the actual capacity can be used) which is 7 minutes per batch versus the average time of 8 minutes per batch. The quality when the paper is wasted during setup time.

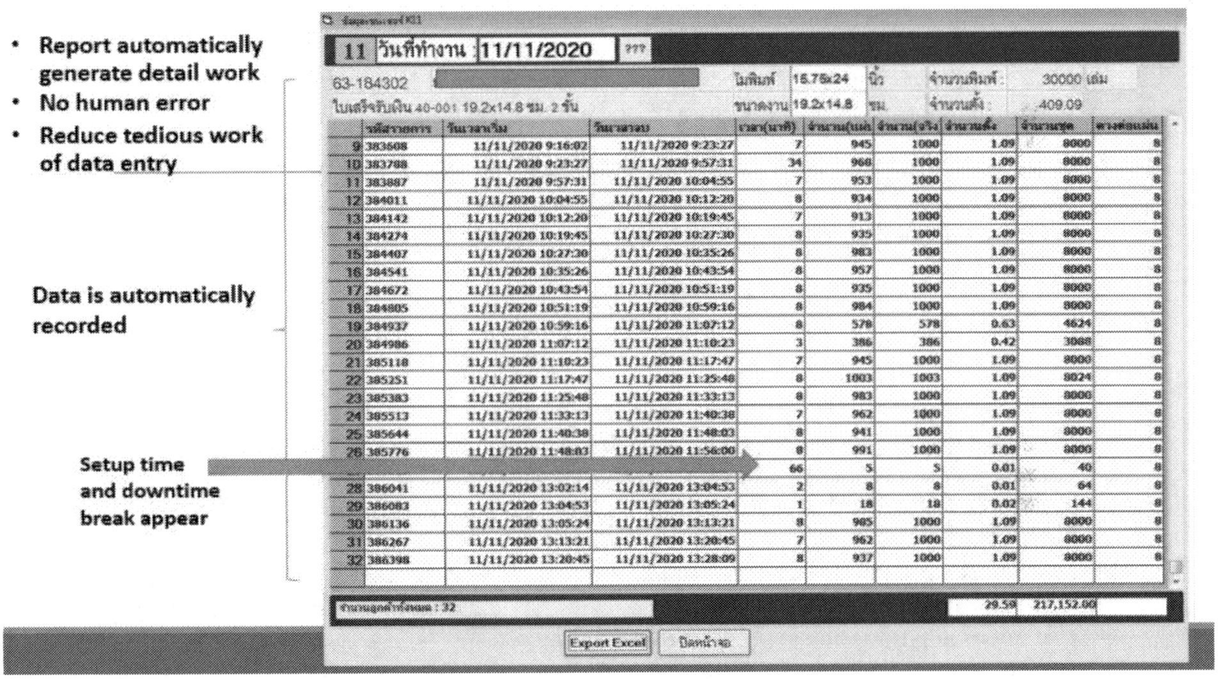

Figure 6. An automatically generated report where work order and each processed box is record with date and time and duration.

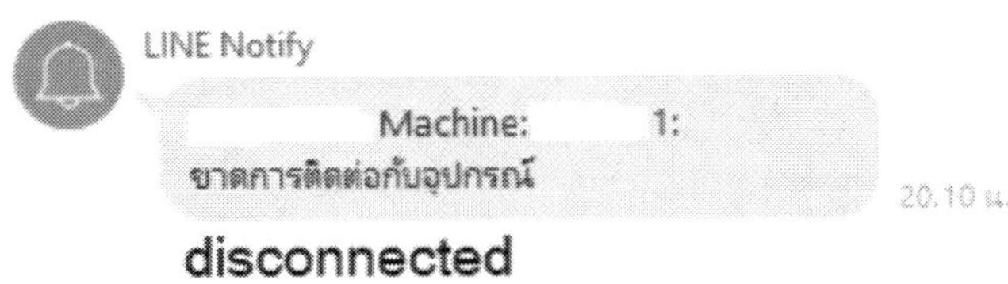

Figure 7. Alert and notification through Line Application.

6. Results and discussion

As the managing people and workers are looking at the same numbers. As the data is accurately logged with actual time of the events. The supervisor and workers know exactly of when and what and then be able to record why and how to solve each incident on-time. The following measures and policies have been proposed to further utilize the real-time and historical data:

- Worker must try best effort to fix the problem and prevent the problem. Since the incident can be seen the worker must act on the timely manner. Also the manager will be notified and take action right away. The planning department can adjust the plan accordingly without having to wait till the end of the day.
- Historical data can be used at regular meeting to monitor and improve setup time.
- Root because analysis can be used to find out the actual cause of the break down.

The direct and indirect outcome from this IoT implementation includes:

- There is no need to have a dedicated person to record and do quality control or record the data such as speed, quantity and wastes as the data can be obtained automatically. The person can help in other more important task such as improving availability and setup time.
- Manager, planner and worker can act on time.
- Historical data can be used to evaluate short-term and long-term performance and can be used to evaluate key performance index of workers and team.

Since speed and quality is a long-term improvement and will require better machine adjustment. The availability factor can be improved by better management of jobs scheduling and planning. Hence in this work we compare the availability before and after the improvement is imposed.

Table 1. Comparison between availability before and after the improvement.

	Before (minutes)	After (minutes)
Total Time	28800	28800
Idle Time	7200	4200
Availability	75%	86.85

The improvement is around 15% compared due the measures. Even though the IoTs monitoring system is not the direct cause of increase in the effectiveness, it is rather a tool that enables sharing and analysing as well as automating the process of gathering data. This frees up resources and allows continuous improvement based on the real-time and actual data.

Finally, the overall success in this demonstration is due to a couple reasons stated below:

- Dedication and support from business owner and management team
- Strong in-house software system that is ready to accept IoTs plug-ins
- Long-term process improvement and long-term IT development who is able to learn new technology as well as technician team who support PLC connectivity

Each IoT unit costs under 1,000 baht, hence, the investment in the project is consider low compared to off-the-shelf IoT solutions and may not be readily suitable to the business flow and may need some customization. This however does not limit the company to integrate in such system in the future.

7. Conclusion

In this paper, we have shown a case study of a printing manufacturer who adopted IoT technology for machine monitoring. The technology transfers and training is also done on-site along with the actual analysis and design while working as a team from the top down to bottom organization allows the adoption successful. The IoTs system is integrated into the main production management software. Usage of such data as a monitoring tool for process improvement resulted in 10-15% improvement of overall equipment effectiveness (OEE).

References

[1] Popkova, Elena G., Ragulina, Yulia V., Bogoviz, Aleksei V 2019 Fundamental Differences of Transition to Industry 4.0 from Previous Industrial Revolutions, Industry 4.0 *Industrial Revolution of the 21st Century* pp 21-29

[2] Hugo K., Markus H., Mario K., Christian R. 2017 Transition towards an Industry 4.0 State of the LeanLab at Graz University of Technology, *Procedia Manufacturing* vol 9 pp 206-213

[3] C. Baur, D. Wee, Manufacturing's next act, McKinsey&Company 2015 Available from: www.mckinsey.com/business-functions/operations/our-insights/manufacturings-next-act

[4] Shohin A., Xun X., Yuqian L., Mauricio A., Juan Pablo V., Benjamin Joa, Yesid V. 2020 IoT-enabled smart appliances under industry 4.0: A case study *Advanced Engineering Informatics* vol 43

Acknowledgments

Authors wishing to acknowledge assistance or encouragement from company, colleagues, special work by technical staff or financial support from TSE and DIP.

Identification of Lime Leaf Diseases with Deep Learning Technique on Android Smartphone

N Thammachot[1*], P Bootkote[1], S Thanimkarn[1] and C Preecha[2]

[1]Curriculum of Agricultural Machinery Technology, Faculty of Agriculture, Rajamangala University of technology Srivijaya, Nakhon Si Thammarat, 80110, Thailand

[2]Curriculum of Plant Science, Faculty of Agriculture, Rajamangala University of technology Srivijaya, Nakhon Si Thammarat, 80110, Thailand

Email address: nasaporn.t@rmutsv.ac.th

Abstract. Leaf plant diseases are difficult to recognize because the symptoms of the disease appear very similar causing farmers to solve problems that don't match the symptoms of the disease. This research presents the creation of a smartphone application for the identification of lime leaf diseases with deep learning technique. This uses 2 sets image data classified by plant disease specialist, 400 images for algorithm training and 100 images for algorithm testing for each disease. Algorithm development using Convolutional Neural Network that imitates the working of human brain neurons. This uses the supervised learning network type. The algorithm was developed based on the TensorFlow Framwork, Google's open source library, and was used by Android Studio software to build smartphone applications in the Android operating system. For use, the algorithm that has been learned through the process of learning is embedded into the developed smartphone application. Then it was tested the application accuracy. The results of a smartphone application in the identification of lime leaf diseases found more than 90% accurate.

1. Introduction

Lime was an economic crop growing almost all over the country. In 2018, 108,779 rai of perennial area, 107,634 rai of a yield area, 152,335 tons of a yield and 1,415 kilograms of a yield per rai. The information of each region was as shown in table 1 [1]. Lime fell into citrus genus that was very useful and valuable because it could be used to cook, to make a beverage and had medicinal properties. Also, limes were high in vitamin C, so limes were popular as a cosmetic ingredient, making it important and needy throughout the year. Not just only the summer every year that affected the production of lime yields, diseases and pests of limes also greatly did. When considering some diseases of limes that appeared on leaves, stems, branches and fruits were very similar, the agriculturists then used the wrong method of treatment. Therefore, the diseases still remained, then causing the epidemic.

Table 1. Lemon planting data and yield for 2018 [1]

	Perennial Area (rai)	Yield area (rai)	Yield (trees)	Yield per rai (kg)
Northern Region	28,124	27,878	38,438	1,379
Northeastern Region	1,872	1,764	459	260
Central Region	71,996	71,342	109,980	1,542
Southern Region	6,787	6,650	3,458	520
Total	108,779	107,634	152,335	1,415

Today, Artificial Neural Network, which imitated the function of the human brain, was widely used specifically for identification, such as for plants [2], fruits [3-4] and plant diseases [5-7]. The results were that in the identification of plant diseases, most of them used effective neural networks and longer learning, so-called deep learning. Deep learning is utilizing neural networks with multiple layers and a large amount of data than traditional ML algorithms. Deep learning can be used to every type of data such as image, text, etc. The popular deep neural network is the Convolutional Neural Network (CNN) [7]. And by comparing CNN's different architectures, such as LeNet, AlexNet, GoogLeNet, and VGG to identify plant diseases, finding that it was more than 90% accurate [6]. CNN was also developed as an application on smartphones to identify plant diseases, such as grape disease [8-9], wheat diseases [5] and citrus disease [10]. However, no development of smartphone applications for lime foliar and citrus disease identification using the CNN was found. Therefore, the researchers appreciated the effectiveness of the algorithm which was developed by using CNN towards lime foliar disease. If the algorithm was used to create an application on a smartphone, it would be a great benefit to lime and citrus agriculturists.

2. Convolutional Neural Network

The strengths of CNN network are the capability of both feature extraction and classification of images, which is unlike the conventional machine learning methods that can only classify and group data. The CNN network consists of convolution layers, max pooling layers, a fully connected layer or a hidden layer, and a classification layer shown in Figure 1. The details are as follows.

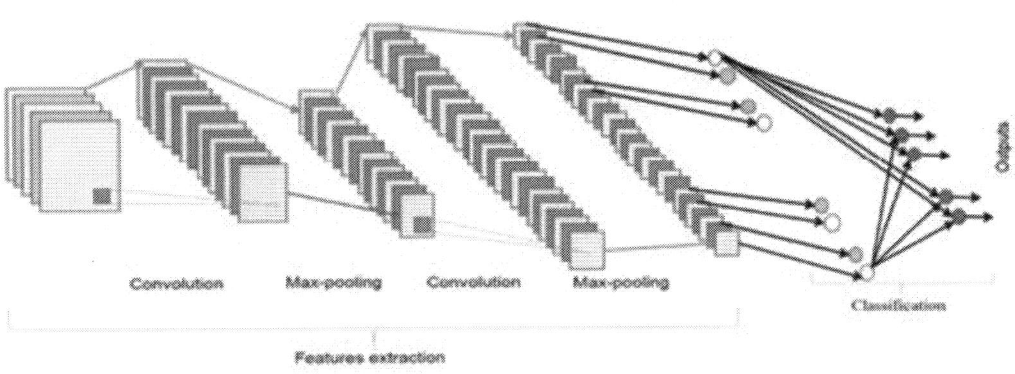

Figure 1. Convolutional Neural Network [11]

(1) Convolutional layer is the result finding of the neuron connected from the local region of the image by reducing the size of the image to a square with the value of the dimension of the image equal to 3 in case of RGB image. The calculation of the convolution layer would make through the sub-region, which is calculated as dot product and kernel. Kernel must be smaller than the image size. The result obtained from the convolution is called Feature Map as shown in Figure 2.

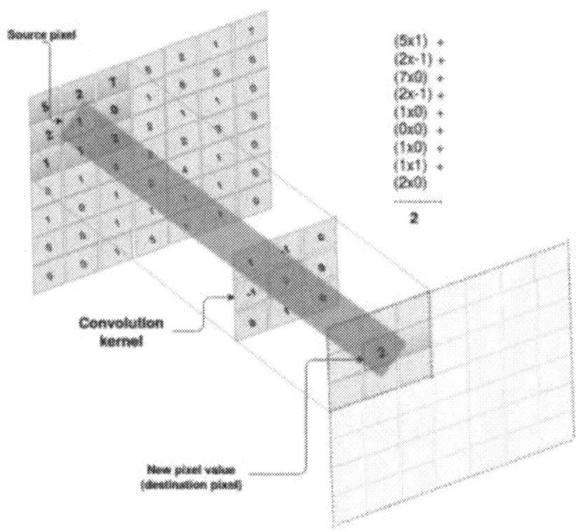

Figure 2. The illustration of the convolutional Layer. The dot production calculation between sub-region of the image and convolution kernel. [2]

(2) Max pooling layer is the intermediate layer between the convolution layers to downsample the feature map. It is divided into local regions, then local regions are divided into p x p / p which are set to be between 2 and 5. Therefore, the largest value in each pool will be chosen as a representative. Then, the system will stride to the local region continuously, for example, stride forwards with the rate of 2 pixels of each move, until meeting the last point of feature map. The calculation of max pooling is shown in Figure 3.

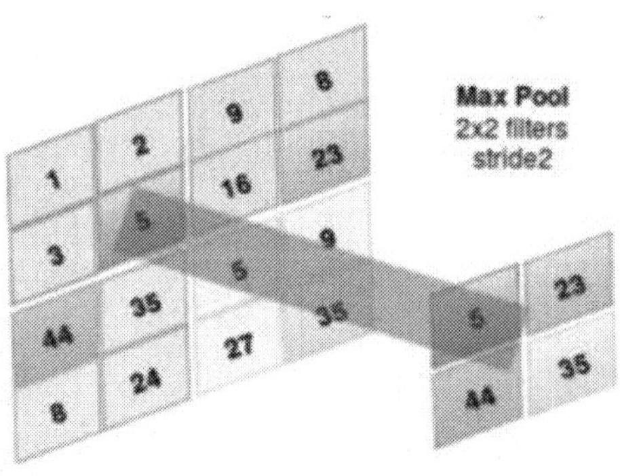

Figure 3. The illustration of max pooling with 2x2 kernel and stride 2. [2]

(3) Fully connected layer is used to classify objects. Every neuron in this layer is completely connected with convolution and pooling layers, also known as the structure of the neural network, in which the result number is the number of groups to be classified.

Inception-v3 was developed from GoogLeNet, the first version of inception deep convolutional architecture, after which it was developed to be Inception-v2, which was about the batch normalization, while Inception-v3 was about factorization ideas which were aimed at reducing the number of connections or parameters without degrading network performance. The details are as follows.

(1) Factorization into smaller convolutions is a replacement of 5 × 5 convolution with 3 × 3 convolutions of 2 layers, resulting in a decrease number of parameters from 25 (5x5) to 18 ((3x3) + (3x3)) parameters. The parameter is reduced by 28% as shown in Figure 4 and the inception module is as shown in Figure 5.

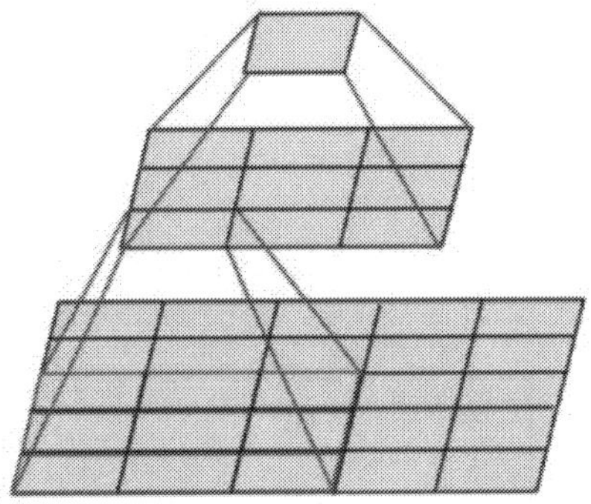

Figure 4. Mini-network replacing the 5× 5convolutions. [12]

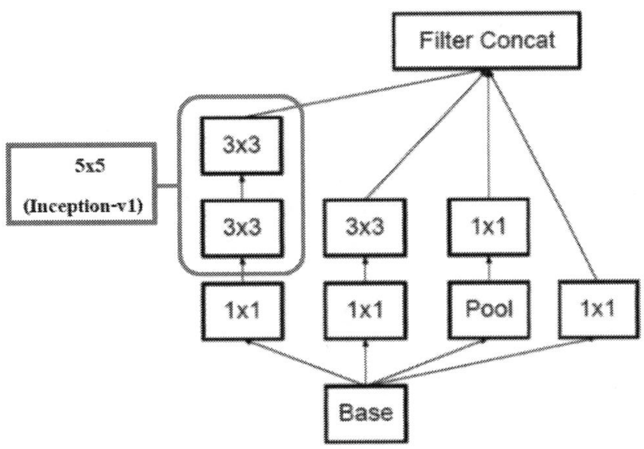

Figure 5. Inception modules where each 5 × 5 convolution is replaced by two 3 × 3 convolution. [12]

(2) Factorization into asymmetric convolutions is the replacement of 3 × 3 convolution with 3 × 1 convolution followed by 1 × 3 convolution, resulting in a 33% decrease in the number of parameters shown in Figure 6. According to the theory, the argument towards the replacement with any n × n convolution by a 1 × n convolution would happen and to be followed by a n × 1 convolution and the computational cost saving increases dramatically as n grows. Inception module was shown in Figure 7.

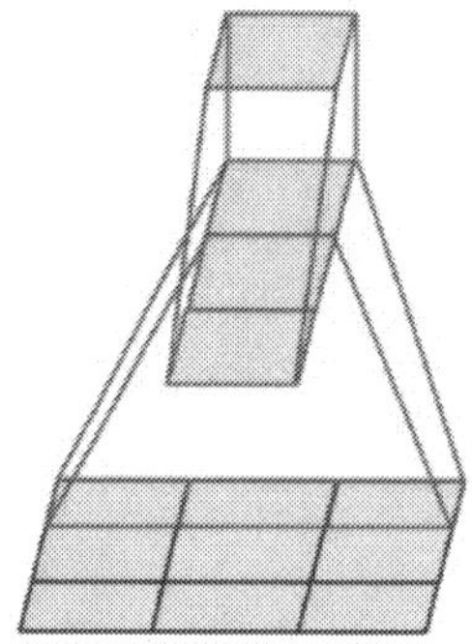

Figure 6. Mini-network replacing the 3 × 3 convolutions. The lower layer of this network consists of a 3× 1 convolution with 3 output units. [12]

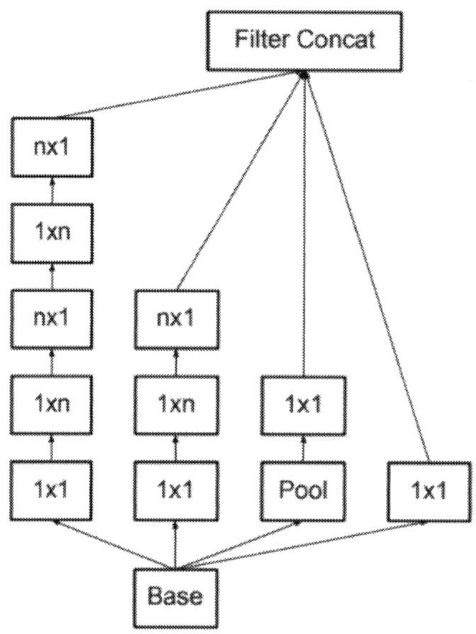

Figure 7. Inception modules after the factorization of the n × n convolutions. [12]

3. Materials and Methods

3.1. Study the data of lime foliar disease, algorithms development, and applications

Explore the lime foliar disease in Nakhon Si Thammarat area to be collected as visual data to develop algorithms for plant disease identification. The images obtained first went through disease classification by plant disease experts. The expertly classified images were divided into 2 data sets: 400 images for algorithm training and 100 images for algorithm testing for each disease.

3.2. Algorithm development

The algorithm was developed based on the TensorFlow Framework, Google's open source library. The development of the algorithm includes the following process: 1) preprocessing by resizing RBG image to 224x224 and converting the visual data into numerical data in the range of 1 to 225 to 3 layers of data including R, G, and B; 2) model creation by Convolutional Neural Network (CNN) by using Inception-v3 architecture because it gave the least error value when comparing with other architectures (12) and the 120 training epochs, the classification output of 6 classes by the number of diseases were all assigned; 3) model evaluation to obtain the training accuracy and the validation accuracy of the test.

3.3. Creation of smartphone application

Creation of smartphone application in the Android operating system using Android Studio software for the application, embedding learning algorithms into smartphone applications. Test applications on smartphones to determine the accuracy and error of identification with 50 images for each disease.

4. Results and Discussion

4.1. The results of the study of the lime disease

From the study, the lime foliar diseases, where were caused by bacteria, such as canker and greening; by fungi, such as greasy spot, sooty mold, root and foot rot; by algae, such as algal spot or red rust; by virus, such as tristeza; and by nutrient deficiency, such as zinc deficiency. The details of the disease are as follows.

4.1.1 Canker disease was caused by Xanthomonas axonopodis pv. citri bacteria. It could occur on the leaves, branches, stems, and fruits of lime. The first staged symptoms were a juicy clear pin-sized circular spot. Then, it got larger and was swollen like sponges with light yellow color. Later, it turned to be a convex with dark brown color, then it was scabbed and crusty like ringworms. The center of lesion was dented and had a pale-yellow halo surrounding the lesion on the leaves. The scabs on the fruits might be larger than on the leaves and might have a rubber flowing out of the scabs, and might cause the fruits to break transversely, starting from the edge of scabs when being exposed to water or heavy rain, and the fruits tent to fall off as it got aged [13-16].

4.1.2 Zinc deficiency disease was caused by zinc deficiency. The first staged symptoms happened with the young leaves. The veins and midribs of the leaves turned yellow and then the symptoms were more obvious as if the veins and the midribs of the leaves were green on the laminas or the yellowish leaves. Severe deficiency would make the young leaves smaller, and the tip of the leaves would be slender and pointed up [14, 17, 18].

4.1.3 Algal spot or red rust disease was caused by Cephaleuros virescens algae. The first staged symptoms were grayish-green spots, with velvet-like fine lines of 0.3-0.5 centimeters, and then the spots turned orange or rusty iron colors. At the stage where the algae produced fully ripe spores, the scabs looked like orange velvet. The symptoms of the branches were similar to those of the leaves, but if there were severe symptoms, the bark would crack. Dry scabs were similar to Canker disease caused by bacteria [16].

4.1.4 Tristeza disease was caused by the Citrus Tristes Virus (CTV). Young leaves turned pale green or alkaline, resembling a mild lack of nutrients. The leaves show a short, transparent condition, the leaves are small, the leaves are usually yellow, pale, or unevenly green. New splinters or dwindling branches often dry to death from the tip of the branch. It's very fruitful, but it's often easy to fall off. The effect is small. The area of the trunk or large branches is not smooth, similar to the trunk or branches, twisting into waves or grooves, many of which are long parallel along the trunk or branches. Diseased trees often grow slower than normal trees, often shabby and eventually die [14].

4.1.5 Greasy spot disease was caused by the Phomopsis citri fungus, which started destroying the leaves from the young age. The characteristics of the disease were small clear spots under the laminas and then the spots became yellow and green or brown convex and then became bigger brown or black convex spreading all over the leaves. Oily and non-irritation to the hands, feeling like stain of the saliva mixed with betel, spat by betel chewers, were the present characteristics of the convex. The diseased leaves turned yellow and fell off before they should [14, 16]

4.1.6 Sooty mold was caused by the Capnodium citricola and Meliola spp. fungus. It could occur on leaves, branches and fruits. It was found to be a stain or black flaky of black fungus. When rubbing or scraping, it came off as dirty flakes, causing the green part of the plant was not able to get sunlight for normal photosynthesis. For large branches, it was often found the cracks and rubber flowing [16-18]

4.1.7 Greening disease was caused by fastidious bacteria. The symptoms were that the young leaves turned yellow with green vein, which was like the leaves having zinc deficiency. The leaves were smaller, narrower, pointed up, and yellow or green with rash all over the leaves. New ruptures were reduced and dry ingress from the tip of the branches were found. The size of the fruits was smaller and the fruits were easy to fall off. The shell color was uneven when the fruits were ripe. The root system was not strong. An unusual incident of less branching of the capillary root was detected. Diseased trees often grew slowly and were susceptible to other diseases, shabby and eventually dying [14, 16].

4.1.8 Root and foot rot were caused by Phytophthora parasitica fungi and might be caused by water trapping. The foliar symptoms were the burning leaves, both in young to fully grown leaves, and dark brown spots were found around the centers and the edges of the leaves. The tips of the leaves were brown and juicy. The scabs were quickly enlarged. The young leaves got burnt and fell off [16, 18].

4.2 Image of the lime foliar disease

From the field trip to Nakhon Si Thammarat Province in order to collect the images of lime disease, with adequate images for the algorithm development it was found the following six most common foliar diseases; (a) canker, (b) zinc deficiency, (c) tristeza, (d) algal spot or red rust, (e) greasy spot, and (f) sooty mold. The disease symptoms were as shown in figure 8.

Figure 8. Examples of symptoms of the lime foliar disease.

IOP Conf. Series: Materials Science and Engineering 1163 (2021) 012003 doi:10.1088/1757-899X/1163/1/012003

4.3Results of algorithmic development

According to Figure 9, the training of the algorithm since epoch fourth provided an accuracy value of 91.89-100% and the validation accuracy value was 75.66-90.42 %. For the number of algorithm training, it should not be less than 20 epochs (Figure 10).

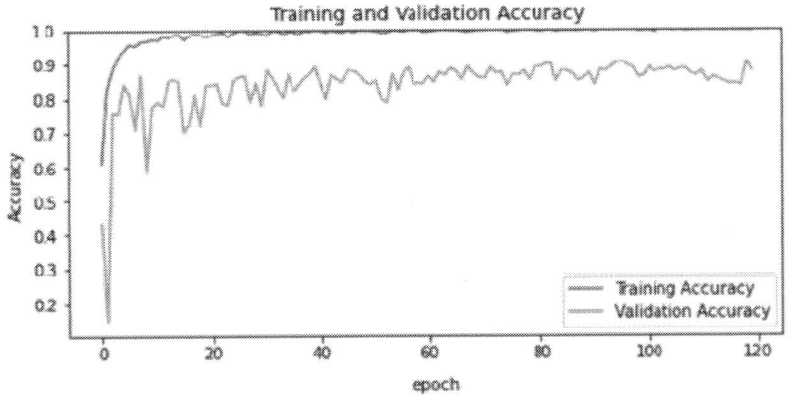

Figure 9. Training and validation accuracy.

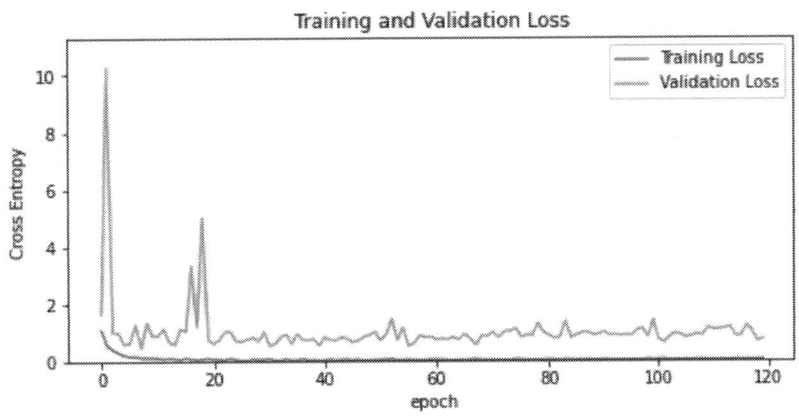

Figure 10. Training and validation loss.

4.4 Results of smartphone application creation in Android operating system.

Developed applications could be installed on Android smartphones. To start, activate the application on the smartphone, then the screen will be displayed as shown in figure 11, where the application would run in real time. Activate your smartphone's camera when the application was opened, and then used the smartphone's camera to the lime leaves to acknowledge the detectable disease. The screen displayed the disease by the percentage that the camera could capture in descending order.

4.5 Results of accuracy in the trial of the developed lime disease identification application.

From the trials using 6 lime disease classification applications with 50 images of each disease, it was found that the accuracy of classification was in the range of 94-98 percent and the mean was 96 percent. Table 2 presents the details.

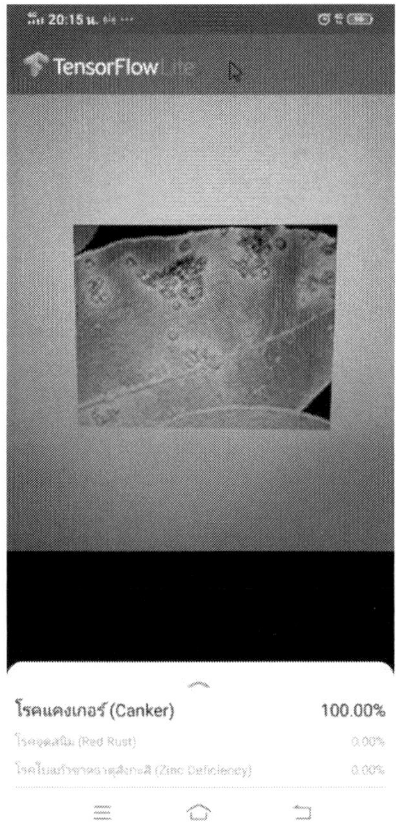

Figure 11. The main screen of the plant disease application.

Table 2. The accurate results in the trial of the developed lime disease classification application.

Disease	Number of images (>80 %)	Number of images (<80 %)	%Accuracy	% Loss
1) Canker	49	1	98.00	2.00
2) Zinc deficiency	47	3	94.00	6.00
3) Tristeza	47	3	94.00	6.00
4) Algal spot or Red rust	49	1	98.00	2.00
5) Greasy spot	48	2	96.00	4.00
6) Sooty mold	48	2	96.00	4.00
Average	48	2	96.00	4.00

5. Conclusions

The algorithm of lime disease identification was developed with the CNN, the Inception-v3 architecture from TensorFlow Framework command set provided a 91.89-100% train accuracy and a validation accuracy of 75.66-90.42%. The accuracy of validation depends on the number of training epochs. When the developed algorithm was used to create a smartphone application in an Android operating system by Android Studio software. The using results of the developed application was more than 90% accurate for all diseases. It can be seen that the developed algorithm is feasible for its improvement by increasing the number of the images or newly detected diseases. This application is very useful for agriculturists to prevent disease from being symptomatic.

References

[1] Office of Agricultural Economics 2018Information of agricultural products "Lime" Retrieved from http://www.oae.go.th/
[2] Sanuksan J and Surinta O 2018 Deep Convolutional Neural Networks for Plant Recognition in the Natural Environment *J Sci Technol MSU* **38**(2) pp 113-124
[3] Wang C, Tang Y, Zou X, SiTu W and Feng W 2017 A robust fruit image segmentation algorithm against varying illumination for vision system of fruit harvesting robot *Optik* **131** pp 626-631
[4] Dorj U O, Lee M and Yun S 2017 An yield estimation in citrus orchards via fruit detection and counting using image processing. *Computers and Electronics in Agriculture* **140** pp103–112
[5] Johannes A, Picon A, Alvarez-Gila A, Echazarra J, Rodriguez-Vaamonde S, Navajas A and Ortiz-Barredo A 2017 Automatic plant disease diagnosis using mobile capture devices, applied on a wheat use case. *Computers and Electronics in Agriculture* **138** pp 200-209
[6] Iqbal Z, Attique K M, Sharif M, Hussain S J, Habib ur Rehman M, and Javed K 2018 An automated detection and classification of citrus plant diseases using image processing techniques: A review. *Computers and Electronics in Agriculture* **153** pp 12-32
[7] Barbedo J G A 2019 Plant disease identification from individual lesions and spots using deep learning. *Biosystems Engineering* **180** pp 96107
[8] N. Petrellis 2017 Mobile Application for Plant Disease Classification Based on Symptom Signatures. *Proceedings of the 21st Pan-Hellenic Conference on Informatics* pp 1-6
[9] Suppatoomsin C and Srikaew A 2018 Imagery Grape Leaf Disease Diagnosis Based on a GA-SASOM Algorithm *The Journal of Industrial Technology* **14**(3) pp 44-61
[10] Petrellis N 2019 Plant Disease Diagnosis for Smart Phone Applications with Extensible Set of Diseases *Appl. Sci* **9** (9) p 1952
[11] Md. Z. Alom et al. 2019 A State-of-the-Art Survey on Deep Learning Theory and Architectures *Electronics* **8** p. 292
[12] Szegedy C, Vanhoucke V, Ioffe S, Shlens J, and Wojna Z 2015 Rethinking the Inception Architecture for Computer Vision
[13] Kositcharoenkul N 2008 Canker disease of citrus plants. Academic documents. *Plant Protection Research and Development Office, Department of Agriculture Bangkok* 82pages
[14] Bhirnuwat A, Thaweechai N, Meerink P, Ratriyat W and Jamsawang J 1999 Regulation of Orange Mite Insect Disease and Management 2nd edition *Department of Plant Pathology Kasetsart University*
[15] Siwakorn N, Puangphaet B, and Songsang P n.d. Using medicinal plants to control grapefruit canker *Plant Pathology Research Group Plant Protection Research and Development Office*
[16] Na Nakorn S, Chaisitpreecha N, Thonglua T, and Intamanee W 2014Production of Pomegranate Siam Pomelo in Pak Phanang River Basin Nakhon Si Thammarat Province *Tropical Fruits and Perennials Research Unit Faculty of Agriculture Rajamangala University of Technology Srivijaya Nakhon Si Thammarat Campus Sai Yai. Somthawee Printing House Nakhon Si Thammarat Province* 62pages
[17] Maneenetr N n.d. Fruit disease and prevention and eradication. Community Agricultural Book Project *Ruengsang printing* Bangkok 72 pages
[18] Chaiprasart P n.d. Lemon Planting "Restoring and Healing Victims by Researching NRCT" National Research Council of Thailand (NRCT)

Acknowledgments
This research was supported by a research scholarship of revenue budget in 2020 of Rajamangala University of Technology Srivijaya.

Effect of pour point depressants and diluents on exergy destruction during pipeline transportation of crude oil

A D Kulkarni[1] and K S Wani[2]

[1] School of Chemical Engineering, Dr.Vishwanath Karad MIT World Peace University, Pune, Maharashtra, India
[2] Department of Chemical Engineering, SSBT College of Engineering and Technology, Bambhori, Jalgaon, Maharashtra, India

E-mail: anand.kulkarni@mitwpu.edu.in

Abstract. Exergy destruction in pipeline flow of crude oils mixed with pour point depressants (PPD) and diluents has been investigated. Ethylene vinyl acetate with vinyl acetate content of 18% and 28% (EVA-18 and EVA-28) and Polymethyl methacrylate (PMMA) with concentrations of 200 ppm, 1000 ppm and 1500 ppm were chosen as the PPDs for this purpose. Four different diluents were considered viz. *n*-hexane, cyclohexane, *n*-butanol and toluene representing paraffins, naphthenes, alcohols and aromatics respectively. The viscosity reduction was found out experimentally. Using these results for a hypothetical pipeline, the effect of API gravity of crude oil, pipeline diameter, Reynolds number, PPD concentration and diluent type on exergy destruction was investigated. These findings would help in making a prudent choice of PPD and diluents for efficient transportation of crude oil through pipelines.

1. Introduction

Pumping of viscous crude oil for long distance transportation results in major energy consumption [1]. It leads to high operating costs and hence overall economic loss. Pumping process is highly irreversible. Detailed second law analysis of this process is highly essential to identify the irreversibilities and reduce the same. These irreversibilities can be due to friction, heat transfer, concentration gradient, mixing or chemical composition. Such irreversibilities can be measured with the help of exergy. Exergy is the maximum amount of useful work that can be obtained in a given environment [2]. The more the irreversibility, greater is the reduction in the exergy and the more the exergy loss, higher is the consumption of energy that drives the process. Hence the exergy analysis of a process gives better insight into the energy consumption and subsequently the economic loss.

Exergy can be classified into physical and chemical exergy. The exergy due to heat, pressure and flow can be considered as physical exergy whereas the exergy of chemical reaction and mixing comes under chemical exergy. Physical exergy is a function of temperature, enthalpy, entropy and hydraulic resistance whereas the chemical exergy is mainly a function of the Net Heating Value (NHV) and the mole fraction of the components. In case of crude oil, these factors are more pronounced due to the changes in wax precipitation behaviour at different ambient conditions. Cheng [3] has studied the effect of wax deposition on physical and chemical exergy of a crude oil pipeline in China.

It has been found that the unavoidable exergy loss in transportation of crude oil is mainly dependent on various factors like the outbound temperature, outbound flow and outbound pressure. Pal [4] has studied the exergy destruction of a surfactant stabilized oil-in-water emulsion. Nyugen [5] has discussed component by component calculation method for more accurate prediction of physical exergies. Using the case study of Yamal-Europe gas pipeline, Chaczykowski [6] has done the exergy analysis of the gas transmission system. Govin [7] has developed the correlations for estimating the chemical exergies of fuels and petroleum fractions. Rivero [8] has developed a method to calculate the exergy of crude oil mixtures. Calculation of exergies is a tedious procedure especially for hydrocarbons. This has been addressed by Abdollahi-Demneh [9] and Montelongo-Luna [10] by developing procedures in process simulators and online tools for calculating the physical and chemical exergies.

Pour point depressants and diluents are generally used to reduce the apparent viscosity of crude oil during transportation through pipelines with lesser pumping power. Proper combination of crude oil and the PPD or diluent helps to reduce the exergy destruction during pumping and improves the economics of the overall transportation process. Literature survey suggests that the exergy analysis of pipelines carrying crude oil mixed with pour point depressants and diluents has not been studied in details. Hence an attempt has been made to perform the exergy analysis of such systems. This work studies the effect of various parameters like the pipeline diameter, Reynolds number, API gravity of crude oil, concentration of the PPDs and type of diluents on the physical exergy for crude oils mixed with PPDs and diluents.

2. Materials and Methods

Crude oils (C_1, C_2 and C_3) obtained from western parts of India were characterized for density at 15 °C, ^{0}API, wax content (%), asphaltene content (%), wax appearance temperature (WAT) (°C) and viscosity at WAT (N s/m^2) as per the standard procedures. The properties are shown in Table 1.

Table 1. Properties of crude oils used in the experimentation

	C_1	C_2	C_3	Method
Density (kg/m^3) at 15^{0}C	875.4	877.0	887.0	ASTM D1298
^{0}API	30.1	29.8	28	Formula
Asphaltene content (%)	12.2	5	7.2	IP 143, ASTM D 6560
Wax content (%)	25.1	35.5	45.7	ASTM UOP 46-85
Pour Point (^{0}C)	18	27	33	ASTM D-97
WAT (^{0}C)	30	43	50	Viscometry
Viscosity at WAT (N s/m^2)	0.0693	0.0554	0.0821	Brookfield DVII+ Pro Viscometer

Two types of pour point depressants (PPD) viz. linear and comb shaped were chosen for study. Ethylene Vinyl Acetate (EVA) was chosen as the linear type having 18% and 28% of vinyl acetate content. The comb type PPD was Poly Methyl Methacrylate (PMMA). The PPDs were dissolved in appropriate solvents viz. xylene for EVA and ethyl acetate for PMMA in required proportions such that their final concentrations in the crude oil were 200 ppm, 1000 ppm and 1500 ppm.

The given samples of crude oils were treated with four different commercially available diluents viz. *n*-hexane, cyclohexane, *n*-butanol and toluene. The concentration of diluents chosen for experimentation was a constant value of 10%.

The viscosities of the above solutions (volume = 50 ml each) were measured. These viscosities were used for the further calculations of exergy.

A hypothetical crude oil pipeline 200 km long was considered to transport 10000 barrels per day of crude oil in an ambient temperature of 15^{0}C. The above samples were considered to be transported through this pipeline. The diameters considered for the pipelines were 0.2032, 0.4064, 0.6096 and 0.7620 m (i.e 8, 16, 24 and 30 inches). The crude oil temperature was maintained at its WAT to avoid gelling and precipitation.

The flow exergy of a stream is a function of physical, chemical, kinetic and potential energy. The kinetic and potential components can be incorporated in the flow exergy calculations which includes pressure and temperature components. According to Chen [11], pressure exergy is a function of temperature and pressure only. They do not depend on the properties of crude oil. Since the flow in the above pipeline is considered at constant temperature maintained at WAT, thermal exergy is neglected. Hence the flow exergy considered here is the pressure exergy. This has been calculated using the following expressions derived by Pal [4] and the results have been discussed in the subsequent sections.

For laminar flow, the exergy destruction is given by

$$Ex_D = \frac{\pi}{2}\left(\frac{T_0}{T}\right)\left(\frac{\mu^3}{\rho^2 D^2}\right)(16\, Re^2) \quad (1)$$

And for turbulent flow, it is given by

$$Ex_D = \frac{\pi}{2}\left(\frac{T_0}{T}\right)\left(\frac{\mu^3}{\rho^2 D^2}\right)(0.079\, Re^{2.75}) \quad (2)$$

Where,
Ex_D = Exergy destructed (J), T_0 = Ambient temperature (K), T = Temperature of the crude oil (K), μ = Viscosity (N s/m^2), ρ = Density (kg/m^3), Re = Reynolds number

3. Results and Discussion

Tables 2 and 3 show the viscosities of crude oils treated with the PPDs at different concentrations and diluents with concentration 10%. These results were used for calculating the exergy using equations (1) and (2).

Table 2. Variation in viscosity due to pour point depressants for different crude oil samples

Crude Oil	PPD Concentration (ppm)	Viscosity (N s/m^2)		
		EVA-18	EVA-28	PMMA
C_1	0	0.0693	0.0693	0.0693
	200	0.0124	0.0143	0.0105
	1000	0.0119	0.0147	0.0095
	1500	0.0117	0.0192	0.0116
C_2	0	0.0554	0.0554	0.0554
	200	0.0182	0.0242	0.0105
	1000	0.0200	0.0356	0.0103
	1500	0.0224	0.0381	0.0102
C_3	0	0.0821	0.0821	0.0821
	200	0.0191	0.0225	0.0137
	1000	0.0194	0.0236	0.0138
	1500	0.0198	0.0251	0.0145

3.1. Effect of pipeline diameter and Reynolds number on exergy destruction

From Figures 1, 2 and 3 it is seen that the larger the pipeline diameter, the lesser is the exergy destruction and it decreases irrespective of the crude oil sample. Also, as the Reynolds number increases, the exergy destruction increases. The lesser diameter pipe has higher Reynolds number and hence the highest exergy destruction. This will result in greater pumping cost.

Table 3. Variation in viscosity due to diluents (10% concentration) for different crude oil samples

	C_1	C_2	C_3
Without diluent	0.0693	0.0554	0.0821
n-Hexane	0.0284	0.0163	0.0225
Cyclohexane	0.0321	0.0180	0.0229
n-Butanol	0.0289	0.0224	0.0323
Toluene	0.0252	0.0159	0.0207

Chen [11] had observed that pressure exergy destruction increases with increase in pressure. It was also observed that the greater the flow, more is the exergy loss. Chen [1] had varied the pipe diameter from 0.168 m to 0.323 m and found that the pressure exergy destructed decreased from 33 to 6 kJ/sec. In the present work, when the pipe diameter is varied from 0.2032 m to 0.7620 m, the exergy destructed decreases from 1.21 kJ to 0.006 kJ. As the flow rate of crude oil is increased, the exergy destructed first decreases and then increases. Thus, the present work yields the similar results as reported by Chen [1], [11]. The greater the irreversibility, higher is the exergy loss.

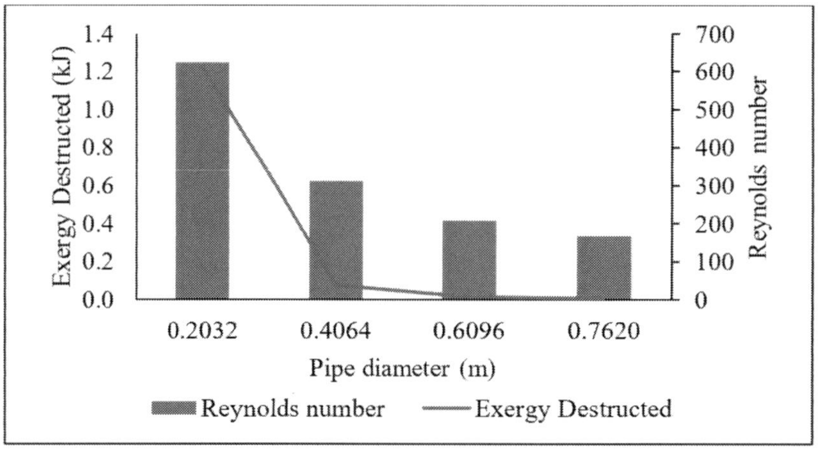

Figure 1. Exergy and Reynolds number vs Pipe diameter for C_1

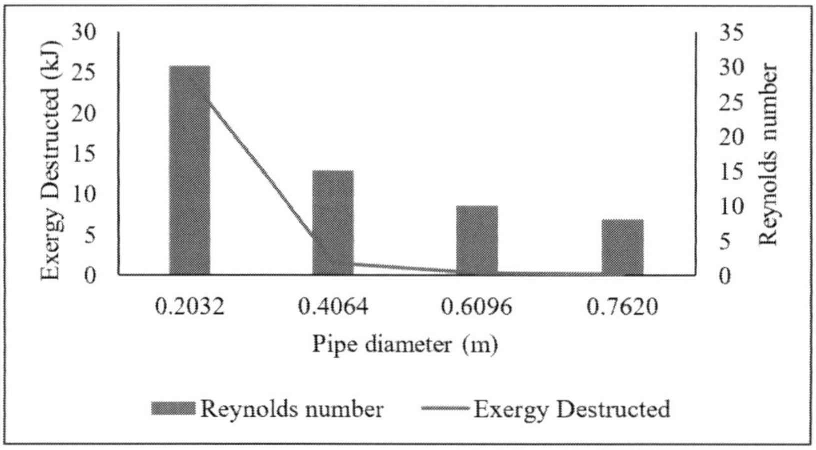

Figure 2. Exergy and Reynolds number vs Pipe diameter for C_2

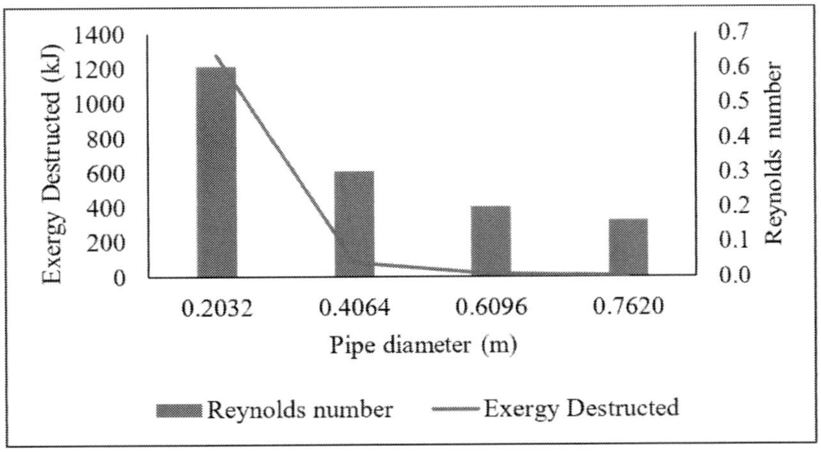

Figure 3. Exergy and Reynolds number vs Pipe diameter for C_3

3.2 *Effect of API gravity on exergy destruction at different pipeline diameters*

Figure 4 shows the variation in the exergy destruction with the API gravity. The exergy destruction increases with decrease in the API gravity. Heavier oil will be more viscous and will generate more pressure drop. This leads to more exergy destruction. Also, as the diameter decreases pressure drop increases further and hence the exergy loss. Thus, a larger diameter pipeline is economical for heavier crude and vice versa. Hence the type of oil is a major factor in determination of exergy destruction.

3.3 *Effect of PPD concentration on exergy destruction*

Figures 5, 6 and 7 show the variation in exergy destruction with changes in PPD concentration for the crude oils C_1, C_2 and C_3. For a given crude oil the exergy destruction decreases with increase in the PPD concentration. But it does not have a linear trend. The exergy destructed decreases with increasing PPD concentration to an optimum value and then starts increasing. The minimum exergy destruction is found to be at around 400 ppm concentration of the PPD. Also, this minimum is more pronounced for PMMA as compared to EVA-18 or EVA-28. The reason for this is as follows. As seen from equations(1) and (2), the exergy destruction is directly proportional to viscosity. It has been seen that reduction in viscosity is more for comb-shaped PPDs like PMMA as compared to linear PPDs like EVA [12-13]. The structure of copolymers, especially the proportion of polar and nonpolar segments, plays an essential role in affecting the rheological properties of crude oil [14]. It has also been found that lower concentrations of PPDs are more effective in reducing the viscosity. Hence, the lower the concentration, better is the viscosity reduction. This is due to the fact that the PPDs associate themselves with wax crystals. An increase in the concentration of PPDs results in a larger network with wax thereby reducing the flow ability [15]. Further the wax to asphaltene ratios for C_1, C_2 and C_3 are 2.05, 7.1 and 6.34 respectively. It is observed that the more is the wax to asphaltene ratio, the more is the exergy destruction. This is in line with the observations of Chen [10] who had observed that waxy crude oil contributed to more exergy loss. This has been attributed to the changes in chemical exergy which results from the wax precipitation. Wax precipitation leads to changes in density, viscosity and heat capacity which in turn changes the chemical exergy.

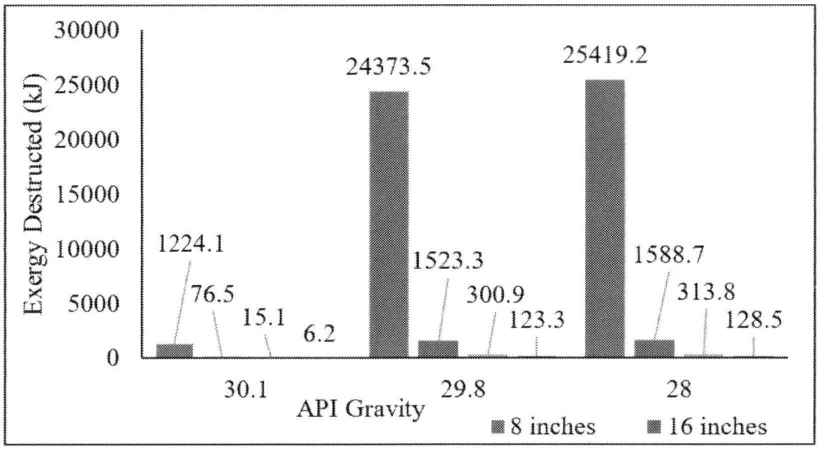

Figure 4. Exergy Destruction vs API gravity for various pipe diameters

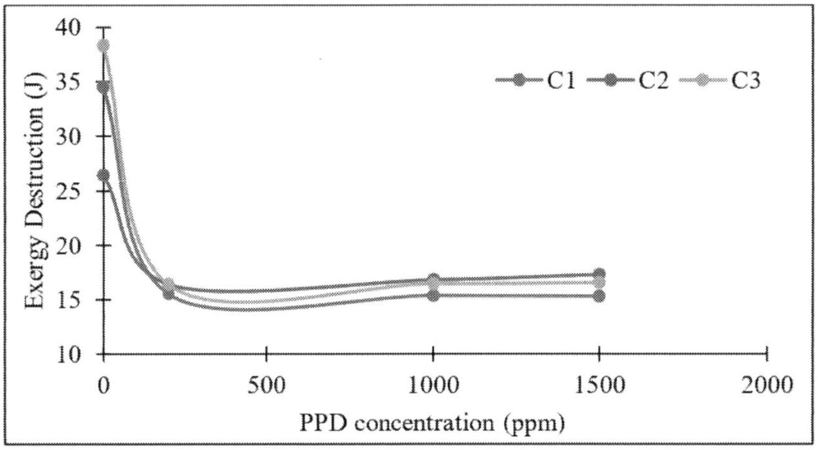

Figure 5. Exergy Destruction vs PPD concentration (EVA-18) for different crude oils

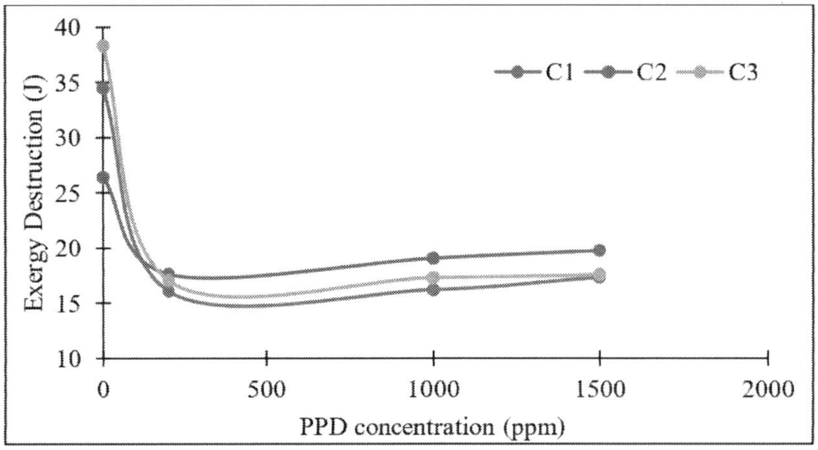

Figure 6. Exergy Destruction vs PPD concentration (EVA-28) for different crude oils

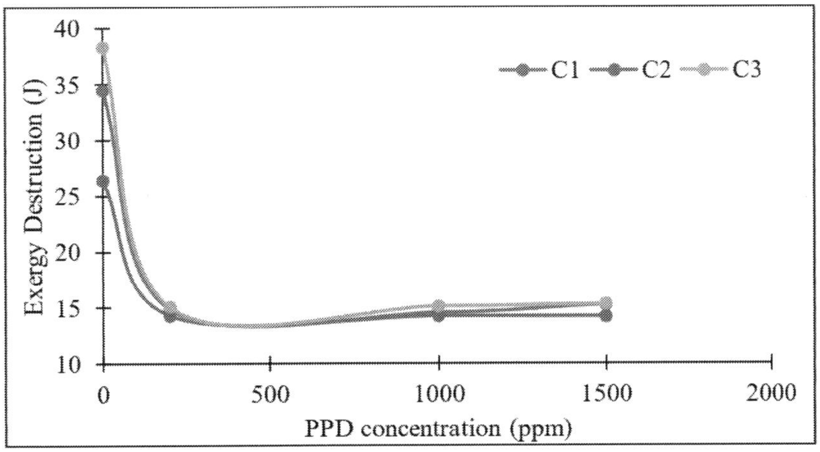

Figure 7. Exergy Destruction vs PPD concentration (PMMA) for different crude oils

3.4 Effect of type of diluent on exergy destruction

Figure 8 shows the exergy destruction for crude oils C_1, C_2 and C_3 mixed with different diluents. It was found that the exergy destruction for toluene was the least. The trend followed with respect to exergy destruction in increasing order was

Toluene < *n*-hexane < Cyclohexane < *n*-butanol.

The reduction in viscosity has been observed in the following order.

Toluene > *n*-hexane > Cyclohexane > *n*-butanol.

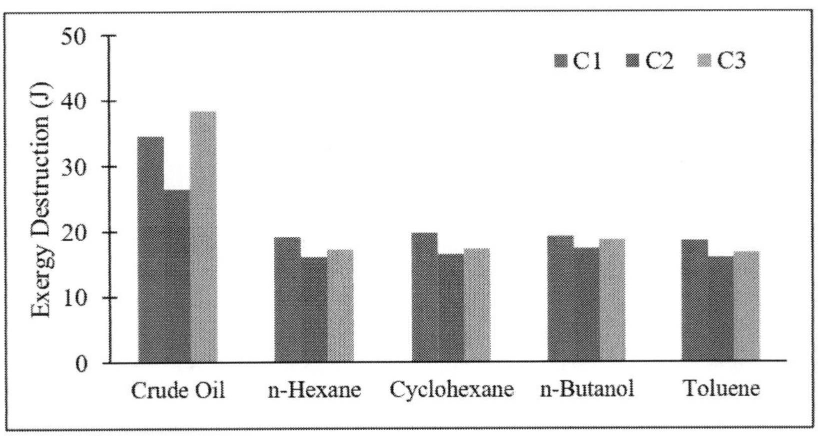

Figure 8. Exergy Destruction for different diluents for different crude oils

The variation of viscosity due to addition of diluents to oil depends on the chemical interactions between the diluents and the oil constituents. These interactions depend on the Hansen Solubility Parameter of the diluents [16]. Hansen has suggested that the solubility parameter comprises of three types of interactions viz. (a) Dispersion (London) forces which correspond to the van der Waals interactions (δ_D) (b) Polar interactions due to charged species (δ_P) (c) Hydrogen bonding (δ_H). The Hansen parameters for the diluents used in the experimentation have been given in Table 4. Toluene has the highest Hansen solubility parameter for dispersion as compared to the other diluents. Moreover, it's polar as well as the hydrogen bonding parameter have moderate values. In case of *n*-hexane, the Hansen parameters representing polar and hydrogen bonding are zero. This prevents its interaction with the polar components of crude oil. Also, no hydrogen bonding is possible. Hence the viscosity reduction takes place by simple dilution effect. The presence of Hansen hydrogen bonding parameter in cyclohexane contributes to the increase in the hydrogen bonding. This results in lesser viscosity reduction than *n*-hexane. Also, the absence of polar part of the Hansen parameter in it, reduces the viscosity reduction as compared to toluene. *n*-butanol gives the least viscosity reduction. Gateau [17] has found that the large hydrogen bonding parameter of alcohols is responsible for their inability to sufficiently reduce the viscosity of crude oil. Hence, amongst all the diluents, toluene contributes to highest viscosity reduction and subsequently less exergy destruction.

Table 4. Hansen Solubility Parameters for the diluents

Diluent	δ_D(MPa$^{1/2}$)	δ_P (MPa$^{1/2}$)	δ_H(MPa$^{1/2}$)
n-hexane	14.9	0.0	0.0
Cyclohexane	16.8	0.0	0.2
n-butanol	16.0	5.7	15.8
Toluene	18.0	1.4	2.0

4. Conclusion

The exergy destruction of crude oil transportation when mixed with pour point depressants and diluents has been studied. It is found that the exergy destruction decreases as the crude oil viscosity decreases. The chemical composition of crude oils especially the wax content and the asphaltene content as well as the type of PPDs and the diluents affect the overall exergy destruction. A wise choice of the PPDs and diluents will give better viscosity reduction and will help in reducing the pumping power and the economics of the transportation process. The present study has not taken into account the chemical exergies of the crude oil, PPDs and diluents. More detailed study involving the mixing effect of these along with the physical exergy needs to be undertaken to get a better insight into the overall exergy destruction.

References

[1] Cheng Q, Yang J, Zheng A et al. 2019 *Entropy* **28** 1–18
[2] Dincer I and Cengel Y 2001 *Entropy* **3** 116–149
[3] Cheng Q, Zheng A, Yang L et al. 2018 *Case Studies in Thermal Engineering* **12** 8–15
[4] Pal R 2014 *Energies* **7** 7602–7612
[5] Nguyen T V, Voldsund M, Elmegaard B et al. 2014 *Energy* **73** 264–281
[6] Chaczykowski M, Osiadacz A J and Uilhoorn F E 2011 *Applied Energy* **88** 2219–2230
[7] Govin O V, Kabo D V V, G and A B 2000 *J. Thermal Analysis and Calorimetry* **62** 123–133
[8] Rivero R, Rendon C and Leodagario M 1999 *International J. of Thermodynamics* **2** 115–123
[9] Abdollahi-Demneh F, Moosavian M A, Omidkhah M R and Bahmanyar H 2011 *Energy* **36** 5320-5327
[10] Montelongo-Luna J M, Svrcek W Y and Young B R 2007 *Asia-Pacific J. of Chem. Eng.* **2** 431–437
[11] Cheng Q, Gan Y, Su W et al. 2017 *Energies* **10** 1956–1975
[12] Chi Y, Daraboina N and Sarica C 2016 *AIChE J.* **62** 4131–4139
[13] Daraboina N, Soedarmo A and Sarica C 2016 *Offshore Technol. Conf.,* 2–5
[14] Xu J, Jiang H, Li T, Wei X, Wang T et al .2015 *Ind. Eng. Chem. Res* **54** 5204–5212
[15] Soni H P, Bharambe D P and Nagar A 2005 *Indian J. of Chem. Technol.* **12** 55–61
[16] Hansen C M 2007 *Hansen Solubility Parameters: A User's Handbook* II ed (CRC Press)
[17] Gateau P, Henaut I, Barre L and Argillier J F 2004 *Oil Gas Sci. Techol.* **59** 503-509

Acknowledgements

The authors wish to acknowledge the support extended by the management and staff of SSBT College of Engineering and Technology and Dr. Vishwanath Karad MIT World Peace University.

IOP Conf. Series: Materials Science and Engineering 1163 (2021) 012005 doi:10.1088/1757-899X/1163/1/012005

Physiochemical properties of vacuum fried durian (*Durio Zibethinus Murr.*) chips combined with microwave heating

C Thongcharoenpipat and R Yamsaengsung*

Department of Chemical Engineering, Faculty of Engineering, Prince of Songkla University, Hatyai, Songkla 90112, Thailand

*Corresponding author's e-mail: ram.y@psu.ac.th

Abstract. This research evaluated the effects of oil temperature, ripeness of durian (*Durio zibethinus Murr.*), and microwave combined heating (MCH) on the physiochemical properties of vacuum fried durian chips. The oil temperature was varied from 90, 100, 110, and 120°C, while the ripeness was determined in term of total sugar (base on day of ripening). In addition, the effect of pretreatment of durian chips by freezing overnight -18°C before vacuum frying (VF) was also investigated.

From this study, it was found that VF at 120°C with the ripeness of day 3 gave the lowest shrinkage, and using MCH produced a higher rate of moisture loss compared to the condition without MCH. Even though it was found that the ripeness of durian lead to more oil absorption, the addition of MCH resulted in decreasing the permeability of oil into the product. Finally, an increase in the frying temperature lead to a slight significant increase in color change ($p<0.05$), while the microwave power intensity did not have a significant effect on the color change ($p<0.05$).

1. Introduction

Durian (Durio zibethinus Murr.) is an attractive and high economic valuable tropical fruit grown widely in South-East Asia [1]. Due to its high nutrients and uniquely unpleasant odor, Durian is known as the "king of tropical fruit" [1]. One of the most important problem occurs in durian season, where there is excessive harvesting which floods the market making it impossible to sell all ripened durians even with high consumer demands. Therefore, it is more economical to process durian in order to extend its shelf-life and add value to the product. Presently, there are several methods including drying and deep-fat frying. Due to its high temperature and oil uptake, which detract the color and nutritive values of the product, alternative methods should be considered.

Drying of fruits and vegetables have been favored for centuries around the world. Not only can they be stored for a long period of time, but dried foods are convenient for carrying and appetizing to consumers. Moreover, drying of fruits and vegetables help to maintain the color and taste similar to the original fresh products meeting the demands of customers. One of the most efficient drying process is vacuum frying. Even though oil is used, the overall oil content is low compared to conventional frying process making it a feasible alternative for many fruit varieties in today's market.

During vacuum frying, products are dried under vacuum pressure, reducing the frying temperature required compared to traditional frying temperature (90-100oC versus 160-180°C), while maintaining high drying rate. The lower frying temperature leads to higher product qualities such as crispness, puffed structure, color, aroma and keeping the flavor of fresh fruits, while minimizing the loss of nutrients [2]. In addition, vacuum frying help to restrict unacceptable chemical reactions, including product oxidation, oxidative degradations, lipid oxidation, browning reactions, and risk of acrylamide formation [3]. Furthermore, oil uptake can be reduced as shown by investigations of Dueik, Robert, & Bouchon [4]; Sothornvit [5] and Yamsaengsung, Yaeed, & Ophithakorn [6].

In recent years, vacuum frying combined with microwave heating have been found to help improve the rate of drying and product qualities even more. Microwaves are produced by an electron tube called a magnetron and they are reflected within the metal interior of the oven where they are absorbed by food. The microwaves excite the water molecules in food to vibrate, generating heat and a driving the moisture out of the food.

While previous researchers have considered the effects of vacuum drying on structural changes of bananas, pineapples, and apples [7], the effects of vacuum frying on the percent thickness expansion and hardness of bananas chips, and the effects of combined microwave techniques for drying of durian chips [9], none have investigated the effect of frying temperature, microwave power, and ripeness on the mechanical and physical properties (in terms of shrinkage, oil absorption, and color) using vacuum frying combined with microwave heating of durian at 90, 100, 110, and 120°C and absolute chamber pressure of 8 kPa. The effect of pretreatment of durian chips by freezing overnight -18°C before vacuum frying was also investigated in determining the most viable alternative in producing a high quality durian chip product.

2. Experimental

2.1. *Materials & Methods*

The schematic diagram of the vacuum fryer used in the present study of Figure 1 depicts the experimental setup comprising of a vacuum fryer, a cooling tower, a water pump, a liquid ring vacuum pump (Model ET32030, Nash Trumbull, CT), and a centrifuge. The vacuum fryer is constructed from stainless steel with a diameter of 400 mm, a height of 300, and a wall thickness of 6 mm. The stainless steel lid of the fryer has a thickness of 8 mm which is installed with generator of microwave (Magnetron) and waveguide controlled the microwave power by power monitor. The cooling tower, water pump and the centrifuge are fabricated and assembled by the Department of Chemical Engineering, Prince of Songkla University, Hatyai, Thailand. The vacuum pump is purchased from Kinetics Engineering Co. (Thailand). After frying, the products are centrifuged for 5 min at 450 rpm to remove excess oil on surface and to minimize oil absorption.

Durian chips (control and pretreatment) were fried at 90, 100, 110, and 120°C and an absolute chamber pressure of 40 mm Hg. Each batch consisted of 100 g of the products and is deep-fried in 13 L of palm oil. The palm oil is replaced after 10 batches of frying to maintain the oil fresh [10]. The sliced durians were fried using three conditions: (1) microwave-vacuum frying (combined microwave power of 640 W); (2) vacuum frying; (3) Pretreatment (frozen at -18°C, 24 hrs) before vacuum frying without MCH. The moisture content of the durian chips were continuously measured at intervals of a minute. The frying experiment were performed until the final moisture content reached 6.0% (w.b). All experiments were performed in triplicates.

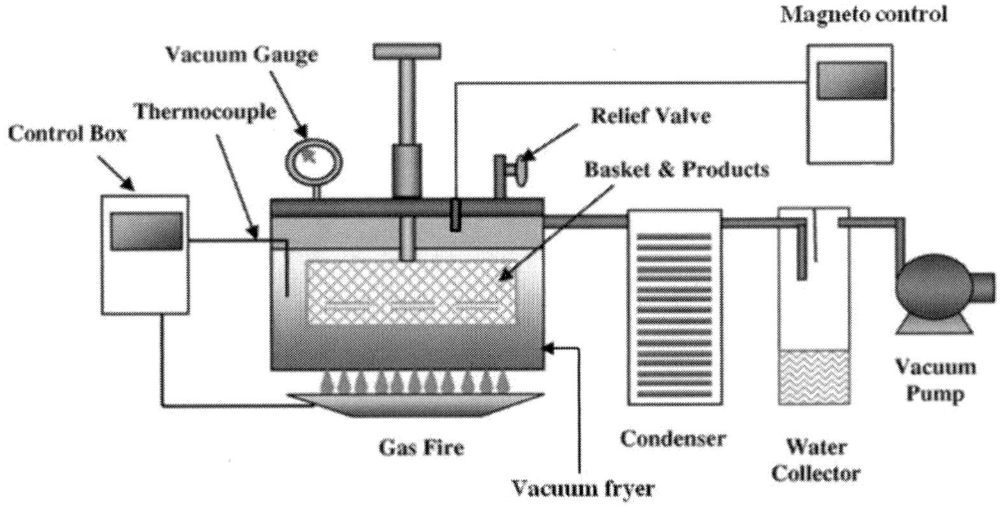

Figure 1 Modified schematic of the microwave-vacuum frying operation [11]

Fresh durians (Cultivated in South of Thailand, cultivar: D159 or 'Durian Monthong' in Thai) were purchased from the local market. The ripeness level in terms of total sugar days of ripening (g glucose/g flesh durian) based on day of ripening (day 1, day 2, and day 3) were used in this study. The fresh durian had the initial moisture content in range of 59.72-71.32% (w.b.) and were fried until the final moisture content of 6.0% (w.b.) was reached. The fresh durian was opened and the flesh was removed and sliced using a blade to 2 - 2.5 mm thickness.

2.2. *Evaluation of frying qualities*

The moisture ratio (MR) of each sample during frying and the frying rate of the sample are determined by the following equations:

$$\mathrm{MR} = \frac{M_{in}-M_e}{M_f-M_e} \tag{1}$$

where: M_0 and M_t = the moisture content (kg water kg dry solid^{-1}) at initial time 0 and frying time t, M_e = the equilibrium moisture content (kg water kg dry solid^{-1}). The equilibrium moisture content (M_e) is assumed to be zero for this experiment because of little value compared to M_0 [12]

2.2.1. Shrinkage The physical qualities of durian chips are evaluated for shrinkage. After each frying time interval, the samples are taken out and new samples are used for each set of frying. An average value of the thickness is obtained from a set of 10 random samples using a digital Vernier Caliper. The percent of shrinkage is described by equation (2):

$$\%\mathrm{Shrinkage} = \left(\frac{D_0-D}{D_0}\right) \times 100 \tag{2}$$

where: D and D_0 are respectively the end-of-each-test and the original diameters (mm).

2.2.2. Color change of fried product The color of the samples are measured using a Hunter Lab color system colorimeter (Juki Model JP100, Japan). The measurements of color change of the samples in each condition are performed by using ten samples and the average value is reported. The color values of fried samples are compared with those of fresh samples and the normalized color changes are then calculated by the following equation [13]:

$$\frac{\Delta L}{L_0} = \frac{L-L_0}{L_0} \tag{3}$$

$$\frac{\Delta a}{a_0} = \frac{a - a_0}{a_0} \tag{4}$$

$$\frac{\Delta b}{b_0} = \frac{b - b_0}{b_0} \tag{5}$$

where: L, a, and b = the lightness, redness and yellowness of the fried sample, respectively, while L_0, a_0, and b_0 represent the initial values of the lightness, redness and yellowness of the sample prior to frying, respectively.

The total color difference (ΔE) was evaluated by using eq. (6) [14]. The results were obtained to express the effect of frying temperature, ripeness of durian, and microwave power on sample quality.

$$\Delta E = \left[\left(L_{ref} - L^*\right)^2 + \left(a_{ref} - a^*\right)^2 + \left(b_{ref} - b^*\right)^2\right]^{\frac{1}{2}} \tag{6}$$

where: L_{ref}, a_{ref} and b_{ref} = the lightness, redness, and yellowness of the initial value of the sample before frying, respectively, while L*, a*, and b* = the lightness, redness and yellowness of the fried sample, respectively.

2.3. *Effect of ripeness (days of ripening)*

Effect of ripeness, or the amount of sugar content, on the degree of drying rate, shrinkage and ability of oil absorption are also studied. Samples were allowed to ripen for 1, 2, and 3 days. Fresh samples are evaluated the total sugar content by Modified Phenol Sulfuric Method [15]. The total sugar content is determined from soluble sugar, oligomeric, and polymeric sugar in original samples and is calibrated using a glucose standard curve.

2.4. *Oil content of durian chips*

Approximately 2 g of fried durian samples were weighted into soxhlet for solvent extraction (using hexane). Total oil content of durian chips were determined by using the Soxhlet extraction apparatus AOAC Official Method 972.28 [16]. All the procedures are carried out in duplicate and mean value reported.

2.5. *Statistical Analysis*

A duplicate of the entire experiments and the mean values with standard deviations are reported. The experimental data are analyzed using an analysis of variance (ANOVA). Duncan's multiple range test is used to establish the multiple comparisons of the mean values; mean values are considered at 95% confidence level ($p = 0.05$). A statistical program SPSS (SPSS software for Windows, SPSS Inc., USA) is used to perform all statistical calculations [8].

3. Results and discussion

3.1. *Influence of process variables*

After vacuum frying, changes in the moisture ratio of durian chips during the vacuum frying at different conditions and processes are presented in Figure 2. The moisture content of the durian samples was decreased from the initial value of durian ripening day 1 of 71.3% (w.b.), durian ripening day 2 of 66.3% (w.b.), and durian ripening day 3 of 59.7% (w.b.) to less than 6.0% (w.b.). At the starting of vacuum frying, the moisture ratio of the durian chips diminished rapidly. The drying kinetics of durian chips provided this similar process to the vacuum frying (VF) combined with microwave heating (MCH) and pretreatment procedure before vacuum frying. The effects of frying temperature on the vacuum frying of durian day 1 (shortest drying) are shown in Figure 2a). The frying time could reach to the required moisture content faster with an increase in frying temperature owing to a greater driving force of heat and mass transfer at the higher frying temperature. However, using the microwave power to combine with vacuum frying enhanced the deduction of the moisture ratio, probably the quick raised of the water temperature in the material come to water boiling point [17] that consisted with those on apple pomace [18], mint leaves [19], and beetroot [20]. Furthermore, Pretreatment method (Freezing at -18°C overnight) prior to vacuum frying also provided more increasing drying rate and leading to the reach of required final content of moisture earlier because of the molecule of water was in an ice-crystal state from freezing and could be evaporated readily compared with VF (control) as following Figure 2b).

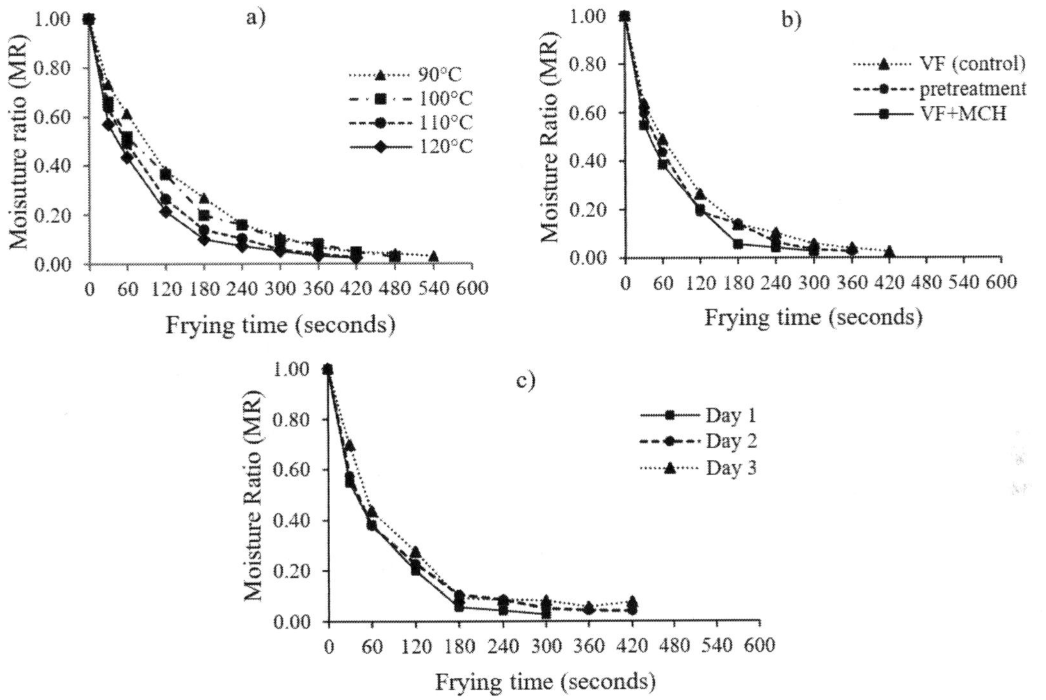

Figure 2 Moisture ratio versus frying time (seconds) of vacuum fried durian chips in different frying conditions of a) vacuum frying of durian day 1 at different frying temperatures, b) vacuum frying of durian day 1 at different processes at 110°C (shortest drying without burning the chips (120°C), c) vacuum frying combined with microwave heating at different ripeness of durian at 110°C.

3.2. Shrinkage

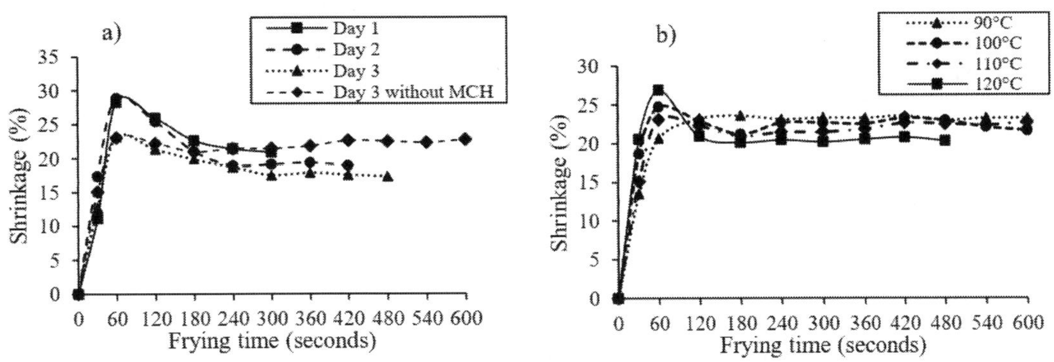

Figure 3 Percent of shrinkage versus frying time (seconds) of vacuum fried durian chips in different frying conditions of a) vacuum frying combined with microwave heat (MCH) at different

ripeness of durian at 110°C (shortest drying without burning the chips (120°C), b) vacuum frying of durian ripening day 3 (least shrinkage) at different frying temperatures.

Figure 3a) and 3b) present the percent of shrinkage as a function of frying time at different ripeness of durian during vacuum frying combined with microwave heating (MCH) compare with vacuum frying without MCH (control) and the percent of shrinkage as a function of frying time at different frying temperature of durian ripening day 3 durian vacuum frying (control). From the Figure 3a), at the beginning of frying, durian chips shrank rapidly and became less shrinkage after then because of significant shrinkage in the product at the first minute of frying pinpointing rapid water loss. However, after a few minutes, the degree of shrinkage slightly decreased which probably have result from the formation of the crust and leading the gaseous vapor expands as it tries to expose [8]. Xanthopoulos, Yanniotis, & Lambrinos [21] suggested the case hardening may be produced in the products having high sugar content which related the lower percent of shrinkage when performed the frying with the higher ripeness of durian (Day 3 has much more sugar content than day 2 and day 1, respectively.). In addition, the vacuum frying combined with MCH produced less percent of shrinkage beside VF (control) at the same ripeness as a result of during microwave heating, the molecules of water absorbed microwave energy and vaporized inside the product which could have resulted the puffing attribute [17]. Meanwhile, it was found that the percent of shrinkage of fried durian chips at the higher temperature were significantly lower than the product that were performed at the lower frying temperatures, likewise the studies of Swasdisevi, Devahastin, Nganchum, & Soponronnarit [13] and Panyawong & Devahastin [22], for bananas, (Wu, Orikasa, Ogawa, & Tagawa [23] for eggplants and Yan et al. [24] for banana, pineapple and mango mentioned that at the highest drying temperature, the shrinkage was finally the least as following the figure 3b).

3.3. Oil content of durian chips

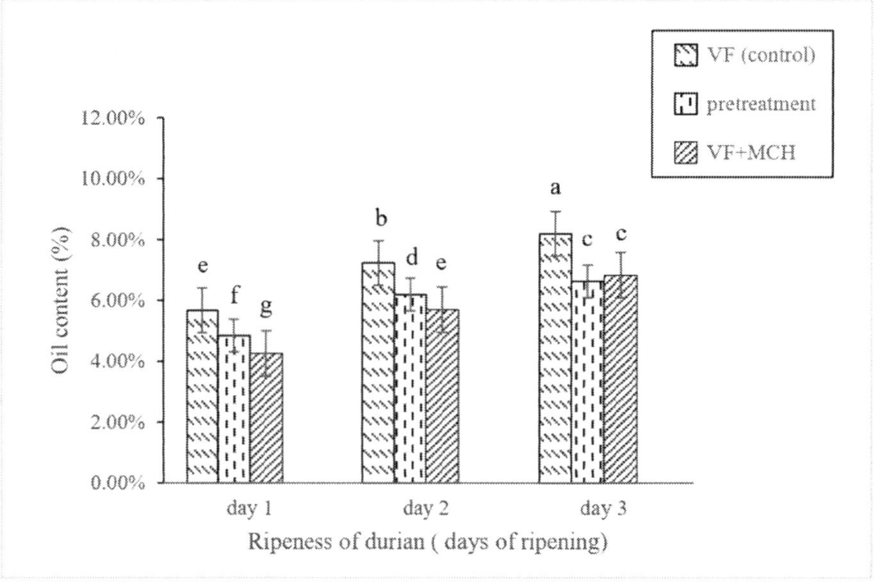

Figure 4 Percent of oil content of vacuum fried durian chips in different frying conditions on durian ripening day 1, day 2 and day 3 at 110°C.

Figure 4 presents the percent of oil absorption at 110°C of vacuum fried durian chips by VF (control), pretreatment prior to VF and VF+MCH at the different ripeness of durian. It could be expressed that more ripeness of durian leading the oil content of durian chips significantly increased ($p<0.05$), agreeably the study of Diamante [25] mentioned that the riper the fruit caused more oil absorption in the vacuum-fried chips. Besides, it was seen in the Figure 4 that the vacuum fried sample that were performed by combined with MCH always presented significantly lower ($p<0.05$) oil content compared to VF and pretreatment prior to sole vacuum frying. At the vacuum frying of durian ripening day 3, the oil content of the samples fried in VF+MCH (6.83%w.b.) was lower compared to VF samples (8.19%w.b.) as the results of the change of microstructural and the deduction of frying time thanks to the MCH [26]. Similar result was presented on edamame with vacuum frying assisted by ultrasound and microwave heating [27].

3.4. Effect of ripeness (days of ripening)

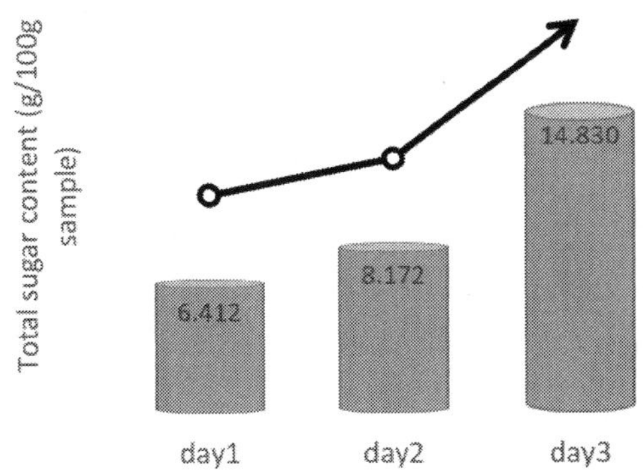

Figure 5 Total sugar content of each durian ripening based on days of ripeness.

The effects of total sugar content based on the ripeness of durian (days of ripening) on degree of drying rate is presented in Figure 2c). From the Figure 2c), the highest degree of drying rate that reached to the required moisture content of ≤6%w.b. occurred for product from day 1 (shortest drying) compared to day 2 and day 3, respectively because of the different initial moisture content and the change of microstructural. In addition, the case hardening may be produced in the products having high sugar content (took more frying time to reach the required moisture content) which related the lower percent of shrinkage when performed the frying with the higher ripeness of sample. Moreover, From the Figure 4, vacuum frying of durian day 3 had more oil absorption than durian day 2 and day 1, respectively due to different microstructure and lower of frying time that occurred in the lower ripeness of the sample.

3.5. Color change of fried durian

Table 1. Mean value for color change for three different conditions of durian ripening day 3.

Conditions	Temperature (°C)	Frying time (seconds)	L*	a*	b*	ΔE
VF (control)	90	600	75.00±1.48^{a}	-1.45±0.17^{a}	27.58±0.91^{a}	13.45±1.65^{a}
VF+MCH	90	480	76.52±0.42^{a}	-1.47±0.13^{a}	25.03±1.59^{a}	13.68±1.27^{a}
Pretreatment	90	540	74.11±0.78^{a}	-1.65±0.03^{a}	27.07±0.86^{a}	14.50±0.95^{a}
VF (control)	100	600	75.42±1.69^{a}	-1.66±0.03^{b}	22.40±1.91^{b}	16.21±2.49^{a}
VF+MCH	100	480	71.15±1.53^{ab}	-1.08±0.08^{a}	28.27±0.90^{a}	16.61±1.61^{a}
Pretreatment	100	480	69.11±1.99^{b}	-1.28±0.14^{a}	29.19±0.27^{a}	18.28±1.97^{a}
VF (control)	110	600	69.56±0.65^{a}	-1.20±0.13^{b}	27.90±1.27^{b}	18.27±0.97^{a}
VF+MCH	110	480	68.19±1.02^{a}	-0.40±0.29^{a}	31.89±0.88^{a}	18.65±1.07^{a}

Pretreatment	110	480	66.64±0.94[a]	-0.70±0.13[ab]	31.98±1.10[a]	20.22±0.98[a]

[a] VF = vacuum frying, VF+MCH = vacuum frying combined with microwave heating, Pretreatment = freezing prior to sole vacuum frying.

[b] Mean value ± standard error of the mean.

[c] The different superscript letters (a-b) in the same column express significant differences within different frying conditions (*p*<0.05).

Pretreatment (freezing at -18°C) prior to vacuum frying and vacuum frying combined with microwave heating of durian chips had no significant difference (p>0.05) on the lightness (L*-value) [17] and the total color difference (ΔE) that are presented in table 1. Nevertheless, the redness (a*-value) and yellowness (b*-value) of vacuum fried (control) durian chips were obviously noticed when they were performed with vacuum frying (control) even though at different frying temperatures. After vacuum frying, the lightness was clearly decreased from the L*, a* and b* of fresh durian of 86.63±0.53, 1.36±0.07 and 33.67±0.58, respectively may be as the result of more frying time that made them slightly darker [27] compared to other conditions that took less frying time to reach the required final moisture content. From the results in Table 1 present the vacuum frying combined with microwave heating could retain the natural color of durian chip as much as vacuum frying (control) and Pretreatment prior to frying, therefore, microwave heating assisted frying does not have any significant effects on the color change of durian chips whereas the higher temperature provided more change of color in L* and a* (darker and more greenness) and decreased in b* (less yellowness) including the total change color (ΔE) that had more value following the tendency of temperature.

4. Conclusions

From the experimental results, the drying rate of durian chips depended on the frying temperature, frying method, and ripeness of durian. It was found that the frying time to reach the required moisture content was faster with an increase in frying temperature meanwhile using MCH provided a higher rate of moisture loss compared to the condition without MCH. Furthermore, Pretreatment method (Freezing at -18°C overnight) prior to sole vacuum frying also provided more an increase in drying rate. Moreover, the ripeness of day 3 resulted in the lowest percent of shrinkage of durian chips; however, the highest oil absorption due to the longer frying time. Finally, vacuum frying with MCH did not whereas significantly affect the change in color of the durian chips, the higher temperature did.

Acknowledgments

The authors acknowledge the Graduate School Research Fund from the Prince of Songkla University and Research and Researcher for Industries Fund for the financial support. We thank the Department of Chemical Engineering, Faculty of Engineering of Prince of Songkla University for quality testing of fried durian chips.

References

[1] Subhadrabandhu S and Ketsa S 2001 *Durian King of Tropical Fruit* (Newyork: Daphne Brasell Associate Ltd.) p 178

[2] Jaya S and DAS H A 2003 A vacuum drying model for mango pulp *Dry. Tech.* **21** 1215-34

[3] Ashraf Z Hamidi Z and Sahari M A 2012 Evaluation and characterization of vacuum drying of date paste *J. Agri. Sci. & Tech.* **14** 565-575

[4] Dueik V Robert P and Bouchon P 2010 Vacuum frying reduces oil uptake and improves the quality parameters of carrot crips *Food Chem.* **119** 1143-49

[5] Sothornvit R 2011 Edible coating and post-frying centrifuge step effect on quality of vacuum-fried banana chips *J. Food Eng.* **107** 319-325

[6] Yamsaengsung R Yaeed S and Ophithakorn T 2017 Vacuum frying of fish tofu and effect

on oil quality usage life *J. Food Proc. Eng.* 1-9

[7] Junlakan W Yamsaengsung R and Tirawanichakul S 2013 Effects of vacuum drying on structural changes of banana slice *ASEAN J. Chem. Eng.* **13** 1-10

[8] Yamsaengsung R Ariyapuchai T and Prasertsit K 2011 Effects of vacuum frying on structural changes of bananas *J. Food Eng.* **106** 298-305

[9] Paengkanya S, Soponronnarit S and Nathakaranakule A 2015 Applications of microwaves for drying of durian chips *Food & Bio. Proc.* **96** 1-11

[10] Ophithakorn T and Yamsaengsung R 2003 Oil absorption during vacuum frying of fish tofu *In Proc. from the PSU-UNS Int. Conf. 2003, Energy and Environment* (Hatyai Songkla Thailand)

[11] Yamsaengsung R and Rungsee C 2003 Vacuum frying of fruits and vegetables *In the 13th Ann. Conf. of Thai Chem. Eng. & Appl.* (Nakhon Nayok) p 1-11

[12] Zakipour E and Hamidi Z 2011 Vacuum drying of some characteristics of some vegetables *Iranian J.Chem. and Chem. Eng.* **30** 97-105

[13] Swasdisevi T, Devahastin S, Ngamchum R and Soponronnarit S 2007 Optimization of a drying process using infrared vacuum drying of Cavendish banana slices *Songklanakarin J. Sci. & Tech.* **29** 809-816

[14] Ahmad Tarmizi A H Niranjan K and Gordon M 2013 Physico-chemical changes occuring in oil when atmospheric frying is combined with post-frying vacuum application *Food Chem.* **136** 902-908

[15] Dubois M Gilles K AHamilton J K Rebers P A and Smith F 1956 Colorimeter method for determination of sugars and related substances *Anal. Chem* **28** 350-356

[16] AOAC 1995 *Official methods of analysis (16th ed.)*, Association of Official Analytical Chemists (Washington DC: AOAC)

[17] Bai-Ngew S Therdthai N and Dhamvithee P 2011 Characterization of microwave vacuum-dried durian chips *J. Food Eng.* **104** 114-122

[18] Wang Z Sun J Chen F Liao X and Hu X 2007 Mathematical modeling on thin layer microwave drying of apple pomace with and without hot aie pre-drying *J. Food Eng.* **80** 536-544

[19] Therdthai N and Zhou W 2009 Characterization of microwave vacuum drying and hot air drying of mint leaves (Mentha cordifolia Opiz ex Fresen) *J. Food Eng.* **91** 482-489

[20] Figiel A 2010 Drying kinetics and quality of beetroots dehydrated by combination of convection and vacuum-microwave methods *J. Food Eng.* **98** 461-470

[21] Xanthopoulos G Yanniotis S and Lambrinos G 2010 Study of the drying behaviour in peeled and unpeeled whole figs *J. Food Eng.* **97** 419-424

[22] Panyawong S and Devahastin S 2007 Determination of deformation of a food product undergoing different drying methods and conditions via evolution of a shape factor *J. Food Eng.* **78** 151-161

[23] Wu L Orikasa T Ogawa Y and Tagawa A 2007 Vacuum drying characteristics of eggplants *J. Food Eng.* **83** 422-429

[24] Yan S H Yin Y P Li W Y Liang T B Wu Y H Geng Q H and Wang Z L 2008 Effect of high temperature after anthesis on starch formation of two wheat cultivars differing in heat tolerance *Ace Ecologica Sinica* **28** 6138-47

[25] Diamante L 2008 Vacuum fried jackfruit: effect of maturity, pre-treatment and processing on the physiochemical and sensory *In Annual Scientific Meeting of the Nutrition Society of Australia* (New Zealand)

[26] Su Y Zhang M Zhang W Liu C and Adhikari B 2018 Ultrasonic microwave-assisted vacuum frying technique as a novel frying method for potato chips at low frying temperature *Food & Bio. Proc.* **108** 95-104

[27] Mojaharul Islam M, Zhang M, Bhandari B and Guo Z 2019 A hybrid vacuum frying process assisted by ultrasound and microwave to enhance the kinetics of moisture loss and quality of fried edamame *Food & Bio. Proc.* **118** 326-335

International Conference on Engineering and Technology (ICET 2021) IOP Publishing
IOP Conf. Series: Materials Science and Engineering 1163 (2021) 012006 doi:10.1088/1757-899X/1163/1/012006

An Investigation of Factors Affecting Measurement Accuracy of Nanoparticles Real-time Sensor using Corona Discharge integrated with Quartz Crystal Microbalance

Nichakran Vichayarom[1,2], Kata Jaruwongrangsri[3], Perapong Tekasakul[1,4] and Racha Dejchanchaiwong[1,2*]

[1]Air Pollution and Health Effect Research Center, Prince of Songkla University, Hat Yai, Songkhla 90112, Thailand

[2]Department of Chemical Engineering, Faculty of Engineering, Prince of Songkla University, Hat Yai, Songkhla 90112, Thailand

[3]Nano-Electronics and MEMs laboratory, National Electronics and Computer Technology Center, Pathumthani, 12120 Thailand

[4]Department of Mechanical Engineering, Faculty of Engineering, Prince of Songkla University, Hat Yai, Songkhla 90112, Thailand

*racha.d@psu.ac.th

Abstract. A nanoparticles real-time sensor in the present study was designed using corona discharge integrated with quartz crystal microbalance (QCM) or Nano-QCM detector. We investigated the following factors affecting the measurement accuracy of the detector; needle to plate distance, flow rate, and number of particles on QCM. DC high voltage power supply of 7 kV was used to maintain the positive corona voltages difference in the charger. The mass sensitivity of 10 MHz AT-cut QCMs coated by plate rigid gold film were investigated. The monodisperse aerosol particles smaller than 300 nm were initially used for the testing of the system. They were charged with negative corona voltages and then deposited on the surface of QCM electrode. Sauerbrey equation was used to describe the frequency-mass relationship at the QCM surface. Results show that the above factors significantly affect the accuracy of the detector measurement. The particles can be collected on a QCM electrode using the partial breakdown high voltage (reach to breakdown points) in terms of needle-to-plate corona discharge. The optimal needle-to-plate distance is 6 mm with maximum mass change (frequency shift) on QCM electrode of 0.607μg/min. The QCM collection efficiency of nanoparticles on the QCM electrode surface using corona discharge technique can be as high as 95% of total particles. Collection efficiency of $PM_{0.1}$ is under investigation.

1. Introduction

Air pollution is a one of significant problems in Thailand. The problem impact on many sides in the society including economy, transportation and, especially health and environment. Development of new technics and knowledge is necessary for real time measurement and monitoring of air quality. United State Environmental Protection Agency (USEPA) defines the national ambient air quality standards (NAAQS) and classifies the particle in term of particulate matter (PM) in 2 sizes including course particle (PM_{10}) and fine particle ($PM_{2.5}$) (EPA, 2018). Recently, the classification of the particulate matter is also including the ultrafine particle or $PM_{0.1}$ or nanoparticles. Although the nanoparticles are not widely concerned and defined in air quality criteria, they are harmful and have greater impact to human than PM_{10} and $PM_{2.5}$ (Herder, et al.,1986; Oberdorster, et al., 2004; Jaques & Kim, 2000; Chen, et al., 2016)

The particle mass concentration measurement devices were improved from several techniques used for example (1) Electrical Low Pressure Impactor (ELPI) - particles are discharged in a corona charger and classified by impactor and are detected by high sensitive aerosol electrometer (Järvinen, Aitomaa, Rostedt, Keskinen, & Ojanperä, 2014). (2) Electrical Aerosol Detector (EAD) - particles are discharged by unipolar diffusion charger. The charged particles are detected by high sensitivity aerosol electrometer (Li, Chen, & Tsai, 2009). (3) Real-Time QCM-MOUDI Impactor - application of Micro-Orifice Uniform Deposition Impactor (MOUDI) and real-time mass detection based on Quartz Crystal Microbalances (QCM) sensors (Chen, Romay, Li, & Naqwi , 2016). (4) Nanosampler or PM0.1 sampler - cascade impactor which operate on the principle of inertial impaction (Furuuchi, et al., 2010). (5) Personal nanosampler or PM0.1 personal sampler – to use inertial filter (IF) which cut off particle size by the principle of inertial impaction (Thongyen, et al., 2015). (6) Monitoring system consisted of electrostatic PM mass monitor (PM_{10} and $PM_{2.5}$ particles) – particles are classified by impactor and discharged in a corona charger and then are measured electrically by PM detector (Intra, et al., 2019a; 2019b; 2020; 2021). For the particle mass concentration measurement device, real-time measurement techniques have been recently developed and gained more acceptance (Li, Chen, & Tsai, 2009; Järvinen, Aitomaa, Rostedt, Keskinen, & Ojanperä, 2014; Thongyen, et al., 2015; Chen, Romay, Li, & Naqwi , 2016; . However, a low cost and online particulate monitoring system for measuring the mass concentrations of ambient $PM_{0.1}$ is limited, especially Quartz Crystal Microbalance (QCM) technique. Therefore, the unipolar needle-plate corona charger technique integrated with QCM for real time nanoparticle ($PM_{0.1}$) mass concentration measurement should be developed and field tested

In the present study, the investigator is interested to use the unipolar needle-plate corona charger technique integrated with Quartz Crystal Microbalance (QCM) for real time nanoparticle ($PM_{0.1}$) mass concentration measurement or Nano-QCM Detector. Personal nanosampler combined with inertial filter which has been developed by researchers from Kanazawa University (Furuuchi, et al., 2010; Otani, Eryu, Furuuchi, Tajima, & Tekasakul, 2007) was used to classify the particle size before passing through the corona charger and measured by QCM. The unipolar needle-plate corona charger is the most common technique to produce high ion concentrations. This technique is mostly used in discharge particle field due to high charging efficiency and simple structure (Alonso, Martin, & Alguacil, 2006; Choi & Kim, 2007; Qi, Chen, & Greenberg, 2008; Intra P., 2012). The charged particles move and deposit on QCM following the principle of electrostatic. QCM could measure highly sensitive mass balance up to nanogram to microgram level changes by frequency oscillation. When addition mass is deposited on the QCM surface, the frequency shift occurs. The difference of frequency can determine in mass per unit area (Sauerbrey, 1959) for real time monitoring.

2. Theoretical and principle

2.1 The corona discharge of needle to plate

The DC steady corona current and voltage relationship for point-to-plane geometry is given by (Townsend, 1914)

$$I = AV(V - V_0) \tag{1}$$

where I is the corona discharge current (A), V is the applied voltage (V), V_0 is the corona onset voltage (V) and A is the dimensional constant depending on the inter-electrode distance, the needle electrode radius, the charge carrier mobility in the drift region and other geometrical factors.

The mathematical model for microscopic point-to-plane coronas in the steady-state regime was developed for prediction of the relationship between corona current and voltage the in gaseous media depending on the distance between the needle tip and the plane by (Henson, 1981).

$$I = \left(\frac{2\pi K\varepsilon}{\alpha}\right)\left[F\left(\frac{\delta}{\alpha}\right)\right]^{-2}(V - V_0)^2 \qquad (2)$$

where δ is the minimum corona glow radius (m), α is the distance between the needle tip and the plane (m), ε is permittivity ($C^2/N.m^2$) K is a dimensional constant and F is a polynomial function of $\frac{\delta}{\alpha}$.

The critical electric field strength, E_c (V/m), for corona onset prediction is given by the Peeks's equation. The equation considers the effects of temperature and pressure, but the effect of air humidity was not included.

$$E_c = m_V E_0 \delta\left(1 + \frac{k}{\sqrt{\delta r}}\right) \qquad (3)$$

where

$$\delta = \frac{P}{P_r} \cdot \frac{273+T_r}{273+T} \qquad (4)$$

m_V is the roughness coefficient of surface of conductor (1 for smooth cylindrical electrodes), r is the radius of curvature of the needle tip (m), E_0 is the electric field strength onset (V/m), k is the empirical constant for cylindrical geometry (normally 0.308 $cm^{1/2}$), δ is the relative air density factor (kg/m^3), T_r is the absolute temperature of room air (K), P_r is the normal atmosphere pressure (bar), and T and P are the operating temperature (K) and pressure of the air (bar). For Peek's equation, the standard reference atmospheric condition is adopted, i.e. P_r=101.3 kPa and T_r=20°C. At normal temperature and pressure conditions, the corona onset voltage, V_0, can be approximated by

$$V_0 = m_v E_0 \delta\left(1 + \frac{k}{\sqrt{\delta r}}\right) r \ln\left(\frac{S}{r}\right) \qquad (5)$$

where S is the distance between the needle electrode and nozzle (m). The air relative humidity is not used to estimate in this equation.

The modified equation which includes the air relative humidity around conductor is determined by (Yawootti et al, 2015)

$$V_0 = m_v E_0 \delta\left(1 + \frac{k}{\sqrt{\delta r}}\right) r \ln\left(\frac{S}{r}\right)\left[1 + \left(5.76 - \frac{1.63}{0.69\sqrt{\delta r}+0.21}\right)\right] \cdot \frac{P_w}{P} \cdot H \qquad (6)$$

and

$$P_w = 611 \times 10^{7.5T/(237.3+T)} \qquad (7)$$

where H is the air humidity, P_w is the partial pressure of saturated water vapor (bar).

The QCM collection efficiency of nanoparticles on the QCM electrode surface was calculated by

$$Collection\ efficiency = \left(1 - \frac{N_{out}}{N_{in}}\right) \times 100 \qquad (8)$$

where N_{in} is particle number concentration of inlet QCM prototype (particle/L). N_{out} is particle number concentration of outlet QCM prototype (particle/L) which are measured by optical particle counter (OPC).

2.2 Quartz Crystal Microbalance (QCM)

Sauerbrey (1959) discovered the relationship between mass absorption on the surface of quartz and offset of its frequency in gas phase. When the particles deposit on the electrode surface, the frequency decreases from the fundamental resonance frequency. The frequency shift (Δf) can be defined by Sauerbrey's equation:

$$\Delta f = -\frac{2f_0^2}{\sqrt{\rho_q \mu_q}} \cdot \frac{\Delta m}{A} \tag{9}$$

where Δf is the measured resonant frequency change (Hz), f_0 is the intrinsic frequency of the quartz crystal (Hz) which depends primarily on plate thickness and cut crystal face, A is the active crystal area or electrodes area (cm^2), ρ_q is the density of quartz (2.643 g/cm^3), μ_q is the shear modulus of quartz for the AT-cut crystal (2.947×10^{11} g/cm s^2), and Δm is the mass changing (g).

3. Experimental Setup

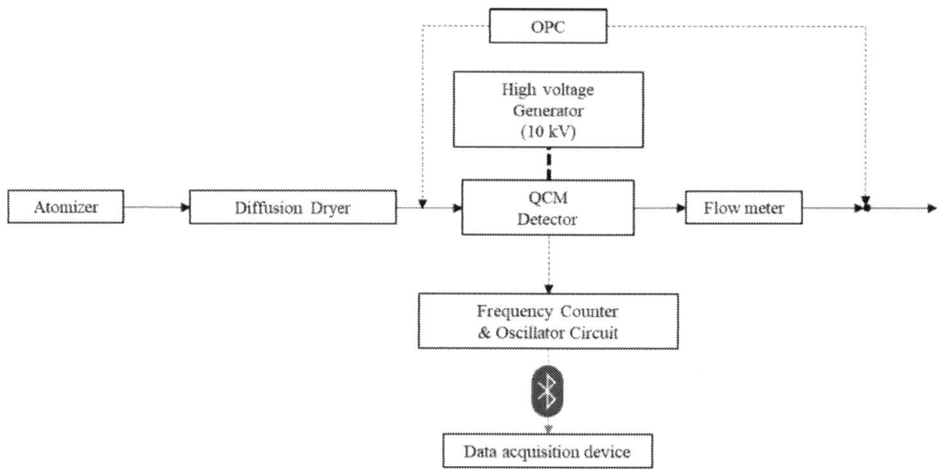

Figure 1 Experimental setup for QCM Prototype.

The experimental setup for investigating the optimum value of flow rate and distance between tip to plate of QCM prototype is shown in Figure 1. The 300 nm monodisperse polystyrene latex (PSL) particles generated by an atomizer flow through a diffusion dryer. The outlet relative humidity is between 45-50%. The particles were charged by 7 kV corona discharge and induced to the QCM electrode. The frequency shift was monitored by a frequency counter and oscillator circuit. The data were transferred to the data acquisition circuit by using Bluetooth communication. Particle concentration was measured by an optical particle counters (OPC) in term of particle number concentration.

IOP Conf. Series: Materials Science and Engineering 1163 (2021) 012006 doi:10.1088/1757-899X/1163/1/012006

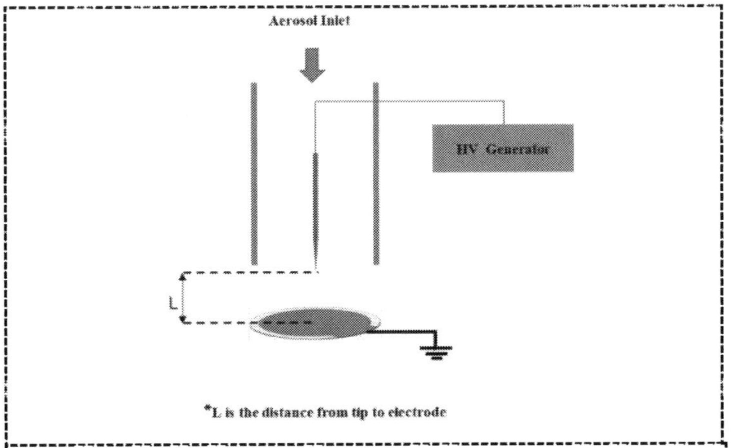

Figure 2 Configuration of QCM prototype.

Figure 2. shows the configuration of QCM prototype. The variation of the distance from tip to plate represents variation of electric field strength which effects the charging efficiency of QCM prototype. The distance was varied from 6-20 mm. Air flow rate inlet was 4.67 lpm. The frequency shift was reported from the data acquisition device. The inlet aerosol flow rate was varied from 1.25-3 lpm by using the flow meter.

4. Results and Discussion

4.1 The relationship between distance from tip to plate and frequency shift.

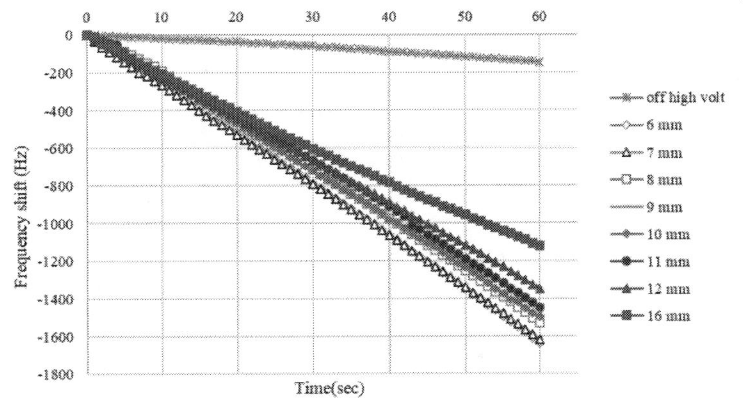

Figure 3 The relationship between distance from tip to plate and frequency shift from 6 to 16 mm.

The relation between distance from tip to plate and frequency shift is shown in Figure 3. The frequency shift represents the particle concentration which is induced on electrode. Frequency shift is inversely proportional to the distance from tip to plate. At the distance below 6 mm, breakdown occurs. Therefore, the distance of 6 mm is the optimal value for this condition. The sensitivity of the QCM when 300 nm particles were used is represented by the shift when the high voltage supply was turned off and it is shown in Figure 3. It decreases from 0-100 in 60 s.

4.2 The relationship between aerosol flow rate and frequency shift.

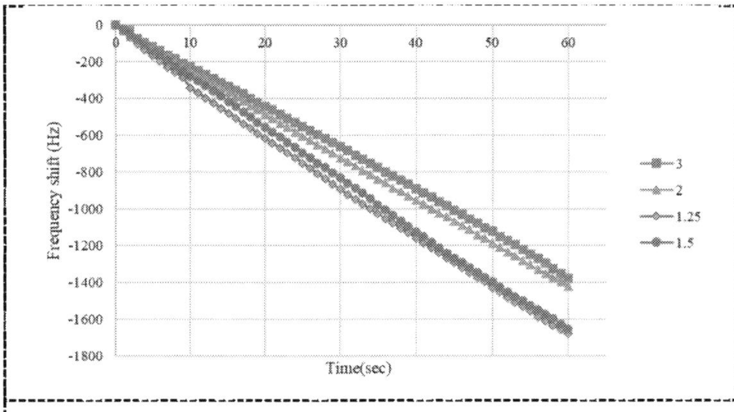

Figure 4 The relationship between aerosol flow rate and frequency shift from 1.25-3 lpm.

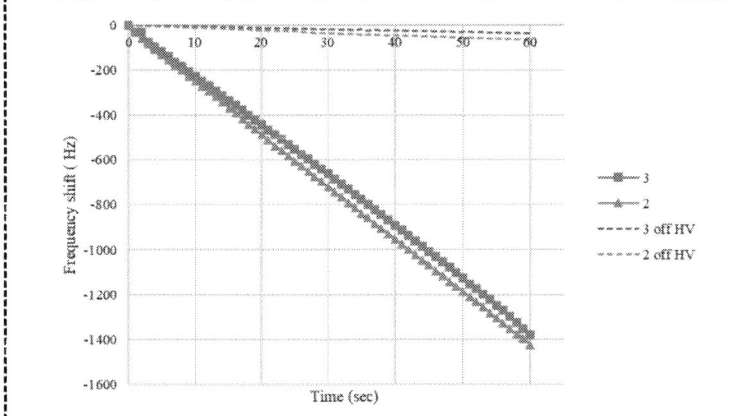

Figure 5 The comparison affecting of frequency shift between on and off high voltage for 2 and 3 lpm air flow rate.

The relation between aerosol flow rate and frequency shift is shown in Figure 4. The highest frequency shift occurs at 1.25 and 1.5 lpm. The higher air flow rate will lead the high concentration of aerosol in the charging zone (between tip to plate) and the aerosol was charged with less efficiency. The suitable values of the flow rate are between 1.25-1.5 lpm. Figure 5 shows the comparison affecting of frequency shift when the high voltage was turned on and off for the flow rate of 2 and 3 lpm. The result is obviously different between 2 and 3 lpm when on and off high voltage. Thus, the air flow rate has a large effect to the induction of aerosol on the QCM electrode.

Moreover, the conversion of frequency shift to particle number concentration and QCM collection efficiency of 6 mm distances were calculated by Eq (9) and Eq (8), respectively. Results shows that the mass change or particle number concentration on QCM electrode is 0.607 μg/min. The collection efficiency of QCM integrated with the unipolar needle-plate corona charger technique was found to be 95% of total particles for 300 nm particles. This is consistent with the study of real-time detection of particles by using electrical detection of charged particles (Intra, et al., 2019a; 2019b; 2020; 2021)

5. Conclusion

The distance from tip to plate and the aerosol air flow affect the induction of particles on a QCM electrode. In this configuration of QCM prototype, optimal needle-to-plate distance is 6 mm with maximum mass change (frequency shift) on QCM electrode of 0.607 μg/min. The QCM collection efficiency of nanoparticles on the QCM electrode surface using corona discharge technique can be as high as 95% of total particles.

References

Alonso, M., Martin, M. I., & Alguacil, F. J. (2006). The measurement of charging efficiencies and losses of aerosol nanoparticles in a corona charger. *Journal of Electrostatics*, 203-2014.

Chen, M., Romay, F. J., Li, L., & Naqwi , A. (2016). A novel quartz crystal cascade impactor for realtime aerosol mass distribution measurement. *Aerosol Science and Technology*, 971-983.

Chen, R., Hu, B., Liu, Y., Xu, J., Yang, G., Xu, D., & Chen, C. (2016). Beyond PM2.5: The role of ultrafine particles on adverse health effects of air pollution. *BBA - General Subjects*, 2844-2855.

Choi, Y., & Kim, S. (2007). An improved method for charging submicron and nano particles with uniform charging performance. *Aerosol Science and Technology*, 259-256.

EPA. (2018, November). (United States Environmental Protection Agency) Retrieved October 2019, from https://www.epa.gov/pm-pollution/particulate-matter-pm-basics#PM

Furuuchi, M., Eryu, K., Nagura, M., Hata, M., Kato, T., Tajima, N., . . . Otani, Y. (2010). Development and performance evaluation of air sampler with inertial filter for nanoparticle sampling. *Aerosol and Air Quality Research*, 185-192.

Henson, B. L. (1981). A space-charge region model for microscopic steady coronas from points. *Journal of Applied Physics*, 709-715.

Herder, J., Gebhart, J., Rudolf, G., Schiller, C. F., & Stahlhofen, W. (1986). Deposition of particle in the human respiratory tract in the size range 0.005-15 um. *Aerosol Science and Technology*, 811-825.

Intra, P. (2012). Corona discharge in a cylindrical triode charger for unipolar diffusion aerosol charging. *Journal of Electrostatics*, 136-143.

Intra, P., Wanusbodeepaisarn, P., & Siri-achawawath, T. (2019a). Experimental Study of Charging Efficiencies and Losses of Submicron Aerosol Particles in a Cylindrical Tri-Axial Charger. *Iranian Journal of Electrical and Electronic Engineering*, 401-410.

Intra, P., Siri-achawawath S. (2019b). Development of an online particulate monitoring system for measurement of the mass and number concentrations and size distributions of ambient PM10, PM2. 5 and sub-400 nm particles. *Songklanakarin Journal of Science & Technology*, 42 (6), 5-30.

Intra P., Wanusbodeepaisarn, P., Siri-achawawath, T. (2020). Evaluation of the Performance in Charging Efficiencies and Losses of Ultrafine Particles Ranging in Sizes from 15 to 75 nm in a Unipolar Corona-based Ionizer. *Journal of Electrical Engineering & Technology*. 1-12.

Intra, P., Wanusbodeepaisarn, P., & Siri-achawawath, T. (2021). Performance evaluation of a corona discharger for unipolar charging of submicron aerosol particles in the size range of 20 – 300 nm. *Indian Journal of Science and Technology*.

Jaques, P. A., & Kim, C. S. (2000). Measurement of total lung deposition of inheled ultrafine partocle in healty men and women. *Inhalation Toxicology*, 715-731.

Järvinen, A., Aitomaa, M., Rostedt, A., Keskinen, J., & Ojanperä, J. Y. (2014). Calibration of the new electrical low pressure impactor (ELPI+). *Journal of Aerosol Science*, 150-159.

Li, L., Chen, D. R., & Tsai, P. J. (2009). Evaluation of an electrical aerosol detector (EAD) for the aerosol integral parameter measurement. *Journal of Electrostatics*, 765-773.

Oberdorster, G., Sharp, Z., Cox, C., Atudorei, V., Elder, A., Gelein, R., & Kreyling, W. (2004). Translocation of inhaled ultrafine particles to the brain. *Inhalation Toxicology*, 437-445.

Otani, Y., Eryu, K., Furuuchi, M., Tajima, N., & Tekasakul, P. (2007). Inertial classification of nanoparticles with fibrous filters. *Aerosol and Air Quality Research*, 343-352.

Qi, C., Chen, D. R., & Greenberg, P. (2008). Performance study of a unipolar aerosol mini-charger for a personal nanoparticle sizer. *Journal of Aerosol Science*, 450-459.

Sauerbrey, G. (1959). Verwendung von schwingquarzen zur wägung dünner schichten und zur mikrowägung. *Zeitschrift für Physik*, 206-222.

Thongyen, T., Hata, M., Toriba, A., Ikeda, T., Koyama, H., Otani, Y., & Furuuchi, M. (2015). Development of PM0.1 personal sampler for evaluation of personal exposure to aerosol nanoparticles. *Aerosol and Air Quality Research*, 180-187.

Townsend, J. S. (1914). XI. The potentials required to maintain currents between coaxial cylinders. *The London, Edinburgh, and Dublin Philosophical Magazine and Journal of Science*, 83-90.

Yawootti, A., Intra, P., Tippayawong, N., & Rattanadecho, P. (2015). An experimental study of relative humidity and air flow effects on positive and negative corona discharges in a corona-needle charger. *Journal of Electrostatics*, 116–122.

This Page Intentionally Left Blank.

IOP Conf. Series: Materials Science and Engineering 1163 (2021) 012006 doi:10.1088/1757-899X/1163/1/012006

Design of Fixture for Welding

Chanada Naksri [1], Somchai Chuchom[2],Supapan Chaiprapat[3]

[1]Benjamarachutit Pattani School , Pattani 9400
[2]Faculty of Engineering, PSU, HatYai, Songkhla 90110
[3]Faculty of Engineering, PSU, HatYai, Songkhla 90110

Abstract. Defects commonly found in welding include distortion, overlapping, warping, and misalignment. Welders probably observe more frequent problems as such if lacking proper fixturing devices. Existing fixture designs are tedious to use, and they could induce even more defects. In this study, four design alternatives of a welding fixture were initially conceptualized from a survey of 25 experienced welders before the best one was selected based on reliability, rigidity, and ease of use. The design was made of SS400 steel because of its exceptional mechanical and thermal properties. A structural analysis was performed using the finite element method to evaluate the strength and stability of the design while in use. Experimental results revealed that this fixturing device could withstand forces exerted vertically from screw clamping. It could also tolerate high temperature from a welding arc and restrain the workpiece as deformed by the heat. This novel design will minimize set-up time, chances of defect formation, and a need for post-welding quality improvement.

Keywords: weld-attaching device, fixture, Finite Element Analysis (FEA), Analytic Hierarchy Process (AHP)

1. Introduction

A jig is a device used to locate the workpiece, while a fixture is designed to support and restrict workpiece movements during operation [1] [2]. On many occasions, fixtures also perform both locating and restraining functionalities[3]. They must be designed in such a way that allows a convenient, reliable, accurate, and quick set-up procedure, more importantly leaving as little workpiece deformation as possible. Jig-fixture, if properly designed, can help save a tremendous cost [4] [5] [6] [7]. Principles of fixture design typically consider the size and shape of a workpiece, material, processes, and economic impacts. The design must also take into account ease of use. Although it provides secure restraining, fixture operational complexity increases setting and handling times that eventually results in less productivity [8] [9]. Simplicity is, therefore, among important considerations the designer must be aware of.

In welding practice, material shrinkage and deformation are common problems resulting from exposure to a welding arc's high temperature. Various devices in the market, such as modular mechanical screws and clamps, magnetic devices, or jigs are used to restrain a workpiece at its nominal position and orientation. One often finds it difficult to accurately keep the workpiece in position while welding using the magnetic-powered tools. The magnetic force obviously cannot withstand the internal forces induced by heat, leading to the workpiece displaced from its intended position and orientation. Using the modular mechanical screws and clamps to orient the workpiece at its right angle is laborious and time-consuming. A device to assist in holding the workpiece while being welded is needed to minimize the defects that can visibly be detected: 1) distortion, 2) overlapping, 3) warping, and 4) misalignment. Besides, the device shall allow a welding operator to put the workpiece in place in less time conveniently. It can also function as a training device for novice welders, or technical students, who lack skills and experiences.

In arc welding operation practice, tools are essential to get the job done correctly [10] fast, safe[11]. Also, the operation duration is the result of the cost [12]. The most common problem encountered in arc welding work is workpiece warpage due to the heat of the welding operation, welding twists and stresses under different mechanical boundary conditions, and welding distortion and residual stress at normal position. The study results showed that the mechanical boundary conditions mainly affected the weld [13]. Beginners and experienced welders encounter problems with welding and twisting. This is caused by the expansion and contraction of the weld metal and the adjacent metal during the heat treatment of the welding process. According to past research studies, it is found that most of the welding heat will cause the workpiece to warp and deform. Some studies use jig-fixture to solve the problem [4] with a finite element program to study the workpiece quality after welding using welding jigs. The result can reduce the problem of warping or deformation of the workpiece.

2. Materials and Methods

2.1 Study theories and research related to design a device to hold the welding work

2.1.1 Metal welding and the principle of Shield Metal Arc Welding (SMAW)

2.1.2 The design of the jig and fixture consists of

1) The positioning and support of the pieces
2) Principle of jig fixture
3) Principle of basic jig fixture

2.1.3 Types of steel

2.1.4 Applying the Analytic Hierarchy Process (AHP)

2.1.5 Analyze the thermal properties of the designed fixtures

2.2 Studying the preliminary welding problems

2.2.1 Survey of the problem arising while performing manual arc welding was conducted using questionnaires, created and verified by welding experts from the institute for skill development, comprising 35 people who issued a test certificate for skill standards (Certificate); 6 people, 19 welding teachers, five welders, one welder student, and four business owners.

2.2.2 Analyzing the results from surveying the problems from welding operators to bring the problems that arise with the workpiece in the welding to be used to design and construct welding equipment.

2.3 Design of welding fixture

Welding fixture design focuses on holding the workpiece to resist the heat shrinkage in the welding work. Quick and convenient to use, including cost-effective use of materials and equipment and designed to reduce the problems of welding that can be visually tested, including 1) distortion, 2) overlapping, 3) warping, and 4) misalignment by designing 4-part sub-component workpieces with SOLIDWORKS program as follows:

1) Five types of a clamping mechanism (a gripping mechanism)
2) Three types of sub-base characteristics to hold the workpiece
3) Two types of the angle rotation plate
4) Three types of keypad or touchpad

2.4 Setting the importance of a subsection of equipment using AHP

The priority will be graded by three experts, with the qualifications of the specialist comprise

1) The qualification must not be lower than the Diploma in industrial metal welding or industrial arts
2) Work experience related to the job at least ten years

3) Welding inspector teachers in the public or private sector.
4) A person who has passed the labor skill standard test manual metal and welder branch level 1

2.5 Assembled grips from selected sub-parts to assess the satisfaction of the experts in 5 areas:

1) Equipment performance
2) Clamping time
3) Equipment cost
4) Equipment size and weight
5) Equipment construction time each issue has a full score of 100

2.6 Determining materials for welding aids

Material designation for welding aids will consider the production cost and the purchasing ability of users as the main. In addition, it must also consider the strength and durability to use. The most popular material is SS400 grade steel.

2.7 Analyze the thermal properties of the welding fixtures according to the design obtained

Using the clamping device simulation while clamping the workpiece at the steel pipe and square tube according to the size as shown in Table 1, which is the size available in the market. Moreover, it is a galvanized steel material clamped at an angle of 45, 90, and 180 degrees. In the analysis, the welding temperature is 10,000 ° F, or 5537.78 ° C. Type of edge weld connectors are determined as groove weld, single-sided, and weld specifications are specified according to American standards. E60 welding electrodes were used to determine the temperature, temperature gradient, and the heat flux on workpieces and fixture.

Table 1. Size of steel pipe and square tube used for testing

STEEL PIPES (INCH)	SQUARE TUBE (INCH)
1/2	3/4
3/4	1
1	1-1/4
1-1/4	1-1/2
1-1/2	2
2	2-3/8
2-1/2	2-1/2
3	3

3. Results and Discussion

3.1 Results of the preliminary study of welding problems

From the study of connection problems, it was found that the most common problems that can be visually tested are 1) distortion, 2) overlapping, 3) warping, and 4) misalignment. If there is a device that helps gripping while working, it can reduce the problem.

Table 2. Questionnaire results.

Welding Problem		Steel square					Steel Pipe				
		No problem	Problem encounter level			Total (person)	No problem	Problem encounter level			Total (person)
			little	moderate	very			little	moderate	very	
BUTT JOINT	 **Figure 1.**										
1. Joint Not equal	 **Figure 2.**	4	3	8	20	35	3	5	10	17	35
	Number of people experiencing the problem (%)	11	9	23	57	89	9	14	29	49	
2. The workpiece is curved.	 **Figure 3.**	3	7	6	19	35	4	4	9	13	35
	Number of people experiencing the problem (%)	9	20	17	54	91	11	11	26	37	
3. The workpiece may move out of the desired position when starting the arc welding.	 **Figure 4.**	9	2	12	12	35	2	8	8	17	35
	Number of people experiencing the problem (%)	26	6	34	34	74	6	23	23	49	
4. The weld line has cracks.	 **Figure 5.**	21	4	4	6		19	4	6	6	35
	Number of people experiencing the problem (%)	60	11	11	17		54	11	17	17	

Welding Problem		Steel square					Steel Pipe				
			Problem encounter level			Total (person)		Problem encounter level			Total (person)
		No problem	little	moderate	very			little	moderate	very	
The welding work is assembled at a 90 degree angle by 45 degree notched workpieces.	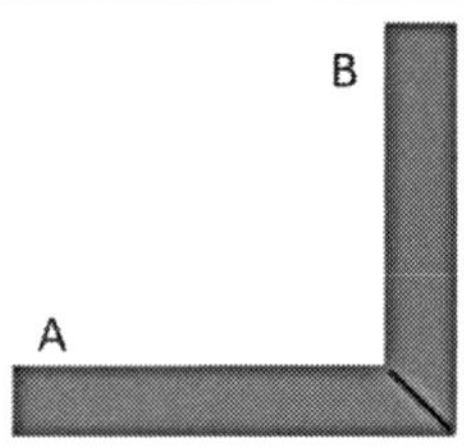 **Figure 6.**										
1. The workpiece is tilted or twisted.	Type 1 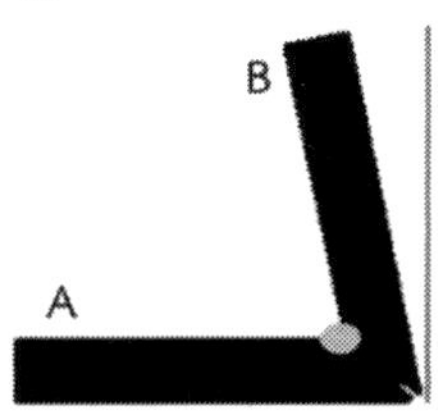 **Figure 7.**	7	2	11	15	35	1	3	10	21	35
	Number of people experiencing the problem (%)	20	6	31	43	80	3	9	29	60	
	Type 2 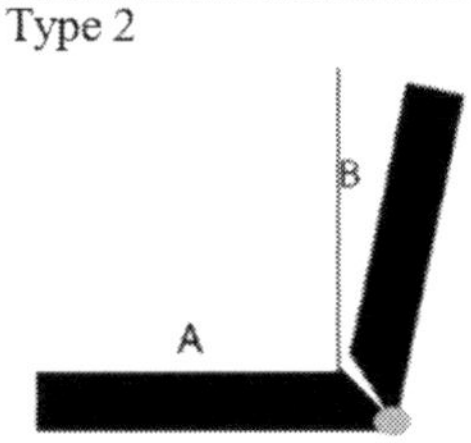 **Figure 8.**	3	6	9	17	35	1	3	14	17	35

	Number of people experiencing the problem (%)	9	17	26	49	91	3	9	40	49	
Welding Problem		Steel square					Steel Pipe				
			Problem encounter level					Problem encounter level			
		No problem	little	moderate	very	Total (person)	No problem	little	moderate	very	Total (person)
2. The weld line has cracks.	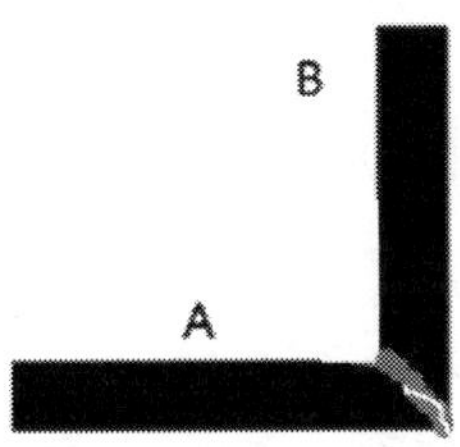 **Figure 9.**	22	4	5	4	35	22	4	6	3	35
	Number of people experiencing the problem (%)	63	11	14	11		63	11	17	9	
3. Joint Not equal	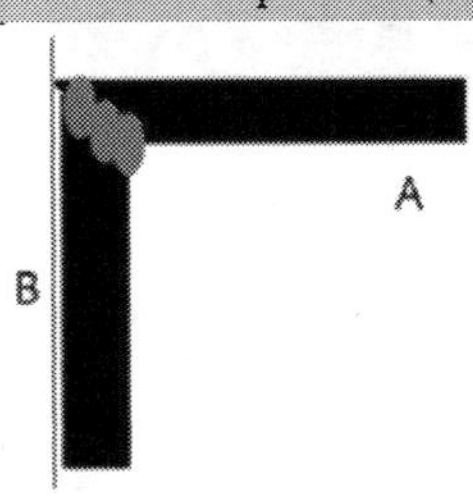 **Figure 10.**	7	6	14	8	35	4	7	17	7	35
	Number of people experiencing the problem (%)	20	17	40	23	80	11	20	49	20	

4. The workpiece may move out of the desired position when starting the arc welding.	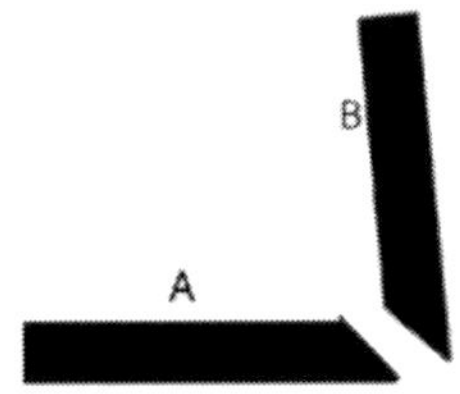 **Figure 11.**	7	6	13	9	35	2	7	11	15	35
	Number of people experiencing the problem (%)	20	17	37	26	80	6	20	31	43	
Welding Problem		Steel square					Steel Pipe				
			Problem encounter level					Problem encounter level			
		No problem	little	moderate	very	Total (person)	No problem	little	moderate	very	Total (person)
T-JOINT 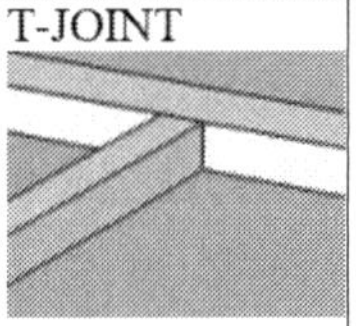	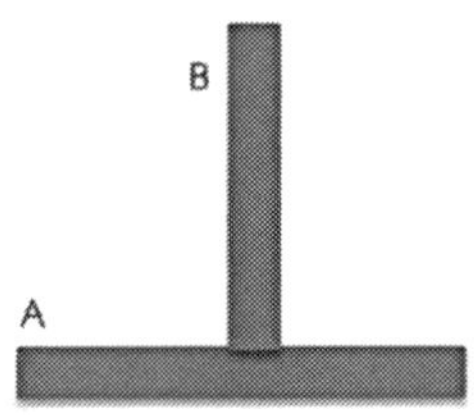 **Figure 12.**										
1. The workpiece is tilted or twisted.	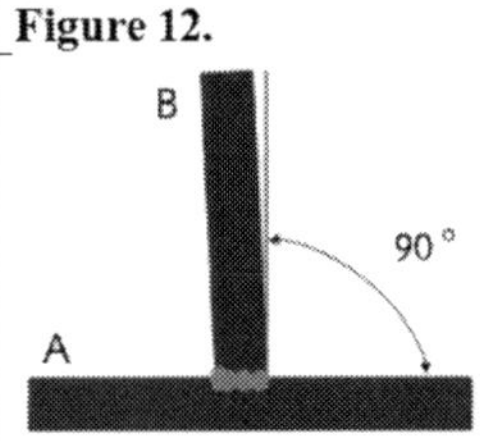 **Figure 13.**	4	8	12	11	35	1	5	13	16	35
	Number of people experiencing the problem (%)	11	23	34	31	89	3	14	37	46	

IOP Conf. Series: Materials Science and Engineering 1163 (2021) 012007 doi:10.1088/1757-899X/1163/1/012007

2. The weld line has cracks.	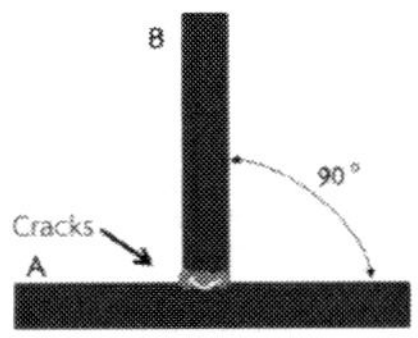 **Figure 14.**	19	6	7	3	35	18	7	6	4	35
	Number of people experiencing the problem (%)	54	17	20	9		51	20	17	11	
Welding Problem		Steel square					Steel Pipe				
			Problem encounter level					Problem encounter level			
		No problem	little	moderate	very	Total (person)	No problem	little	moderate	very	Total (person)
3. Joint Not equal	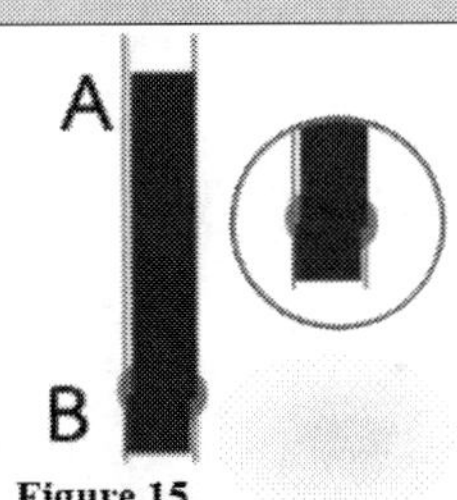 **Figure 15.**	5	5	17	8	35	1	8	14	12	35
	Number of people experiencing the problem (%)	14	14	49	23	84	3	23	40	34	
4. The workpiece may move out of the desired position when starting the arc welding	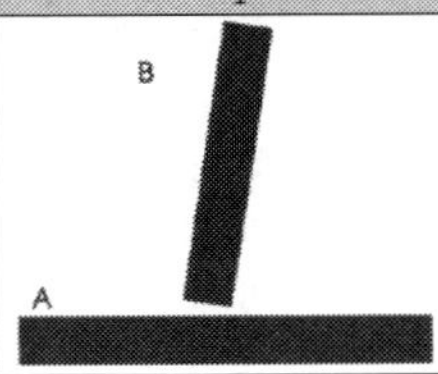	8	7	13	7	35	4	5	13	13	35

	Figure 16.										
	Number of people experiencing the problem (%)	23	20	37	20	77	11	14	37	37	

From the table, it can be seen that the Practitioner will encounter most of the problems that may arise from the visual inspection. 1) The workpiece is tilted or twisted accounted for 86 percent 2) Joint Not equal accounted for 84 percent 3) The workpiece is curved accounted for 81 percent, and 4) The workpiece may move out of the desired position. When starting the arc welding with the workpiece accounted for 77 percent

3.2 The results of the design of the welding fixture

As a result of the welding problem, the welding fixture sub-component was designed and shown in Table 3.

Table 3. shows the sub-component of the device

Sub-component	Type	
1) Characteristics of clamping mechanisms (a gripping mechanism) (A)	**(A) (1) spring**  **Figure 17.**	**(A) (3) toggle clamp**  **Figure 18.**
	(A) (2) spiral 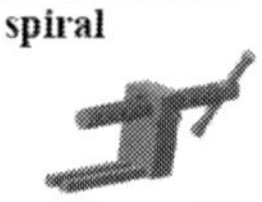 **Figure 19.**	**(A) (4) magnet**  **Figure 20.**
2) Sub-base characteristics to hold the workpiece (B)	**(B) (1) unperforated base** 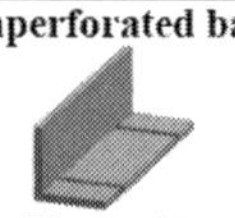 **Figure 21.**	**(B) (3) only edge**  **Figure 22.**
	(B) (2) perforated base 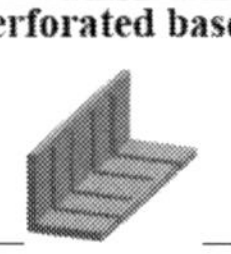	

	Edge **Figure 23.**	**Circular** **Figure 24.**
3) Adjustable rotating plate (C)	**(C) (1) grooved** 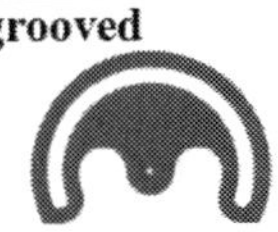 **Figure 25.**	**(C) (2) not grooved**  **Figure 26.**
4) Keypad or touch pad (D)	**(D) (1) attached with fixture**	**(D) (2) separate**

Fixture component selection criteria used the AHP selection process, fixture design, and system design. The jig is a mnemonic device. For convenience, the focus is 1) efficiency, 2) speed of use of the equipment, and 3) equipment weight and cost. The results of the design of welding jigs Selected components are 1) Characteristics of clamping mechanisms (a gripping mechanism) (A) (2) spiral because the gripping force can be controlled 2) Sub-base characteristics to hold the workpiece (B) (2) perforated base circular lightweight, non-slip surface 3) Adjustable rotating plate (C) (1) grooved The shape is more robust 4) Keypad or touchpad (D) (2) separate it can modify the equipment.

3.3 The results of prioritizing the sub-component by AHP

The prioritize the sub-component by AHP all four parts (A-D) found that the first sub-component, clamping mechanisms (A), the experts give top priority to the spiral (0.496) and toggle clamp (0.265), spring (0.130), and magnet (0.109) orderly. In the second sub-component, sub-base characteristics (B), the experts give top priority to the perforated base (0.601) next, unperforated base (O.245), and having an only edge (0.170). In the third sub-component, adjustable rotating plate (C), the experts give top priority to grooved one (0.750) and not grooved one (0.250). The fourth sub-component, the keypad or touchpad (D), the experts give top priority to separate one (0.833) and next attached with the fixture (0.167), which summarized the sub-components significance as shown in table 4.

Table 4. shows expert-graded sub-components

(A) Pattern 2	**(B) Pattern 2**	**(C) Pattern 1**	**(D) Pattern 2**
 **Figure 27.**	 **Figure 28.**	 **Figure 29.**	**Separate**

(A) Pattern 2 Characteristics of clamping mechanisms (a gripping mechanism)
(B) Pattern 2 Sub-base characteristics to hold the workpiece
(C) Pattern 1 Adjustable rotating plate
(D) Pattern 2 Keypad or touch pad

3.4 Model of sub-component fixture that has selected
According to the satisfaction survey of 25 users, the model with the highest score was type 3, fixture vertically circular perforated base with 384 points, followed by type 4 fixture vertically perforated base with 379 points. Type 1 fixture horizontally circular perforated base 345 points, and Type 2 fixture horizontally and fixture with a square gripping 326 points, indicating that Pattern 3 as shown in Figure 30. Is the selected one according to analyze the thermal properties.

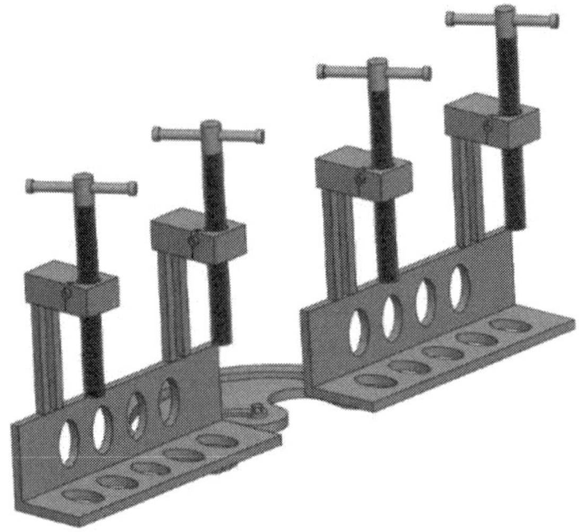

Figure 30. Welding fixture

3.5 Results of thermal analysis of welding fixture according to model
From the thermal analysis of the welding fixture in both steel pipe and square tube at angles of 45, 90, and 180 degrees, it was found that the results were in the same direction, with the same welding temperature at the highest level (Max.) 5538 ° C which was close to the set temperature of 5537.78 ° C. The temperature gradient and the heat flux value increased with the size of the workpiece grew. Moreover, from the thermal analysis figure, it can be concluded that the welding heat does not affect the workpiece clamping device, as shown in Figure 31. Simulation of welding is electric welding. Simulate heating temperature at 10,000 ° F or 5537.78 ° C and clamping at 45, 90, and 180 angles using E60 welding electrodes. It may not be used for other welding types because the heat may change the jig may not be stable.

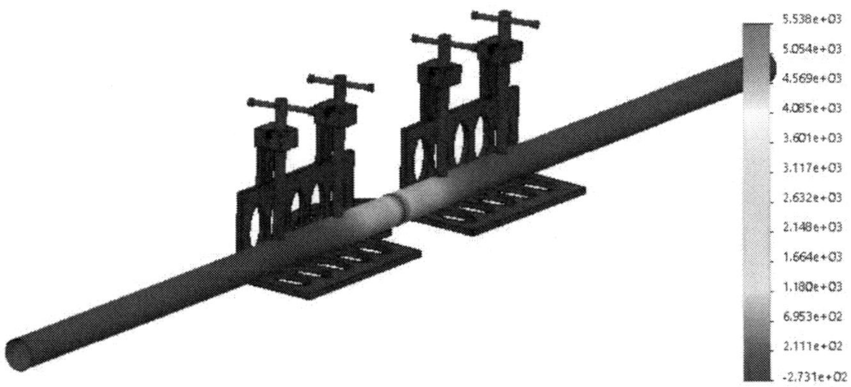

Figure 31. Thermal analysis result

4. Conclusion

This research aims to design a welding fixture-that can adjust angled using AHP to obtain the most suitable welding fixture assembly for the application. The thermal properties analyzed by Finite Element Analysis (FEA) were also used, it was found that the designed welding fixture was not affected by heat during welding. As a result, the fixtures can hold the workpiece efficiently as expected. Device results, especially Jig fixture designed to press the workpiece from the vertical with a twist the clamping force of 7-9 pounds can hold the workpiece to resist the shrinkage of welding and does not cause the workpiece to be deformed or damaged, which can be obtained from the design principles of the Jig fixture. For other research approaches or to further carry out this research, the end of gripping should be taken into account. Equipment weight, and by combining the device with a practical welding table, can perform more versatile and ergonomic work.

5. Acknowledgments

This thesis has been successful because people join hands so thank them all. Thank the advisor, Associate Professor Somchai Chuchom and Assistant Professor Dr. Supapan Chaiprapat, Professor from Department of Industrial Engineering Faculty of Engineering Prince of Songkla University. They both give useful advice to improve the research together with checking, correcting the mistake until this research becomes entirely correct.

6.References

[1] J. D. Suthar, K. M. Patel, and S. G. Luhana, "Design and Analysis of Fixture for Welding an Exhaust Impeller," Procedia Engineering, vol. 51, pp. 514–519, Jan. 2013.

[2] B. Vijaya Ramnath, C. Elanchezhian, S. Rajesh, S. Jaya Prakash, B. M. Kumaar, and K. Rajeshkannan, "Design and Development of Milling Fixture for Friction Stir Welding," Materials Today: Proceedings, vol. 5, no. 1, Part 1, pp. 1832–1838, Jan. 2018.

[3] A. Das, P. Franciosa, and D. Ceglarek, "Fixture Design Optimisation Considering Production Batch of Compliant Non-Ideal Sheet Metal Parts," Procedia Manufacturing, vol. 1, pp. 157–168, Jan. 2015.

[4] H. Wang and Y. (Kevin) Rong, "Case based reasoning method for computer aided

welding fixture design," Computer-Aided Design, vol. 40, no. 12, pp. 1121–1132, Dec. 2008.

[5] C. Yurci, A. Akdogan, and M. N. Durakbasa, "Investigation of Effects (Welding Sequence, Fixturing, Welding Points) on Distortions after Spot Welding for Determining Individual and Cumulative Tolerances," IFAC-PapersOnLine, vol. 49, no. 29, pp. 30–35, Jan. 2016.

[6] P. Dong, Y. P. Yang, Z. Cao, and J. Zhang, "Effects of Welding Procedures on Formability: A Finite Element Study," SAE Transactions, vol. 108, pp. 638–652, 1999.

[7] S. A. David and T. DebRoy, "Current Issues and Problems in Welding Science," Science, vol. 257, no. 5069, pp. 497–502, 1992.

[8] R. Weck, "Review Lecture: The Need for New Welding Processes," Proceedings of the Royal Society of London. Series A, Mathematical and Physical Sciences, vol. 356, no. 1684, pp. 1–23, 1977.

[9] Y. Zhang and Y. Wang, "The influence of welding mechanical boundary condition on the residual stress and distortion of a stiffened-panel," Marine Structures, vol. 65, pp. 259–270, May 2019.

[10] "Weld Distortion | Lincoln Electric," lincolnelectric. [Online]. Available: http://www.lincolnelectric.com/en-us/support/welding-how-to/Pages/weld-distortion-detail.aspx. [Accessed: 17-May-2019].

[11] I. Feddal, A. Khamlichi, and K. Ameziane, "Resistance to buckling of a stiffened panel under axial compression; Effect of imperfections resulting from welding process," Procedia Manufacturing, vol. 32, pp. 921–927, Jan. 2019.

[12] N. Ma, H. Huang, and H. Murakawa, "Effect of jig constraint position and pitch on welding deformation," *Journal of Materials Processing Technology*, vol. 221, pp. 154–162, Jul. 2015.

[13] J. Ordieres, E. Rodríguez, A. Bayón, J. Caixas, A. Barbensi, and P. Guglielmi, "Improvement of manufacturing jigs design for reduction of welding distortion in Vacuum Vessel PS1 through finite element analysis," *Fusion Engineering and Design*, Mar. 2019.

IOP Conf. Series: Materials Science and Engineering 1163 (2021) 012008 doi:10.1088/1757-899X/1163/1/012008

Determination of feasible sites for wind farm using WAsP in Uttaradit, Thailand.

T Khan[1], **J Taweekun**[2,3], **T Theppaya**[2]

[1] Energy Technology Program, Faculty of Engineering, Prince of Songkla University, Hat Yai, Songkhla 90112, Thailand.
[2] Department of Mechanical and Mechatronics Engineering, Faculty of Engineering, Prince of Songkla University, Hat Yai, Songkhla 90112, Thailand.

Email: teri_jv@msn.com

Abstract. Electricity production from the wind plays the potential role to reduce conventional way of electricity generation to reduce the negative impacts on the environment. In this study the data is collected from Uttaradit, the northern province of Thailand. The calculation in this study is based on ten-minutes interval data at 10m observed during the year 2017-2019 by Thai meteorological department. Weibull distribution is applied to extract the monthly and yearly wind in the selected area by using wind atlas analysis and application program (WAsP). Moreover, in WAsP a stimulation was created to observe the comparison of potential of wind energy between three windiest sites located in the study area. The results generated show that the mean wind speed is 2.35m/s and the gradient wind direction from north and south are quite distinct. The annual energy production calculation of three sites in Uttaradit by WAsP estimates about 2125.815MWh, 3379.621MWh and 3369.235MWh respectively by using Bonus 300kW Mk III wind turbine. This paper suggests and supports the suitable sites and wind turbine for the electricity generation in Uttaradit. Hence, it will not only reduce the stress on conventional methods but also create awareness regarding the potential of wind energy.

1. Introduction

Rapidly growing economy, expanding urbanization, new infrastructure demands such as mass rapid transit systems and campaigns for the thriving automobile sector to invest in electric vehicles have raised energy demand [1]. On the other hand, consumption of energy at the current level would deplete the reserves of oil and natural gas gradually [2]. Thailand is turning to alternative and indigenous energy sources because of high energy dependency ratios and exposure to commodity market price volatility [3]. By the rising unpopularity of coal-fired plants that made this pivot hardened with remarkable social opposition to coal projects in the south. Over the last decade the installed capacity of renewable energy doubled, mainly driven by biomass and hydropower. Since 2018, renewables shared 15% of the total consumed power in Thailand. The installed capacity of wind energy is still one-fifth of the 3,002MW target set by Alternative Energy Development Plan-2036 [4].In Thailand approximated

754MW is the overall capacity of wind power installed in Thailand, gained from the northern, north-eastern and southern parts of Thailand [5].

The annual energy production and how the value of the energy produced compares to other sources of energy are the crucial factors in siting a wind farm or wind turbine also known as wind plant or wind park [6]. The data available at meteorological stations throughout the world is of little use during prediction of wind power potential and estimation of energy production from wind farms. In order to build the full potential wind farm, there must be an accurate and precise study on the data for the regarding area [6].

Northern Thailand is geographically characterized by several mountain ranges, which continue from the Shan Hills in bordering Myanmar to Laos, and the river valleys which cut through them [7]. Though like most of Thailand, it has a tropical savanna climate, its relatively high elevation and latitude contribute to more pronounced seasonal temperature variation, with cooler winters than the other regions of Thailand.

The Uttaradit province is selected in this study for certain reasons, firstly the huge inhabited region which is required for the suitable wind farm site as shown in figure 1. Secondly a study has been done by Maejo University, Thailand to find the wind potential in Uttaradit province in the year 2011 [8]. Moreover, Energy Department of Thailand is planning to decrease the use of conventional ways of energy production and increase the clean and sustainable methods of electricity generation in Uttaradit province [9].

According to official statement the total annual consumption of electricity in Uttaradit is 482,900,653kWh which is allocated to three sectors, residential (204,396,557kWh), industrial (201,361,347kWh), government sector (72,301,132kWh) and others (4,841,618kWh) [10].

In this study, the wind resource map and evaluation of the suitable site in Uttaradit region is created by WAsP software. This software is developed by the Riso National Laboratory, Denmark to forecast the climate and study the power generation by wind farms [11]. WAsP is used in several countries by the researchers to develop wind atlas and wind resource maps including Lithuania, Bangladesh, Fiji, Portugal and South Korea [12].

This research will help to allocate the most suitable site for the wind farm by using WAsP software. Potential energy generation by turbine will provide sufficient power in Uttaradit, thus this will help to achieve the target set by the AEDP-2036.

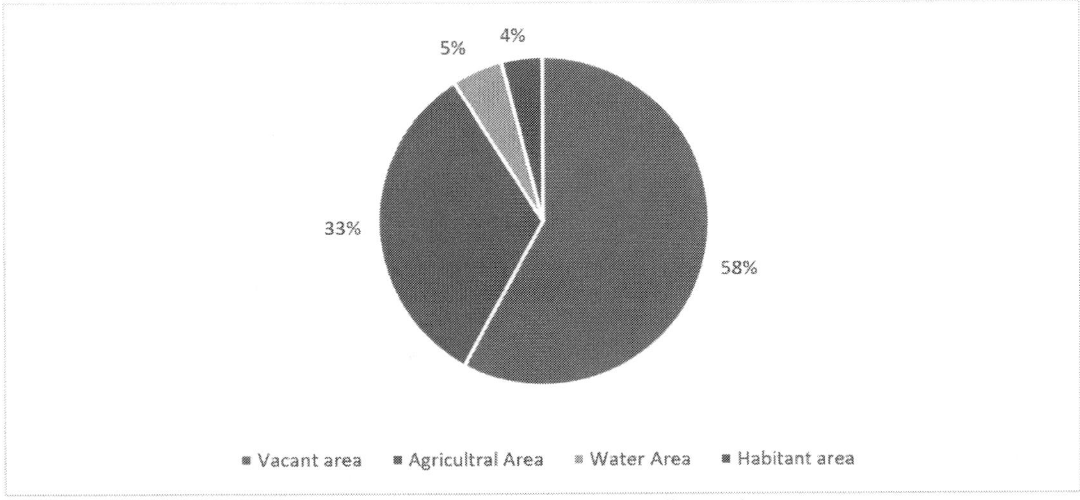

Figure 1. Land division of Uttaradit province.

2. Methodology

Wind data of 3 years, which includes wind speed and wind direction **is** processed and analysed to get the information on monthly and annually basis. The wind data is obtained from the meteorological stations of Uttaradit provinces. Two significant Weibull parameters, shape parameters 'k' and scale parameter 'c' will be measured by using wind speed and wind direction data, obtained from Thai meteorological department. Wind Atlas Analysis and Application Program (WAsP software) is used to make monthly and yearly wind profile and to calculate average speed (m/s) and power density P (W/m^2) for the selected site. Moreover, determination of the most suitable kind of wind turbine for the selected area to study the efficiency of wind energy in terms of mean wind speed and annual energy production.

2.1. Location

Uttaradit is located in the upper northern part of Thailand bordered with Laos. It is about 488km north of the capital city, Bangkok. The wind data including wind speed and wind direction is obtained from Uttaradit Weather observing station site positioned: 17°37'00.0"N 100°06'00.0"E; at 10m above the ground level anemometer height.

The mean wind speed per hour in Uttaradit has little variation throughout the year. Normally the windier months are from February to August and calmer are from September to January. However, the mean wind direction per hour changes over the course of the year. Figure 2 depicts the mean wind speed of Thailand (left side) and study area (right side) [13].

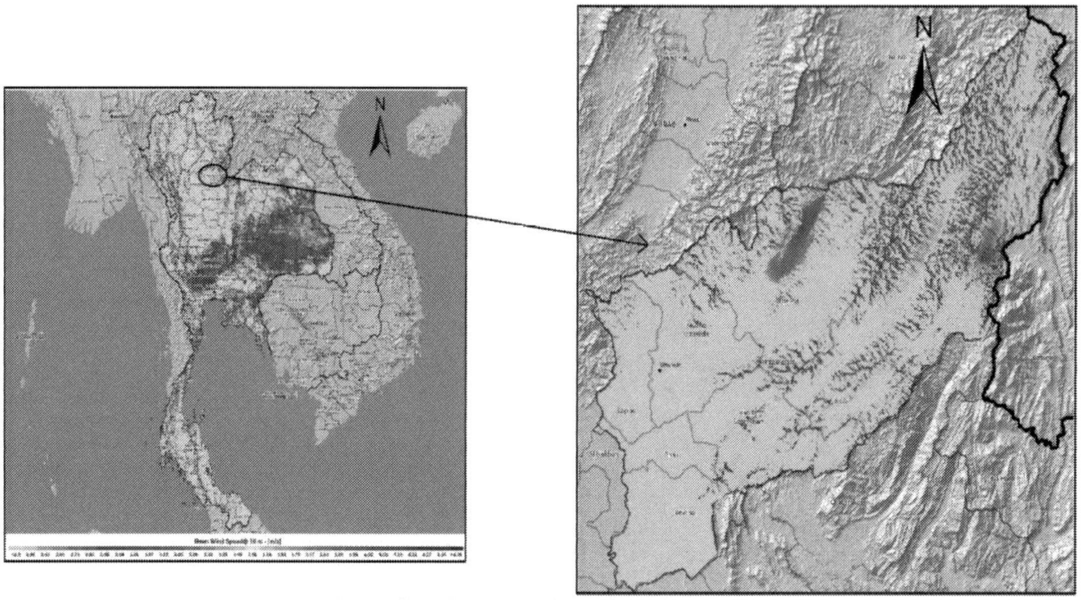

Figure 2. Overview of study area. (adapted source: Global Wind Atlas).

2.2. Wind Data

The Wind data of 3 years starting from 01.01.2017 to 31.12 2019 recorded at Uttaradit weather observing station which includes wind speed and wind direction. The data is recorded repeatedly every 10 minutes at the height of 10m [14]. Firstly, this data is checked properly, then the combined data is separated into two excel sheets categorically as wind speed, wind direction for each year.

Figure 3 represents the the wind rose and histogram. Wind rose shows that the mostly wind direction is from south and north. Moreover, in histogram it can been seen that the mean speed represented as "U" is 2.35m/s and power destiny represented as "P" is 20W/m^2.

Estimated calculation is based on Weibull probability distribution function (PDF) by finding the values of k-shape and scale parameters [15].

To characterise wind behaviours the two parameter Weibull distribution is regularly connected because it appears a great representation of wind data.

The Weibull distribution function is expressed as [16].

$$P_{(v)} = \frac{k}{v}\left(\frac{v}{c}\right)^{k-1} exp\left\{-\left(\frac{v}{c}\right)^{k}\right\} \tag{1}$$

Where $P_{(v)}$ represents the frequency of incidence of wind speed, v. The scale factor c (m/s) shows the average wind speed and the dimensionless shape factor k indicates the shape and width of the distribution. Hence, the parameters c and k determine the Weibull distribution. The cumulative Weibull distribution, $P_{(v)}$, which gives the probability of the wind speed greater than the value, v, is expressed as:

$$P_{(v)} = exp\left\{-\left(\frac{v}{c}\right)^k\right\} \quad (2)$$

The values after the calculation for the shape and scale parameters are 1.60 and 2.6m/s represented as "k" and "A" respectively. The value of shape parameter shows the regular winds focus around an average value. Scale parameter value shows the possibility for selection of the site for the windfarm.

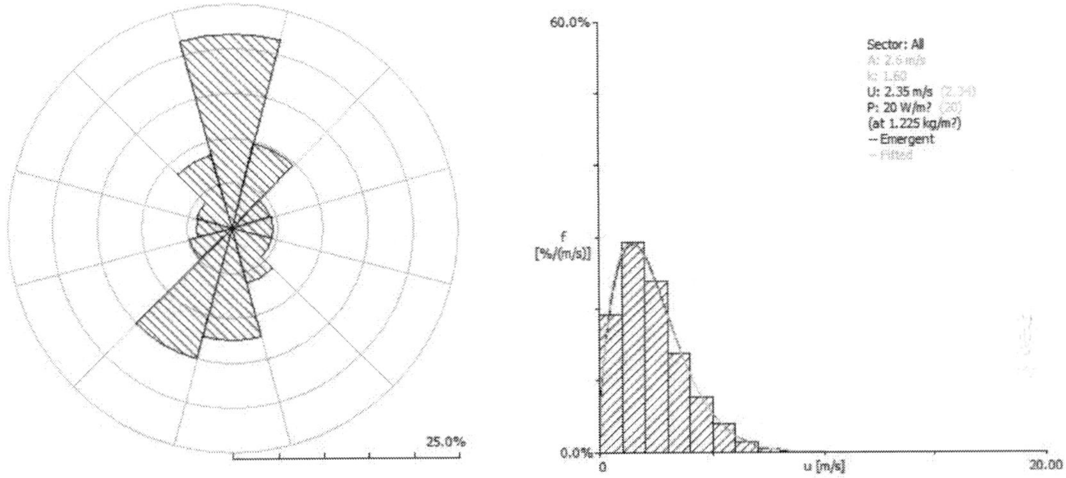

Figure 3. Histogram and wind rose of Uttaradit station.

Three years wind data measured at 10m height as mentioned earlier was processed by WAsP (Wind Analysis and Application Program) developed by the Riso National Laboratory with the collobroration of Danish Metoerological Organization and Danis Technical University, to forecast and generalize wind climate for the suitable site in Uttaradit province. WAsP software typically calculates following areas:

- Calculation of energy yield for single wind turbines and wind farms.
- Calculation of wind farm efficiency.
- Wind resource and turbulence mapping.
- Complex terrain sites.
- Onshore and offshore wind farms.
- Calculation of wind conditions for IEC site assessment for individual wind turbines in a wind farm, e.g. mean wind speed, wind shear, ambient turbulence, extreme wind and wind flow inclination.

The exponential contour map is extracted from Shuttle Radar Topographic Mission(SRTM) data linked with WAsP Map Editor. The extracted map is 90 km from west to east and 60 km from north to south while the station is in the center point. Furthermore details are also derived such as the roughness and, highest and lowest point of the selected area. The surface terrain generated by WAsP software obtained from SRTM and Global Wind Atlas (GWA) of the selected area are classifed into 5 roughness parametters showed in table 1. Surface roughness indicates the concrete picture of the desiered area for the wind study.

Table 1. Land cover division along with surface roughness lengths (Source: WAsP Map Editor)

Land Cover Class Name	Roughness Length (m)
Water Surface	0.0000
Grassland, savannas or lichens/mosses	0.0300
Croplands and Shrubland	0.1000
Mosaic natural vegetation / cropland	0.3000
Forests	1.5000

Processed wind data and the countour map were infused together to create the wind resource map with the total area of about 4400km^2 and 2000m resolution. Wind resource map of the selected area can be seen in figure 4. The range of the wind speed value from 1.36m/s to 5.82m/s is also mentioned in the figure produdced by WAsP.

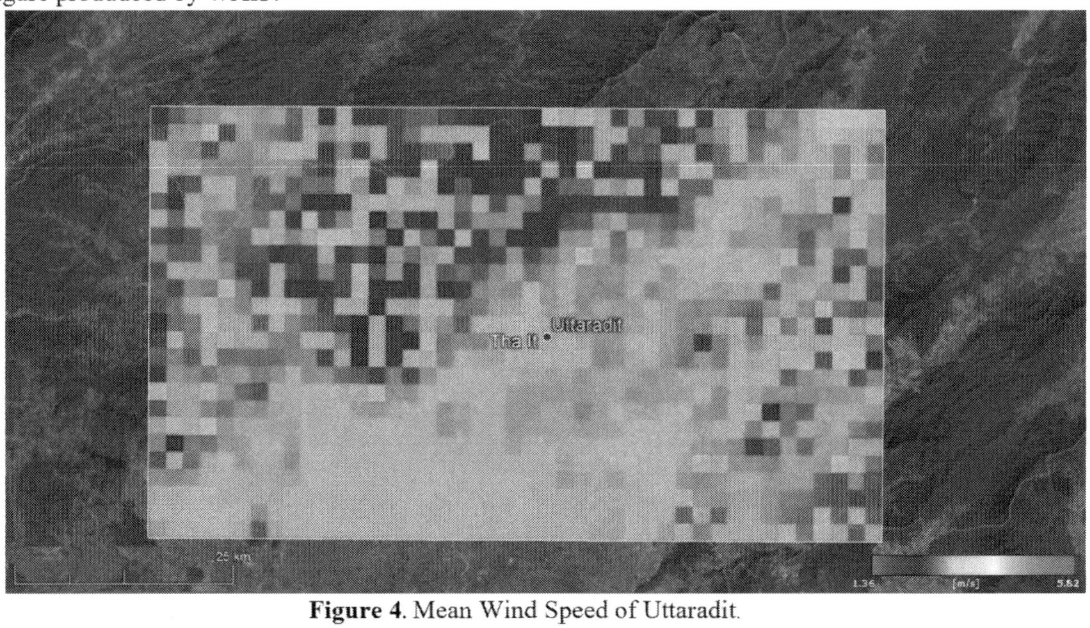

Figure 4. Mean Wind Speed of Uttaradit.

Following the creation of wind resource map of the specific area, three suitable sites for the wind farm were selected for the simulation to aquire the maximum annual energy production(AEP) in accordance with the most feasible sites as shown in figure 5.

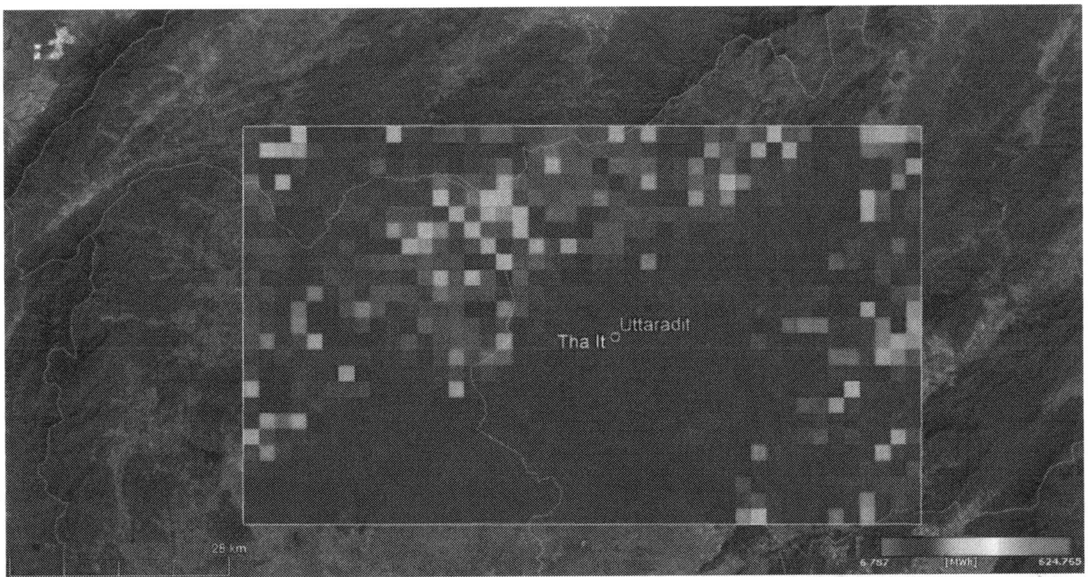

Figure 5. Estimated Annual Energy production of Uttaradit station.

2.3. *Wind Farm and Turbine*

In this study, three sites for the wind farm were selected while cosidering annual energy production.. Each wind farm consists of 8 wind turbines at the 30m hub height. Bonus 300kW Mk-III wind turbine generator is used in this wind farm simulation manufactured by LM Glasfieber/Bonus. The power curve of this wind turbine indicates the expected electrical power output which is 305kW at speed of 14m/s as displayed in figure 6.

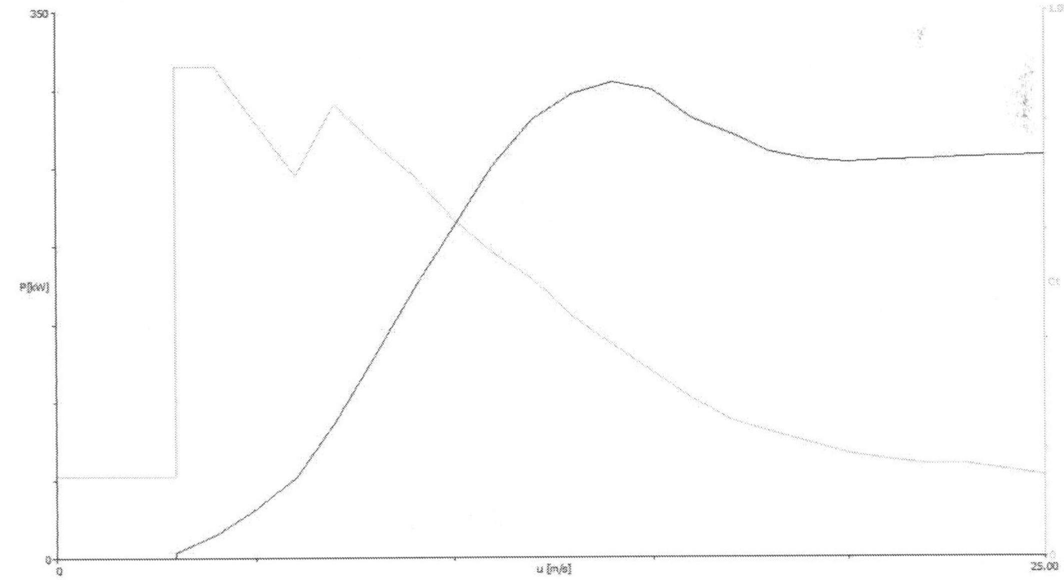

Figure 6. Power curve of Bonus MK III (300 kW) wind turbine.

The turbines installed in each site showed the different results due to the roughness of the area, as the selected sites are in mountainous region. Wake losses generated by WAsP also differs in each turbine as they are not positioned at the same height. Three different sites including 8 turbines each, suitable for the wind farm are marked in figure 7.

Figure 7. Sites selection for the wind farm.

3. Results and discussion

The net annual energy production (AEP) is calculated after deduction of losses. Table 2 shows least AEP and capacity factor which is 2125.815MWh and 10.8% respectively in site number 1, among other selected sites. The maximum mean speed and power density of the site 1 are 4.17m/s and 92W/m^2. RIX represents ruggedness index in the WAsP that helps as an objective measure of the extent of steep slopes in an area.

Table 2. Statistical analysis of site 1.

Variable	Total	Mean	Minimum	Maximum
Total gross AEP [MWh]	2312.601	289.075	174.019	378.568
Total net AEP [MWh]	2125.815	265.727	165.346	344.828
Proportional wake loss [%]	8.08	-	4.46	10.87
Capacity factor [%]	10.8	-	6.5	14.2
Mean speed [m/s]	-	4.17	3.47	4.69
Mean speed (wake-reduced) [m/s]	-	4.03	3.39	4.54
Air density [kg/m^2]	-	1.107	1.102	1.114
Power density [W/m^2]	-	93	53	125
RIX [%]	-	-	13.0	15.4

Table 3 depicts site number 2 having more potential as compared to site number 1. Site number 2 shows the increase in AEP and capacity factor which is 3379.621MWh and 16.4% respectively. The maximum mean wind speed and power density of site 2 are 4.86m/s and 166W/m^2.

Table 3. Statistical analysis of site 2.

Variable	Total	Mean	Min	Max
Total gross AEP [MWh]	3515.660	439.458	245.262	557.527
Total net AEP [MWh]	3379.621	422.453	233.830	534.725
Proportional wake loss [%]	3.87	-	1.86	4.66
Capacity factor [%]	16.4	-	9.2	20.9
Mean speed [m/s]	-	4.86	3.80	5.50
Mean speed (wake-reduced) [m/s]	-	4.77	3.70	5.39
Air density [kg/m^2]	-	1.127	1.123	1.135
Power density [W/m^2]	-	166	77	231
RIX [%]	-	-	16.5	19.1

Finally, table 4 illustrates site number 3 with AEP and capacity factor which is 3369.235MWh and 16.1%. respectively.

The maximum mean speed and power density of the site 3 are 4.93m/s and 153W/m^2.

Table 4. Statistical analysis of site 3.

Variable	Total	Mean	Min	Max
Total gross AEP [MWh]	3447.385	430.923	191.964	596.542
Total net AEP [MWh]	3369.235	421.154	187.480	585.537
Proportional wake loss [%]	2.27	-	1.84	2.78
Capacity factor [%]	16.1	-	7.2	22.3
Mean speed [m/s]	-	4.93	3.61	5.78
Mean speed (wake-reduced) [m/s]	-	4.88	3.56	5.73
Air density [kg/m^2]	-	1.103	1.097	1.114
Power density [W/m^2]	-	153	59	233
RIX [%]	-	-	12.9	19.2

The results show the sites number 2 and 3 are suitable for the maximum AEP among the selected region of Uttaradit province.

4. Conclusion

The assessment of the wind energy potential in the selected region of Uttaradit province, northern Thailand is studied by utilizing WAsP software. The mean wind speed of the region is 2.35m/s at the height of 10m above the ground level. The promising result at 10m height are used for the calculation in WAsP in other studies as well such as Venezuela and Algeria. In Alegria the annual mean speed shifts from 1.2 to 6.3m/s at 10m [17,18].

Moreover, the wind resource map spotted three suitable sites for the windfarm. Among three sites, the site number 2 showed the maximum AEP, while equipped with 8 Bonus 300kW Mk-III wind turbine generators. The net AEP calculated for the site number 2 is 3379.621MWh and capacity factor is 16.4%. Finally, the region has competent potential in the accommodation of a wind farm installation such as accessibility, low population density, vacant space, local government policies and zero protected zone. This will also reduce the consumption of conventional ways of power generation.

References

[1] K. Kim, H. Park, H. Kim, Real options analysis for renewable energy investment decisions in developing countries, Renew. Sustain. Energy Rev. (2017). https://doi.org/10.1016/j.rser.2016.11.073.

[2] S. Chingulpitak, S. Wongwises, Critical review of the current status of wind energy in Thailand, Renew. Sustain. Energy Rev. 31 (2014) 312–318. https://doi.org/10.1016/j.rser.2013.11.038.

[3] EPPO, Summary of Thailand power development Plan 2012 - 2030 (PDP2010: Revision 3), 2012. https://www.egat.co.th/en/images/about-egat/PDP2010-Rev3-Eng.pdf.

[4] AEDP, Alternative Energy Development Plan: AEDP2015, 2015. http://www.eppo.go.th/images/POLICY/ENG/AEDP2015ENG.pdf.

[5] J. Waewsak, C. Chancham, S. Chiwamongkhonkarn, Y. Gagnon, Wind resource assessment of the southernmost region of Thailand using atmospheric and computational fluid dynamics wind flow modeling, Energies. (2019). https://doi.org/10.3390/en12101899.

[6] M. Djamai, N. Kasbadji Merzouk, Wind farm feasibility study and site selection in Adrar, Algeria, in: Energy Procedia, 2011: pp. 136–142. https://doi.org/10.1016/j.egypro.2011.05.016.

[7] K. Sasujit, N. Dussadee, Evaluation of Wind Energy Potential and Electricity Generation in Northern of Thailand, Naresuan Univ. J. Sci. Technol. 24 (2014) 41–54.

[8] M. University, Study of possibility and preliminary environmental effects to construction of wind power plants in suitable wind sources, Chiang Mai, 2012.

[9] T. Ministry of Energy, Energy Plan of Uttaradit, Uttaradit, 2019.

[10] Uttaradit Province, Development Plan of Uttaradit Province, Uttaradit, 2019.

[11] N.G. Mortensen, D.N. Heathfield, O. Rathmann, M. Nielsen, Wind Atlas Analysis and Application Program: WAsP 11 Help Facility, 2014.

[12] U. Yilmaz, F. Balo, L.S. Sua, Simulation Framework for Wind Energy Attributes with WAsP, in: Procedia Comput. Sci., 2019. https://doi.org/10.1016/j.procs.2019.09.076.

[13] World Bank, Technical University of Denmark, Global Wind Atlas, Https://Globalwindatlas.Info/. (2020) 28782.

[14] V. Warudkar, P.K. Sharma, S. Ahmed, Analysis of a terrain characteristic using WAsP and windPRO, in: Energy Procedia, 2019: pp. 1223–1228. https://doi.org/10.1016/j.egypro.2019.01.311.

[15] J. Waewsak, T. Chaichana, C. Chancham, M. Landry, Y. Gagnon, Micro-siting wind resource assessment and near shore wind farm analysis in Pakpanang district, Nakhon Si ฑ Thammarat province, Thailand, in: Energy Procedia, 2014: pp. 204–215. https://doi.org/10.1016/j.egypro.2014.07.071.

[16] F.C. Odo, S.U. Offiah, P.E. Ugwuoke, Weibull distribution-based model for prediction of wind potential in Enugu , Nigeria, Adv. Appl. Sci. Res. (2012).

[17] F.M. Gonzalez-Longatt, Wind resource potential in Los Taques-Venezuela, IEEE Lat. Am. Trans. (2015). https://doi.org/10.1109/TLA.2015.7111999.

[18] S.M. Boudia, A. Benmansour, M.A. Tabet Hellal, Wind resource assessment in Algeria, Sustain. Cities Soc. (2016). https://doi.org/10.1016/j.scs.2016.02.010.

Acknowledgements

This work was supported by the grants of PSU-PhD. Scholarship and Energy Conservation and Promotion Fund Office (Project number: RE-2-0037), Prince of Songkla University, Hat Yai, Songkhla 90112, Thailand. The authors acknowledge the help of the Thai Meteorological Department (TMD) for sharing valuable wind data.

Effect of Thai Silk Fibroin on Mechanical Properties of Bioactive Glass Silk Fibroin Hybrid Bone Scaffolds

Ayuth Vejpongsa[1,2], Peerapat Thongnuek[1,2]

[1]Biomaterial Engineering for Medical and Health Research Unit, Faculty of Engineering, Chulalongkorn University, Pathum Wan 10330, Thailand
[2]Interdisciplinary Program of Biomedical Engineering, Faculty of Engineering, Chulalongkorn University, Pathum Wan 10330, Thailand

Email: Peerapat.T@chula.ac.th

Abstract. Bioactive glass (70 mol% Si and 30% Ca; 70S30C) has been used as tissue engineering scaffolds in various research despite its brittleness due to its easy-to-replicate formula, extensive interconnected pores and natural bone bonding property of bioactive glass (BG). Thai silk fibroin (SF) is natural polymer with many desirable properties; biocompatibility, mechanical strength, biodegradability. Using BG-SF hybrid scaffolds could employ the positive effects of both bioactive glass and silk fibroin. Serving as bone scaffolds, BG-SF scaffolds must resist compression and other types of mechanical load depending on the application. This study fabricates scaffolds with various ratios of BG:SF (BG:SF; 90:10, 80:20 with 100:0 as control), and their stress-strain profiles were investigated. As expected, the scaffolds with highest BG content exhibited the curve resembling that of porous foam while the elastomeric stress-strain profile was becoming more evident when the SF content was increased. This study hence demonstrated fine-tuning mechanical properties of the BG using locally enriched SF.

1. Introduction

Bioactive glass (BG) has been widely used in tissue engineering research. This is due to its composition similarity to natural bone minerals, ability to bond with newly formed bone, and the release of ions promoting osteogenesis [1-3]. There are many formulae of BG and numerous additives can be added to enhance certain aspects to suit different applications [4, 5]. 70S30C is one of the easiest formulae to fabricate. It consists of only 2 distinct compositions: silica and calcium (70% mol being silica and 30% mol being calcium) [2, 6, 7]. However, being ceramic in nature leads to having a brittle structure.

Silk fibroin (SF) is also a biomaterial widely used, both in research and clinics. The SF is strong, non-inflammable, and biodegradable. These properties make SF an excellent candidate for various approaches in tissue engineering [8]. In this study, to overcome the brittleness of BG, SF is added to reinforce the silica network formed beforehand to create a more resilient scaffold overall. SF can be crosslinked with BG by (3-glycidyloxypropyl)trimethoxysilane (GPTMS) [9]. The resulting hybrid scaffolds were reported to have enhanced mechanical stability. Briefly, GPTMS is a silane coupling agent with epoxy and methoxysilane group with oxirane ring reacting to amine group forming covalent bond. Hydrolysis of methoxy resulted in pendant silanol groups which condense into Si-O-Si.

70S30C BG usually requires a silica network with calcium incorporated. There are many fabrication methods. In this study, the gel foaming method is used. Gel foaming can be done easily and does not require any specialized equipment [2]. In the final step, the solution containing every composition of the scaffold is agitated vigorously, introducing air bubbles, and inducing gelation. These bubbles will later act as porogens and create an interconnected porous scaffold [10] which is an advantageous property for the scaffold. The porous scaffolds enable cells to communicate, proliferate and differentiate within the scaffolds, ensuring maximum cellular activities [11]. However, porous scaffolds often face similar pitfall: structural integrity. It is a challenge to navigate this limitation while providing adequate porosity for the cells to function properly.

This study focused on the effect the of ratio between BG and SF on the integrity of the scaffolds. Hybrid scaffolds with various ratios of BG:SF were fabricated by gel foaming method. The scaffolds were then characterized.

2. Experimental methods

2.1. *Materials*

Silk cocoons were obtained from The Queen Sirikit Department of Sericulture, Thailand. Tetraethyl orthosilicate (TEOS), lithium bromide (LiBr), n-hexane, and Teepol were purchased from Sigma-Aldrich. Hydrochloric acid (HCl) and Hydrofluoric acid (HF) were purchased from QreC. Calcium chloride ($CaCl_2$) and sodium carbonate (Na_2CO_3) were purchased from Ajax Finechem and GPTMS was purchased from Gelest, Inc.

2.2. *Preparation of silk fibroin solution*

Silk cocoons were dissolved in 9.3 M LiBr solution, but they must first be washed and degummed. The process was described elsewhere [12]. Briefly, silk cocoons were washed with deionized water to clean out debris and larvae. Then, 0.02 M Na_2CO_3 solution was used to boil the cocoon for 20 minutes, this was to remove the glue-like coating called sericin, or degumming [13]. The cocoons were then boiled for the second time after washing with deionized water. After degumming, the fibres were left to dry open-aired for 2 days. The dried fibres were then dissolved in 9.3 M LiBr solution and dialyzed for 3 days in deionized water. The solution was then centrifuged for 20 minutes at 9000 rpm, 4°C to remove remaining debris.

2.3. *Scaffold fabrication*

TEOS was hydrolysed by HCl to and allow to condense to form silica network. Calcium chloride was added, as a network modifier, with molar ratio of Si:Ca of 7:3. SF solution was added last with varying ratios of BG: SF. SF solution was preconditioned by GPTMS before adding to the sol. After mixing all the ingredients, the sol was agitated vigorously with milk froth whipper to induce gelation and introduce air bubble. The sol was agitated until foamed and then transferred to a silicone mould. The foam was left to age for 3 days at 60°C with the cover closed. Later, the cover was removed, and the foam was left to fully dry for another 2 days at 60°C. The scaffolds were then collected and stored in a dry environment at room temperature [2, 6].

3. Characterizations

3.1. *Thermogravimetric analysis*

The scaffolds' organic/inorganic mass ratios and thermal stability were investigated with thermogravimetric analysis. The scaffolds were heated from room temperature to 900°C with a heating rate of 20°C/min under a nitrogen atmosphere.

3.2. *Surface morphology*

Scaffolds were subjected to scanning electron microscope (SEM) with Hitachi SU3500 model to observe scaffold morphology. The scaffolds were gold-coated, and the voltage was 5kV.

3.3. Porosity
To determine the porosity of each scaffold, fluid displacement analysis was conducted. We used n-hexane as the fluid that filled the porous structure [10]. Briefly, the scaffolds were cut into a 5×5×5 mm cubes. The cubic scaffolds were submerged under n-hexane in a closed container and weighted again. The scaffolds were left to rest for 5 minutes and retrieved from the container. The containers were weighed again and calculated using the following equation to determine the amount of n-hexane that occupied the porous cavity:

$$\%\,Porosity = \left(\frac{W_1 - W_3}{W_2 - W_3}\right) \times 100$$

When W_1 is the initial n-hexane weight, W_2 is the weight of n-hexane with scaffold submerged and W_3 is the weight of remaining n-hexane after removing the scaffold.

3.4. Compressive analysis
To determine the scaffolds' compressive strength and stress-strain response, the scaffolds underwent compression test. The analysis was done with Universal Testing Machine (Shimadzu EZ-S) with a 500 N load cell and a crosshead speed of 1 mm/min. The scaffolds were compressed until the strain reached 75% to avoid damaging the machinery.

The data obtained was Force (N) versus Stroke (mm). Manual calculations were done to convert force to stress and stroke to strain.

$$Stress = \frac{Force}{Crosssectional\ area}$$

$$Strain = \frac{Stroke}{Specimen's\ original\ height} \times 100$$

4. Results

4.1. Scaffold fabrication
BG-SF hybrid scaffolds with various ratios of BG:SF were successfully fabricated with gel foaming method. Table 1 shows compositions of scaffold fabricated. (100 uL of Teepol and 200 uL of 5% HF were added to BG100 as surfactant and gelling agent, respectively.)

Table 1. Composition of fabricated BG-SF hybrid scaffold

Composition	BG (wt%)	SF (wt%)
BG100	100	0
BG90SF10	90	10
BG80SF20	80	20

4.2. Thermogravimetric analysis
Scaffolds were subjected to TGA (Fig.1). The results were normalized with the amount of moisture trapped inside, and hence started at 150°C onwards. We found that the ash content decreased as the SF content increased. Indeed, the ash contents are as follows: BG100-74.58%, BG90SF10-59.9% and BG80SF20-53.04%. The weight started decreasing steadily at 150°C for all scaffolds. The scaffolds with SF content experienced major weight loss started at 320°C and returned to steady decline at 400°C until 900°C.

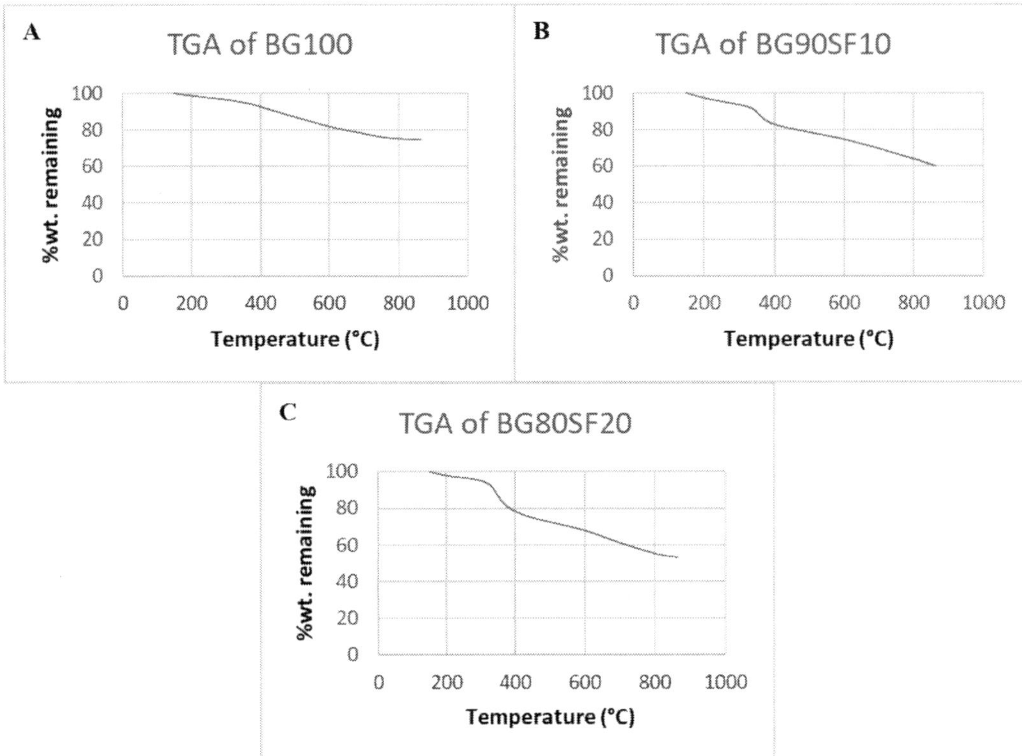

Figure 1. TGA of scaffolds: (A) BG100. (B) BG90-10SF. (C) BG80-20SF.

4.3. Surface morphology

Scaffolds BG100, 90BG10SF and 80BG20SF were subjected to SEM (Fig.2). Overall, the micrographs showed porous structures throughout the scaffolds (Fig.2C and 2E). We found that the BG100, which only contain bioactive glass, had relatively smooth surface (Fig.2A and 2B). On the contrary, the BG90SF10 and BG80SF20 showed rough surfaces and ball-shaped structure (Fig.2D). The surface of BG80SF20 were covered almost entirely in rough patches (Fig.2E and 2F).

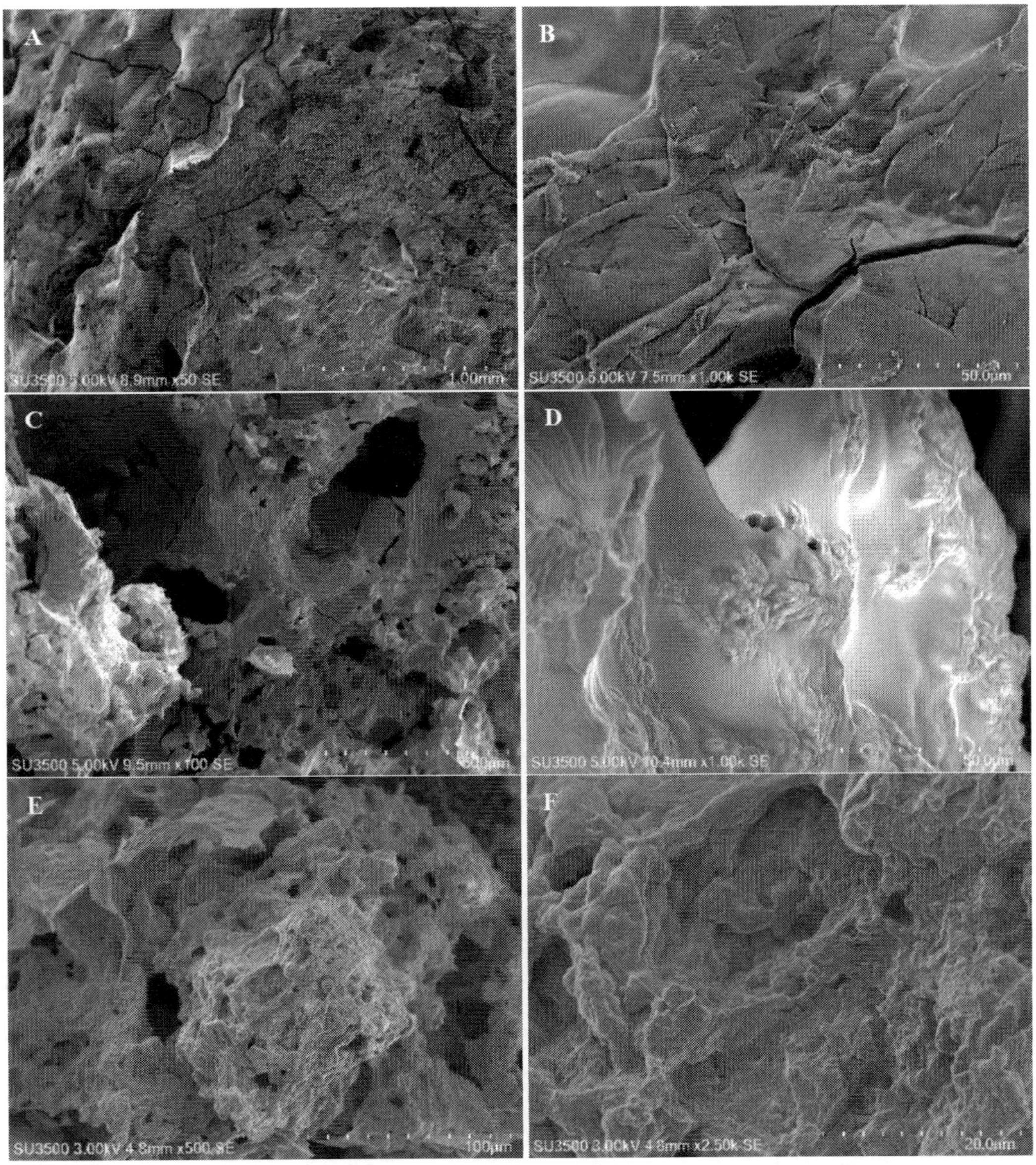

Figure 2. SEM images of scaffolds with different magnifying strength: (A) and (B) are BG100. (C) and (D) are BG90SF10. (E) and (F) are BG80SF20.

4.4. Porosity
The scaffold porosity was determined by fluid displacement (Fig.3). Introduction of SF increased the porosity drastically. The porosity also increased as silk content increases. The increase is in a direct variation trend as the increase from 10%wt SF to 20%wt SF resulted in an increase of 23.88% in porosity.

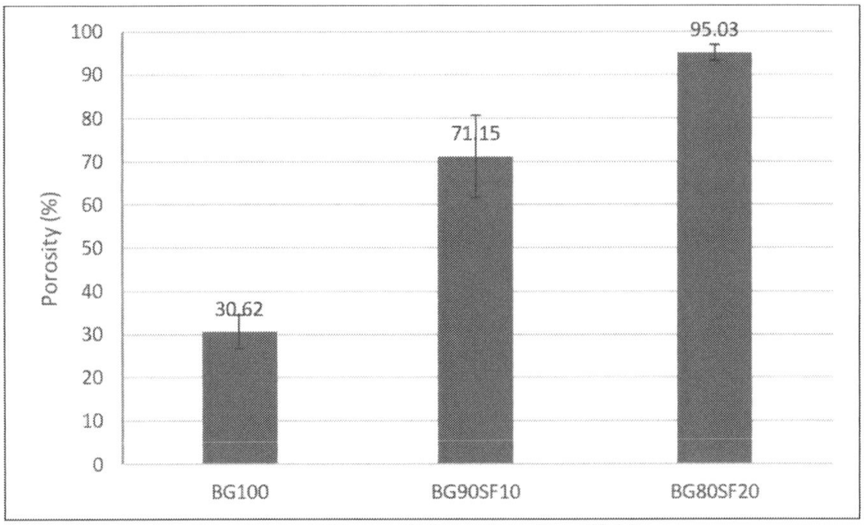

Figure 3. Porosity of scaffolds

4.5. Compression analysis
Scaffold BG100, BG90SF10 and BG80SF20 underwent compression and the results are shown below (Fig.4). The jagged curves in stress-strain profile were seen in BG100 and BG90SF10 (Fig.4A and 4B). However, BG80SF20 showed no jagged curves previously presented (Fig.4C). Young's modulus were calculated from stress-strain graph (Table 2). The yield point of BG100, BG90SF10 and BG80SF20 were 8% strain, 11% strain and 22% strain, and with yield strength of 912 kPa, 39 kPa and 29 kPa respectively.

Table 2. Scaffolds' Young's modulus and yield strength

Scaffold	Young's Modulus	Yield strength
BG100	118.34 kPa	912 kPa
BG90SF10	11.681 kPa	39 kPa
BG80SF20	1.51 kPa	29 kPa

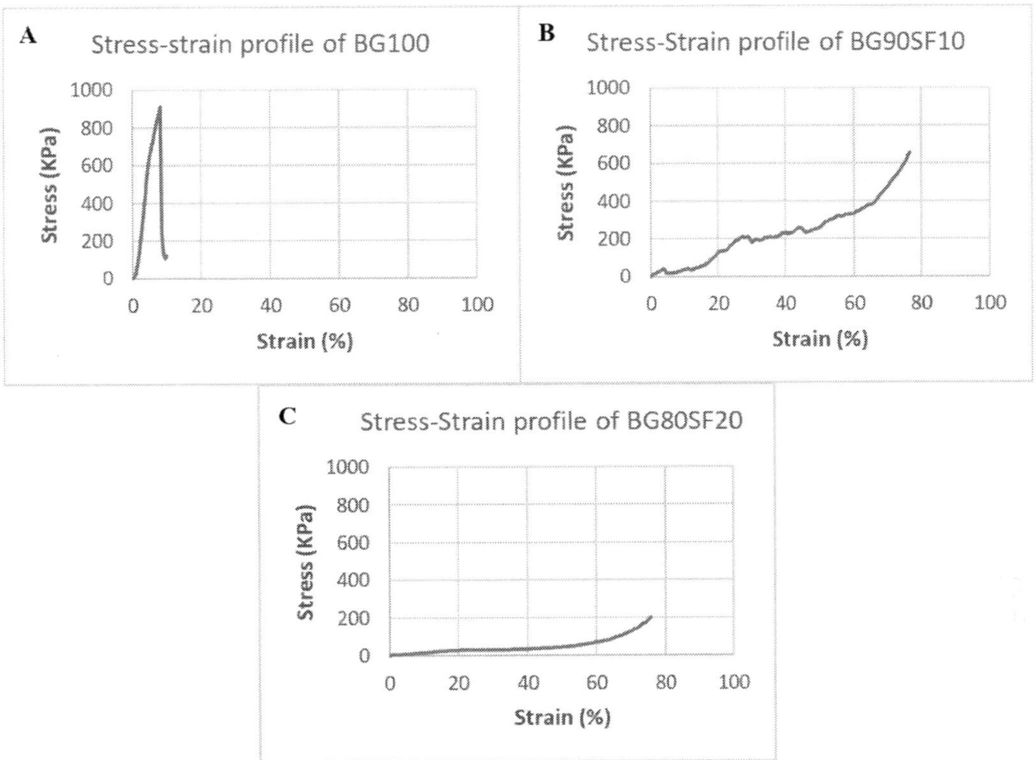

Figure 4. Stress-strain responses of scaffolds: (A) BG100. (B) BG90SF10. (C) BG80SF20.

5. Discussion and conclusion

Normally, SF would disintegrate at a temperature over 300° [14]. However, the results suggested that SF and BG underwent some degree of modification and resulted in higher thermal stability. It is also important to note that SF has been functionalized with GPTMS before fabrication. It has been reported that functionalized SF has a slower degradation rate [9], this could carry over to thermal stability. The ash remaining at 900°C decreased when more SF content was added to the scaffolds. This confirms that SF was successfully incorporated to the bioactive glass.

The scaffolds underwent major weight loss during 200°C – 400°C, which is the temperature range where SF disintegrates [14]. The value of these weight loss of each scaffold corresponded with the amount of SF content in the scaffold. BG100, BG90SF10 and BG80SF20 lost 6.02%, 14.44% and 19.41% of its original mass, respectively. This confirms that the more SF added, the more it exists in the scaffolds.

Hybridizing BG with SF yields drastically different mechanical profiles. BG100 exhibit ceramic-like response to compressive load [15]. Ceramics have low compliance but can withstand high compressive force. The stress-strain response produced a steady and rapid climb only to reach the ultimate strength and sudden drastic drop occur as the scaffold shatters. BG80SF20 can withstand less stress than BG90SF10. The changes in the stress-strain graphs' characteristic can be attributed to the SF content. It was shown that scaffolds with foam structure will slowly fail under crushing load but not entirely due to the failure of struts, causing sudden little drops shown in figure 4B. The overall integrity of the scaffold has not yet failed but some struts have already failed [16]. BG80SF20 (Fig.4C) shows almost linear line of stress-strain response with no sudden drops. This is attributed to crack bridging effect of polymer content covering the surface of the scaffold [17, 18]. Although the ultimate stress was decreased, it should be noted that increasing the SF content can help alleviate the effects of these micro failures of struts. Our results suggest that we may be able to tune the mechanical properties of bioactive glass to suit different applications by changing the polymeric silk fibroin content. This corresponded with SEM images of the scaffolds. BG90SF10 (Fig.2C) shows very little SF deposit on the surface. In contrast, scaffold BG80SF20 (Fig.2E and 2F) shows abundant deposition of SF content or rough surfaces.

The porosity is also a major factor for compressive strength. The increase of SF content also caused the porosity to increase (Fig.3) and naturally, the structure with higher porosity can withstand less compression. This increase of porosity is an effect from SF being a surfactant [19]. The sol that has more surfactant, when undergoes agitation, will be able to generate and retain more air bubbles resulting in more pores. Also, gelation can occur easier with more SF content.

Drastic change in porosity from BG100 to BG90SF10 and BG80SF20 is likely due to removal of Teepol and HF from the sol. The gelation time and bubble stability due to SF solution changed drastically compared to using chemical means.

In conclusion, the BG-SF hybrid scaffolds with varied BG:SF ratios were fabricated. The fabrication is rather straightforward and does not require specialized equipment. The scaffolds were characterized, and this study shows that by increasing SF content we can modify mechanical properties of the scaffolds. The responses of scaffolds to compressive load drastically change when percentage of SF reaches 20% of total scaffold weight. The scaffolds' compliance increases but at the expense of compressive strength. Scaffolds' responses to crushing load also change from foam-like response to compliant material's response possibly due to crack bridging effect of SF content. This could serve as information for future uses regarding micro-failures and scaffold compliance.

References

[1] Hench, L.L., et al., *Bonding mechanisms at the interface of ceramic prosthetic materials.* 1971. **5**(6): p. 117-141.

[2] Martin, R.A., et al., *Characterizing the hierarchical structures of bioactive sol–gel silicate glass and hybrid scaffolds for bone regeneration.* 2012. **370**(1963): p. 1422-1443.

[3] Wilson, J., et al., *Toxicology and biocompatibility of bioglasses.* J Biomed Mater Res, 1981. **15**(6): p. 805-17.

[4] Baino, F., S. Hamzehlou, and S. Kargozar, *Bioactive Glasses: Where Are We and Where Are We Going?* Journal of functional biomaterials, 2018. **9**(1): p. 25.

[5] Vuornos, K., et al., *Bioactive glass ions induce efficient osteogenic differentiation of human adipose stem cells encapsulated in gellan gum and collagen type I hydrogels.* Materials Science and Engineering: C, 2019. **99**: p. 905-918.

[6] Jones, J.R., et al., *Extracellular matrix formation and mineralization on a phosphate-free porous bioactive glass scaffold using primary human osteoblast (HOB) cells.* Biomaterials, 2007. **28**(9): p. 1653-1663.

[7] Midha, S., et al., *Preconditioned 70S30C bioactive glass foams promote osteogenesis in vivo.* Acta Biomaterialia, 2013. **9**(11): p. 9169-9182.

[8] Kundu, B., et al., *Silk fibroin biomaterials for tissue regenerations.* Adv Drug Deliv Rev, 2013. **65**(4): p. 457-70.

[9] Manissorn, J., et al., *Osteogenic enhancement of silk fibroin-based bone scaffolds by forming hybrid composites with bioactive glass through GPTMS during sol-gel process.* Materials Today Communications, 2020: p. 101730.

[10] Alves-Rosa, M.A., et al., *Sulfated zirconia foams synthesized by integrative route combining surfactants, air bubbles and sol–gel transition applied to heterogeneous catalysis.* RSC Advances, 2016. **6**(8): p. 6686-6694.

[11] Zadpoor, A.A., *Bone tissue regeneration: the role of scaffold geometry.* Biomater Sci, 2015. **3**(2): p. 231-45.

[12] Kim, U.-J., et al., *Three-dimensional aqueous-derived biomaterial scaffolds from silk fibroin.* Biomaterials, 2005. **26**(15): p. 2775-2785.

[13] Mondal, M., K. Trivedy, and N. Kumar, *The silk proteins, sericin and fibroin in silkworm, Bombyx mori Linn. - A review.* 2006. **5**.

[14] Kaewprasit, K., et al., *Physico-chemical properties and in vitro response of silk fibroin from various domestic races.* J Biomed Mater Res B Appl Biomater, 2014. **102**(8): p. 1639-47.

[15] Liu, X., et al., *Mechanical properties of bioactive glass (13-93) scaffolds fabricated by robotic deposition for structural bone repair.* Acta Biomater, 2013. **9**(6): p. 7025-34.

[16] Řehořek, L., et al., *Response of 45S5 Bioglass® foams to tensile loading.* Ceramics International, 2013. **39**(7): p. 8015-8020.

[17] Mohamad Yunos, D., O. Bretcanu, and A.R. Boccaccini, *Polymer-bioceramic composites for tissue engineering scaffolds.* Journal of Materials Science, 2008. **43**(13): p. 4433-4442.

[18] Philippart, A., et al., *Toughening and functionalization of bioactive ceramic and glass bone scaffolds by biopolymer coatings and infiltration: a review of the last 5 years.* Expert Rev Med Devices, 2015. **12**(1): p. 93-111.

[19] Cooper, A. and M.W. Kennedy, *Biofoams and natural protein surfactants.* Biophysical chemistry, 2010. **151**(3): p. 96-104.

Optimization of operating parameters for rubber sheet fast drying in a forced-convection rubber smoking room

Pennung Intararak[1,3], Racha Dejchanchaiwong[2,3] and Perapong Tekasakul[1,3*]

[1]Department of Mechanical Engineering, Faculty of Engineering, Prince of Songkla University, Hat Yai, Songkhla 90112, Thailand
[2]Department of Chemical Engineering Faculty of Engineering, Prince of Songkla University, Hat Yai, Songkhla 90112, Thailand
[3]Energy Technology Research Center, Faculty of Engineering, Prince of Songkla University, Hat Yai, Songkhla 90112, Thailand

Email: perapong.t@psu.ac.th

Abstract. A new design of a rubber smoking room for fast drying based on conventional rubber smoking room was tested. Effects of the operating parameters for the fast drying of rubber in a forced-convection rubber smoking room were investigated. Temperature distribution, moisture content, thermal efficiency, specific fuelwood, electricity and energy consumption were evaluated for drying of 1,500 rubber sheets. The suitable operating condition from the experiment is 60°C and 12 m/s for the temperature and velocity. The rubber sheet moisture content was reduced to 0.4% db within 36 hours, i.e. 36 hours less than conventional smoke drying. The variation temperature inside the rubber smoking room was near uniform with a maximum of 8.0°C. The temperatures ranges were 38.9-64.6°C at the front plane, 40.4-65.0°C at the middle plane, and 36.7-62.9°C at the rear plane. A new rubber smoking room can save up to 49.3% in fuelwood consumption as compared to conventional rubber smoking room. Hence, the new rubber smoke drying system is capable of enhancing thermal efficiency and saving energy compared to existing rubber smoking room.

1. Introduction

Natural rubber (NR) is an important agricultural product of Thailand. In 2016, Thailand produced more than 4.25 million tons of NR, accounting for 34.0% of global production. Most of them are exported in several forms. Ribbed smoked sheet (RSS) is one of the major products used in tire industry. The raw rubber sheets are dried in a smoking room where fuelwood is use as a heat source. Smoke from wood burning is also allowed to the smoking room to coat the rubber sheets as a means of product preservation before they are shipped to downstream industries.

Conventional natural-convective smoking rooms take at least 3 days (72 hours) to reduce moisture content (MC) about from 40% to 0.4%. In the drying room temperature is non-uniform and high specific energy consumption is about 1.26 kg/kg dried-rubber sheets [1-2]. Tekasakul and Promtong [2-4] have applied the computation fluid dynamics (CFD) technique to develop a more energy efficient RSS room. Size, position and numbers of gas supply ducts and ventilating lids were redesigned. Test results showed that the average temperature difference between any planes was is 4-7°C and modified RSS room was able to save 66.7% fuelwood consumption as compared to a conventional RSS room with 15.7% thermal efficiency [5]. Dejchanchaiwong, et al. (2017) developed a forced convection smoking room prototype and it could reduce drying time to 48 hours. Energy consumption was reduced by 55% compared to the convention smoking room. Temperature distribution inside the drying room was uniform with largest variation of about 6.75°C [1,5]. Maximum temperature variation in a typical convention smoking room is about 15.0°C. The reduction of the smoking time by 24 h give opportunity to manufactures mostly small business groups of local farmers or rubber cooperatives with daily production capacity of 1-3 metric tons [6]. However, large commercial smoking room with a capacity of 5-7 metric tons may require different design

In the present study, an inlet velocity of 12 m/s is used to compare with results from conventional natural-convective smoking room. Moisture content of fuelwood is about 60% dry. The objectives are to investigate the thermal performance and the unit cost. The RSS should meet the conditions for the class 3 standard [7].

2. Methodology

2.1. Rubber smoking room description

A fast drying rubber smoking room was constructed at the Saikao cooperative (N 7° 10' 32": E 100° 36' 52"), located in the Muang District, Songkhla Province, Thailand. The room dimension is 5 m x 6 m x 2.5 m. A furnace generated heat and smoke supplied to the smoking room by forced convection. It has a dimension of 2.0 m × 2.1 m × 1.4 m and is connected to a 1.2 m diameter cylindrical tube and a blower (4,000 CFM). The flow rate of hot air was controlled by adjusting the speed of the blower motor using an inverter as shown Figure 1

Figure 1. A photo of the fast-drying rubber smoking room at Saikao rubber cooperative.

2.2. Methodology

Experiment was conducted for 1,500 rubber sheets. The rubber sheets in the present study were obtained from Saikao Cooperative. The thickness of rubber sheets was 3-4 mm which was controlled with the constant spacing of milling and followed the RSS standard [7]. Type-K thermocouples were used to measure temperature; 7 positions in the smoking room and 2 positions outside as shown Figure 2. Temperature was recorded by a data logger (Yokogawa, FX112-2-4). Velocity of air was measured at the inlet duct (V1) using a hot wire type anemometer (DIGICON DA-45S). The weight of three rubber sheet samples at Exam1, Exam2 and Exam3 positions in smoking room were measured by weight balance (Shimadzu ELB3000). Fuelwood was measured by a weighing scale (OHAUS Defender, TM 2000-E Series). Maximum temperature in the smoking room was maintained near 60°C. Average heating value of fuelwood used was 6.57 MJ/kg.

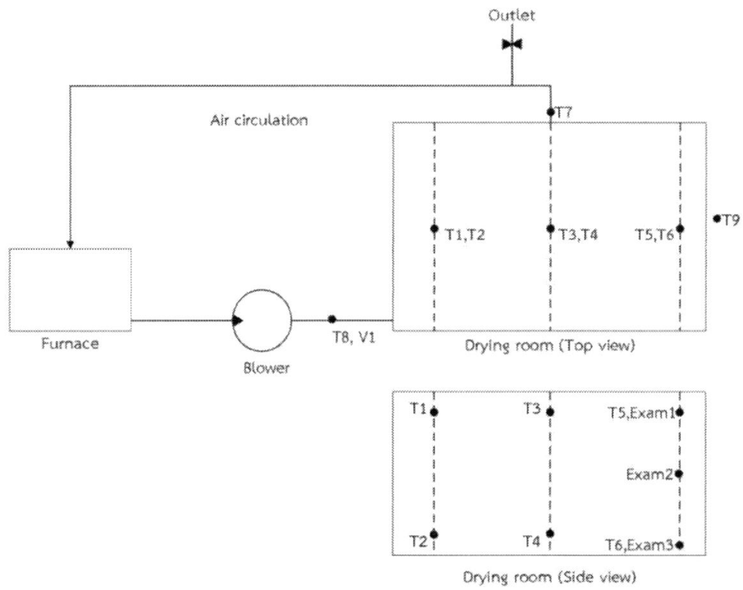

Figure 2. The positions of temperature, velocity and weight of rubber sheet samples measurement in the drying room.

3. Theory

3.1 Moisture content (MC)

Dry basis moisture content is preferable to measure the internal water of agricultural product. Initial moisture content of rubber sheets was about 40% and finalize moisture content was about 0.4%. It can be calculated from [2]

$$\%Md = \frac{W_w}{W_d} x100 \quad (1)$$

where W_w is the weight of water and W_d is the weight of the dried rubber sheet.

3.2 Thermal efficiency and specific energy consumption.

Thermal efficiency and specific energy consumption are indicators for thermal performance of rubber smoking dryer. Thermal efficiency can be calculated from [2]

$$\eta_{dryer} = \frac{m_L L}{(HV)_{wood} m_{wood} + E_e} \quad (2)$$

where m_L is the mass of water evaporated (kg), L is the latent heat of vaporization (kJ/kg), $(HV)_{wood}$ is the heating value of the fuelwood (kJ/kg), m_{wood} is the total mass of fuelwood (kg), and E_e is the total electricity (kJ) [8].

Specific energy consumption (SEC) is the ratio between total supplied and amount of water removed. It can be calculated as [8]

$$SEC = \frac{P_t}{W} \quad (3)$$

where P_t is the total supplied energy to the RSS smoking room (kJ), and W is the mass of water evaporated (kg).

3.3 Cost of energy

Cost of energy takes into account of fuelwood and electricity costs. The unit cost of energy is defined as

$$Cost = \frac{(m_{wood} * m_c) + (E_e * E_c)}{m_{total-dried-rubber}} \quad (4)$$

where m_c is the price of fuelwood (current 0.7 baht/kg), E_e is the price of electricity per unit (current 3.7171 baht/unit), and $m_{total-dried-rubber}$ is the total number of dried rubber sheets.

4. Results and discussion

4.1 Fast drying

The rubber sheets were dried after 36 hours. The temperature ranges were 47.4-64.7°C at the front plane, 52.6- 67.5°C at the middle plane and 46.6-62.0°C at the rear plane, as shown in Figure 2. Temperature distribution in smoke room is uniform with largest variation of 6°C. Average temperature at front, middle, and rear plane is 58.2°C, 62.0°C and 56.0°C, respectively.

Moisture content of 3 rubber sheet samples at different locations is shown in Figure 3. It was reduced from initial MC of 40% to 0.4% in 36 hours. Drying rate in the first 6 hours is the fastest and it is slower until the process is over.

Total fuelwood consumption was 871.0 kg and electricity consumption were 110 kWh. The exhaust hot air recirculation was 0% from 0-6 hour, 50% from 6-24 hour, and 100% from 24-36 hour. Performance of the fast-drying rubber smoking room was evaluated. Thermal efficiency and SEC of fast-drying rubber smoking room are 20.5% and 10.8 MJ/kg of water evaporated, respectively. Details of performance are shown in Table 1. Physical appearance of the rubber sheets used in the experiment is shown in Figure 4. They passed the class 3 standard for RSS.

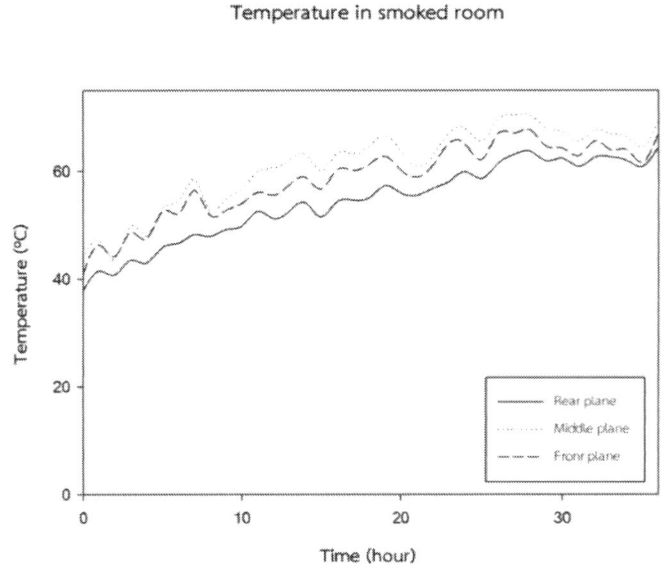

Figure 3. Temperature vs time at front, middle and rear plane.

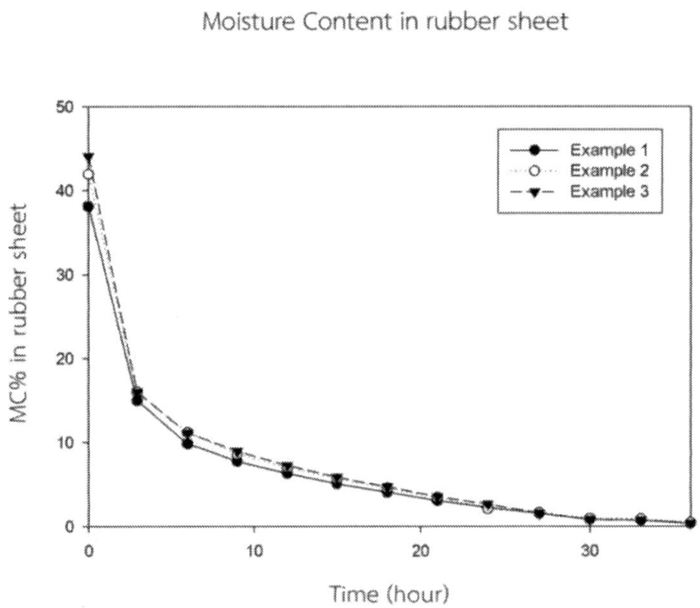

Figure 4. MC during RSS drying in velocity 12 m/s.

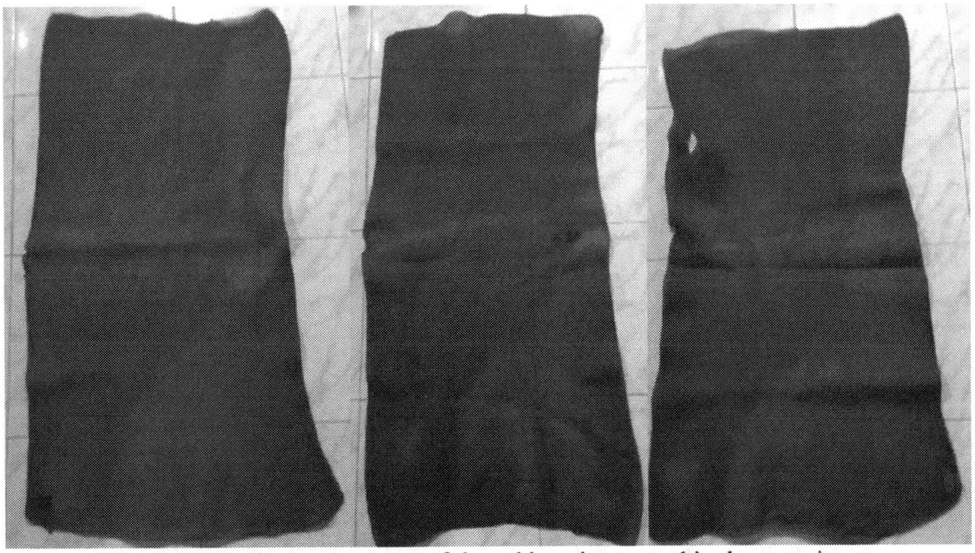

Figure 5. Physical appearance of the rubber sheets used in the experiment.

Table 1. Overall performance of the fast rubber sheet drying.

Item	value	unit
Good quality	1,361	kg
Good quality (%)	100.0	%
Drying time	36	hour
Fuelwood consumption	871.0	kg
Specific fuelwood consumption	0.64	kg/kg dried rubber
Specific Electricity consumption	0.11	kWh/kg dried rubber
Overall SEC	10.8	MJ/kg of water evaporated
eff. dryer	20.5	%

4.2 Comparison between conventional and fast-drying smoking room

Thermal efficiency of the fast drying is 20.5%, about 3 times improvement from the 6.9% of the conventional drying [1]. The specific energy consumption of the fast drying is 10.8 MJ/kg of water evaporated. It is improved from 34.4 MJ/kg of water evaporated for the conventional drying. Unit cost of rubber smoke sheet fast drying is 0.75 baht/dried rubber sheet (calculate at fuelwood price 0.7 baht/kg) in comparison to 0.90 baht/dried rubber sheet of the conventional smoking-room cost*(4)*.. Details of the performance comparison are shown in Table 2.

Table 2. Compairson between conventional rubber smoking room and fastdrying smoking room

Case	Drying time (hour)	SEC (MJ/kg of water evaporated)	η drying (%)	Cost of RSS (baht/dried rubber sheet)
Conventional room	72	34.3	6.9	0.90
Fast drying room	36	10.8	20.5	0.75

5. Conclusion

A fast-drying rubber sheet smoking room is able to maintain a temperature near 60°C and uniform distribution with temperature deviation of 4-6 °C between planes. Moisture content of the rubber sheets is decreased from about 40% to 0.4% in 36 hours so that 36 hours of the conventional drying can be saved. Thermal efficiency of the fast smoking rooms increases from 6.9% to 20.5%. 13.6%, while the SEC is reduced from 34.3 to 10.8 MJ/kg of water evaporated. Cost in RSS production is reduced from 0.90 baht/dried rubber sheet to 0.75 baht/dried rubber sheet.

References.

[1] Dejchanchaiwong R, Kumar A, Tekasakul P. *Performance and economic analysis of natural convection based rubber smoking room for rubber cooperatives in Thailand*. Renew energy. 2019 Mar;**132**:233–42.

[2] Kalasee W, Pongpariyawatee Y, Tantaprapa W, Suwanjarat S, Kirirat P. *Fuel efficiency enhancement in the rubber smoking process of the rubber cooperatives*. 2013.

[3] Tekasakul, P., and Promtong, M. *Energy efficiency enhancement of natural rubber smoking process by flow improvement using a CFD technique*. Applied Energy. 2008. **85(9)**: 878–895

[4] Promtong, M., and Tekasakul, P. *CFD study of flow in natural rubber smoking-room: I. Validation with the present smoking-room*. Applied thermal engineering, 2007. **27(11–12)**: 2113–2121.

[5] Dejchanchaiwong R, Tirawanichakul Y, Tirawanichakul S, Kumar A, Tekasakul P. *Techno-economic assessment of forced-convection rubber smoking room for rubber cooperatives*. Energy. 2017 Oct; **137:152**–9.

[6] TK Technician Limited partnership (2020). Commercial rubber room (Online). Available: http://www.tktechnicianth.com/ Accessed at 2 SEP 2020

[7] Rubber intelligence unit. Standard of smoked rubber. (Online). Available: http://rubber.oie.go.th/box/ELib_Document/3729/) Accessed at 2 SEP 2020

[8] Matiyanon T, Sathitreaungsuk P, Sangworakran S, Boonprasit P. *Design and experiment of an energy-saving combustor equipped with a multi-pass heat exchanger for a high-efficiency rubber-smoked system*. 2014.

Acknowledgement

This research was supported by Natural Rubber Innovation Research Institute, Prince of Single University (Grant No. ENG610281S).

IOP Conf. Series: Materials Science and Engineering 1163 (2021) 012011 doi:10.1088/1757-899X/1163/1/012011

Evaluation of Insecticidal Properties of Four Essential Oils from *Illicium verum*, *Curcuma longa*, *Foeniculum valgare*, and *Ocimum tenuiflorum* against *Spodoptera exigua*.

Hataichanok Passara[1] and Ammorn Insung[2]

[1]Department of Plant Production Technology, Faculty of Agricultural Technology, King Mongkut's Institute of Technology Ladkrabang, Bangkok, Thailand.

E-mail: HataichanokP12@gmail.com

[2]Department of Plant Production Technology, Faculty of Agricultural Technology, King Mongkut's Institute of Technology Ladkrabang, Bangkok, Thailand

E-mail: ammorn.in@kmitl.ac.th

Abstract. Four plant essential oils from *Illicium verum* (Star anise), *Curcuma longa* (Tumeric), *Foeniculum vulgare* (Sweet funnel), and *Ocimum tenuiflorum* (Holy basil) were tested against the 2nd instar larvae of the beet armyworm, *Spodoptera exigua* in order to evaluate their killing efficiency, and antifeedant activity by using leaf dipping bioassay. The results revealed that the essential oils from *Illicium verum* and *Curcuma longa* were the highest effective essential oils in having killing, and antifeedant property against the insect. At 24 hours, they caused 100% mortality with only 0.750% (v/v) concentration and showed the LC50 of 0.438 and 0.413%, respectively. Both essential oils at 0.500% concentration expressed the antifeedant property at 100% within 24 hours. So that, the essential oils of *Illicium verum* and *Curcuma longa* have a potential to be applied as botanical insecticide for *Spodoptera exigua* control management.

1. Background

The quantity and quality of the product are the most important output of agricultural production. The insect pest is always mentioned as the most problematic of crop production. Using the chemical pesticides has been very effective in reducing pest infection and increasing agricultural production and economic growth. However, it has an effect on the environment, reduced agricultural, negatively impacted on agricultural production, and ill-health of human (Pimentel et. al., 1992; Pimentel and Greiner, 1997). Additionally, targeted pests have developed resistance to pesticides resulting in rapid increase in pesticide use. Therefore, an increase in using pesticide for pest control has led to an increase in the poison of many species of economic pests due to the demolition of non target species and natural predators of pests and parasites (Pimentel et al., 1992) and affects the economic system.

Using the synthetic insecticides causes some anxiety about their adverse effect on the environment. In the present, some farmers use the plant extracts for controlling pests. The compound of plant extracts are eco-friendly because of they are easily degraded, safe to non-target organism, and cheaper than imported insecticide (Lee, et. al., 2001).

The beet armyworm, *Spodoptera exigua*, is one of the most destructive pests of agricultural vegetables, fruits and crops because it can destroy as much as 50% of crop product (Toby, 2019). It occurs worldwide because of higher reproductive rate and migration. The larvae feed sociable by initially grinding the surface of the leaf and result to high level of economic agricultural losses. The methods of pest control are chemical, biological, and physical control. It has been heavily depending on the use synthesis pesticide. The long continued long term of problems is increasing corroboration of negative environmental, health of living things and infexible environmental regulation of pesticides (Ozkara, et. al., 2016).

According to the following reports, herbal extracts can be used for pest control. Yinghui, et. al., 2014 showed the essential oil extracted from fructification of *Illicium verum* against the instar larvae and pupa of *Aedes albopictus* and *Culex pipens*. Jia. et. al., 2018 presented the ethanolic extracts of *Curcuma longa* root, the ar-turmerone and 8-hydroxyl-ar-turmerone of *Curcuma longa* against the 4th-instar larvae of *Culex pipens* with high efficiency after 24 h of treatment. Abbas. et. al., 2015 revealed the Arturmerone and curcuminoids of *Curcuma longa* rhizome and leaf against *Aedes aegypti* and *Anopheles quadrimaculatus*. Diara, et. al., 2015 reported the essential oil of *Foeniculum vulgare* displayed strong larvicidal effect against *Aedes aegypti*. L. B. Gende, et. al., 2009 carried out the *Pimpinella anisum* and *Foeniculum vulgare* essential oils against Paenibacillus larvae and both oils presented great similarity in physiochemical properties values and antimicrobial activity. Sarita, et. al., 2017 demonstrated *Ocimum basilicum* leaves against *Aedes aegypti* as a suitable and eco-safe alternate to chemical insecticides as well as Zorica, et. al., 2013 showed the essential oil of sweet basil, *Ocimum basilicum* tested for their antifeedant properties against 2nd instar gypsy moth larvae, *Lymantria dispar*.

Above mentioned implies that plant extract or plant essential oils play an important role in protection methods against insect pests. In this research, the effectiveness of four essential oils, namely *Illicium verum*, *Curcuma longa*, *Foeniculum vulgare*, and *Ocimum tenuiflorum* were evaluated for their insecticidal activities against *Spodoptera exigua* by using leaf dipping bioassay.

2. Methods

2.1. Essential oils preparation

Essential oils from dried flower of star anise *Illicium verum*, fresh rhizome of tumeric, *Curcuma longa*, seed of sweet fennel, *Foeniculum vulgare*, and fresh leaves of holy basil, *Ocimum tenuiflorum* which have been previously reported to contain insecticidal properties against many insect pests were selected to test. All essential oils were purchased from Thai-China Flavours and Fragrances Industy Co. Bangkok, Thailand. Essential oils were diluted in distilled water plus tween-20 to be 0.25, 0.50, 0.75, 1.00, and 1.25% (v/v) concentrations where the control group used was 1.25 % of tween-20 plus distilled water

2.2. Insect samples

The insect culture of *Spodoptera exigua* was maintained in laboratory conditions at 25°C and 12:12 light dark. The second instar larvae were used in this study.

2.3. Bioassay

Leaf dipping method was applied, when 3 CM diameter leaves of Chinese cabbage were dipped in 0.25, 0.50, 0.75, 1.00, and 1.25% concentrations of essential oil and also control group. After drying under shade at room temperature, five treated leaves were placed in circle box. Then, ten individuals of second instar larvae were released in each box. The mortality was observed after 24 h and Chinese cabbage leaf consumption percent was calculated over graph paper.

2.4. Data analysis

The completely randomized design (CRD) with 3 replications was applied. The data obtained were statistically analyzed by via probit analysis for LC50 values.

3. Results and discussion

The essential oils of *Illicium verum* and *Curcuma longa* treatments performed the highest killing rate against the larvae of *Spodoptera exigua* with 0.75% concentration in which caused 100% mortality at 24 hours, those showed the LC50 of 0.438 and 0.413%(v/v), respectively, whereas, the essential oils of *Foeniculum vulgare*, and *Ocimum tenuiflorum* gave the highest killing rate against the larvae of *Spodoptera exigua* with 1.25% concentration in which caused 100% mortality at 24 hours, those showed the LC50 of 0.601 and 0.646%(v/v), respectively (Table1). The essential oils from *Illicium verum* and *Curcuma longa* showed the highest antifeedant property against the larvae at the 0.50% concentration where the essential oils from *Foeniculum vulgare*, and *Ocimum tenuiflorum* at 0.75% concentration presented the highest antifeedant property against the larvae (Table2).

*3.1**Table 1*** The mortality of second instar larvae of *Spodeptera exigua* caused by different plant essential oils and their LC50 values.

Essential oil	Concentration (%), (v/v) Average mortality percentage						LC50
	0	0.25	0.50	0.75	1.00	1.25	
Illicium verum	0 ± 0.00	46.66 ± 0.00	66.66 ± 5.77	100 ± 0.00	100 ± 0.00	100 ± 0.00	0.438
Curcuma longa	0 ± 0.00	53.33 ± 0.00	83.33 ± 5.77	100 ± 0.00	100 ± 0.00	100 ± 0.00	0.413
Foeniculum vulgare	0 ± 0.00	30.00 ± 5.77	53.33 ± 5.77	66.66 ± 0.0	73.33 ± 0.0	100 ± 0.00	0.601
Ocimum tenuiflorum	0 ± 0.00	30.00 ± 4.71	53.33 ± 5.77	56.66 ± 0.0	70.00 ± 0.0	100 ± 0.00	0.646

*3.1 **Table 2*** The antifeedant activity of fresh chinese cabbage leaf after treated with different plant essential oils against *Spodoptera exigua* larvae

Essential oil	Concentrations (%) Leaf feeding area (%)					
	0	0.25	0.50	0.75	1.00	1.25
Illicium verum	100 ± 0.00	8.33 ± 0.00	0 ± 0.00	0 ± 0.00	0 ± 0.00	0 ± 0.00
Curcuma longa	100 ± 0.00	5.56 ± 0.00	0 ± 0.00	0 ± 0.00	0 ± 0.00	0 ± 0.00
Foeniculum vulgare	100 ± 0.00	36.11 ± 0.00	13.89 ± 5.77	0 ± 0.00	0 ± 0.00	0 ± 0.00
Ocimum tenuiflorum	100 ± 0.00	44.44 ± 4.71	19.44 ± 5.77	0 ± 0.00	0 ± 0.00	0 ± 0.00

Our obtained results were similar to the previous other studies, particulary their chemical compound activities such as the tumeric powder from Curcuma longa rhizome and its derivaties caused 10-20% mortality in third instar *Trichoplusia ni* at a very low dose, 10 μg/larva by contact toxicity (Wagner, et. al., 2016). A treatment of 1% ($m.v^{-1}$) acetonic solution of ar-turmerone mixed in an artificial diet produced 58.3% mortality of one-day-old larvae of the fall armyworm, *Spodoptera frugiperda* after 10 days of feeding (Smith, 1797). The compound ar-turmerone caused 100 and 64% mortality of adult brown planthopper, Nilaparvata lugens (Stal., 1854). Curcuminoids of turmeric rhizome powder showed the growth inhibitory activity against the desert locust, *Schistocera gregaria* and tumeric oil showed 10% growth inhibition and 60% nymphal mortality in *Schistocera gregaria* (Forsskal, 1775). The tumeric extract controlled the peach fruit fly and growth inhibition (Abdul, 2011). The 50 $\mu g ml^{-1}$ tumeric leaf extract showed the highest killed activity within 24 h. It may use to control instar larvae of *Aedes aegypti*. This experiment implied an eco-friendly and low-cost approach (Kalyani, et. al., 2017). The essential oil of *Illium verum* demonstrated high larvicidal activity with a LC50 < 18 mgL^{-1}. Eucalyptol and caryophyllene compound of the *Illicium verum* extract showed the least significant activity against mosquito larvae (Athanasios., 2012). The *Illicium verum* showed high insecticidal property against adults of *Tribolium castaneum* and *Sitophilus zeamais* by the fumigant toxicity and repellency (SH Ho., 1997). The essential oil of the *Illicium verum* was also tested against the third instar larvae of *Aedes aegypti* and presented the highest larval mortality with LC50 28.2 μg mL^{-1} and adult mortality with LC50 10.3 μg mg $female^{-1}$ (Diego, 2018). The effectiveness of leaf and flower extracts of *Ocimum sanctum* against fourth instar larvae of *Aedes aegypti* and *Culex quinquefasciatus* was investigated and showed the highest larval mortality (Mohamed, 2008). This interesting result obtained when 99% mortality of *Aedes aegypti* larvae was found when it treated with 37.1 and 52.4 μL L^{-1} of *Foeniculum vulgare* essential oils from Cape Verde and Portugal, respectively. It displayed strong larvicidal effect against *Aedes aegypti* (Diara., 2015). The essential oil of *Foeniculum vulgare* seed showed larvicidal activity caused 90% mortality for the fourth instars larvae of *Culex pipens* at concentration 60 mg/L after 4h (Safia, 2014). So that those two essential oils demonstrated high potential to be used as botanical insecticide. These oils contained insecticidal properties in terms of successfully killing larvae, and antifeedant effect. In the future, field study is still needed.

4. Conclusion

Illicium verum and *Curcuma longa* were the highest effective essential oils in having killing, and antifeedant property against the insect. At 24 hours, they caused 100% mortality with only 0.750% (v/v) concentration and showed the LC50 of 0.438 and 0.413%, respectively. Both essential oils at 0.500% concentration expressed the antifeedant property at 100% within 24 hours.

References

[1] A.J. Rajamma, Sonal Dubey, Sateesha Shivallyborregowda, S.N. Tiwari. 2011. *Natural product research* vol25. Comparative larvicidal activity of different species of Ocimum against *Culex Quinquefasciatus*. pp 1916-22

[2] A. Mohamed Anees. 2008. *Parasitology Research*.vol 103. Larvicidal activity of *Ocimum sanctum* Linn. (Labiatae) against *Aedes aegypti* (L.) and *Culex quinquefasciatu*. pp 1451–1453

[3] A. Ozkara, D Akyil, M Konuk. 2016. *Pesticides*, environmental pollution, and health

[4] Abbas Ali, Yan-Hong Wang, Ikhlas Khan. 2015. *Journal of Medical Entomology* vol 52. Larvicidal and Biting Deterrent Activity of Essential Oils of *Curcuma longa*, Ar-turmerone, and Curcuminoids Against *Aedes aegypti* and *Anopheles quadrimaculatus* (Culicidae:Diptera). pp 979-86

[5] Abdul Rauf Siddiqi, Athar Rafi, Falak Naz, Rafique Masih, Iftikhar Ahmad, Ghulam Jilani. 2011. *Agrotec*. vol35. Effect of *Curcuma longa* extracts on mortality and fecundity of *Bactrocera zonata* (Diptera: Tephritidae). pp 1413-7054

[6] Athanasios C Kimbaris, George Koliopoulos, Antonios Michaelakis, Maria A Konstantopoulou. 2012. *Parasitology research*. vol 111. Bioactivity of *Dianthus caryophyllus*, *Lepidium sativum*, *Pimpinella anisum*, and *Illicium verum* essential oils and their major components against the West Nile vector *Culex pipens*. pp 2403-2410

[7] David Pimentel, H. Acquay, M. Biltonen, P. Rice, M. Silva, J. Nelson, V. Lipner, S. Giordano, A. Horowitz and M. D'Amore. 1992. *BioScience* vol 42. Environmental and Economic Costs of Pesticide Use. pp 750-760

[8] David Pimentel, Christa Wilson, Christine McCullum, Rachel Huang, Paulette Dwen, Jessica Flack,Quynh Tran, Tamara Saltman, and Barbara Cliff. 1997. *BioScience* vol 47. Economic and Environmental Benefits of Biodiversity. pp 747-757

[9] Diara Rocha, Olivia Matos, Maria Teresa Novoa, Ana Cristina Figueiredo. 2015. *Natural product communications* vol 10. Larvicidal Activity against *Aedes Aegypti* of *Foeniculum Vulgare* Essential oils from Portugal and Cape Verde. pp 677-82

[10] Diego Gomes da Rocha Voris, Luciana dos Santos Dias, Joselia Alencar Lima, Keila dos Santos
Cople. 2018. *Environmental Science and Pollution Research* vol 23. Evaluation of larvicidal, adulticidal, and anticholinesterase activities of essential oils of *Illicium verum* Hook. pp 22541-22551

[11] Jia Liu, Min Zhang, Wei-jie Fu, and Feng Hu. 2018. *Journal of Applied Entomology* vol 142. Efficacy of bioactive compounds from *Curcuma longa* L. against mosquito larvae

[12] Liesel B. Gende, Matias D. Maggi, Rosalía Fritz and Martin J. Eguaras, Pedro N. Bailac and Marta I. Ponzi. 2009. *Journal of Essential Oil Research* vol 21. Antimicrobial Activity of *Pimpinella anisum* and *Foeniculum vulgare* Essential Oils Against Paenibacillus larvae

[13] S E Lee, B H Lee, W S Choi, B S Park, J G Kim, B C Campbell. 2001. *Pest Manag Sci* vol 57.
Fumigant toxicity of volatile natural products from Korean spices and medicinal plants towards the rice weevil, Sitophilus oryzae (L). pp 548-53

[14] Yinghui Mu, Yunyun Zhao, Xiaolong Fang, and Yanbo Cheng. 2014. *Bulletin of Environmental*
Contamination and Toxicology vol 92. Metal Pollution in (Cd, Pb, Zn, and As) Agricultural Soils and Soybean, Glycine max, in Southern China

IOP Conf. Series: Materials Science and Engineering 1163 (2021) 012012 doi:10.1088/1757-899X/1163/1/012012

Identification of strategies over tools for static code analysis

Darko Stefanović[1], Danilo Nikolić[2], Sara Havzi[3], Teodora Lolić[4], Dušanka Dakić[5]

[1,2,3,4,5] Department of Industrial Engineering and Engineering Management, Faculty of Techincal Sciences, University of Novi Sad, Novi Sad, 21000, Serbia

E-mail: nikolic.danilo@uns.ac.rs

Abstract. Static code analysis tools are being increasingly used to improve code quality. The source code's quality is a key factor in any software product and requires constant inspection and supervision. Static code analysis is a valid way to infer the behavior of a program without executing it. Many tools allow static analysis in different frameworks, different programming languages, and detecting different defects in the source code. Different strategies of using static code analysis tools are often used, and these strategies are not classified. In this paper, an experiment was conducted on different tools and their use in relation to the standard code review cycle. The identified strategies for using static code analysis tools and the steps required to implement them are presented. When using the tool, users should choose one of the identified strategies to implement following the defined steps for successful implementation.

1. Introduction

Since the beginning of software development as a scientific discipline, software developers have lacked a source code quality assessment method [1].

The source code's quality is a key factor in any software product and requires constant verification and monitoring. Static analysis is used to maintain and improve the quality of the source code. Static analysis of program code has been used since the early 1960s to optimize the operation of compilers [2]. Later, it proved useful for debugging tools, as well as for software development frameworks. A growing number of tools allow static code analysis, many of which allow an analysis of different programming languages [3].

Static analysis tools are used to generate reports and point out certain deviations from the prescribed code quality standards. However, these tools do not allow automatic modifications to the source code. The decision to change the structure of previously written code is up to the software developers. Static code analysis tool helps software developers by, in addition to generating a report, stating the cause of a particular defect, as well as how this defect can be corrected.

As the first step in this research, a systematic literature review in the field of static code analysis was conducted [4], and various tools for static code analysis, programming languages for which they have support, and the types of defects they detect were presented and analyzed. This systematic review of the literature is presented in detail in Section 3 of this paper.

Static code analysis tools are applied directly to the source code of the program [1]. Software developers and other users of these tools use different ways to apply these tools over program source code. This paper aims to identify different ways (strategies) of applying tools for static code analysis over program source code. Furthermore, the paper aims to identify the standard steps of implementing different strategies over static code analysis tools.

IOP Conf. Series: Materials Science and Engineering 1163 (2021) 012012 doi:10.1088/1757-899X/1163/1/012012

The remainder of the paper is organized as follows. Section 2 presents the basic concepts of static code analysis tools; Section 3 describes related work based on a conducted systematic literature review; Section 4 describes the methodology used to conduct the research; Section 5 presents the results, strategies, and implementation steps defined strategies; Section 6 concludes the paper and suggests future research.

2. Theoretical foundations

This section will present how static code analysis is performed, which tools help with this process, and an introduction to domain-specific programming language (DSL) and general-purpose programming language (GPL).

2.1. Static code analysis

The process of static code analysis is useful not only for optimizing the compiler's operation (which was the original purpose) but also for detecting defects and possible shortcomings. In this way, it is possible to create tools that will help developers understand the program's behavior and identify various shortcomings of the program without its execution. The tools used for static code analysis are programs that explain the behavior of other programs [2].

At the outset, it is necessary to consider the types of defects that static analysis can detect. Problems can also arise due to a lack of understanding of what secure programming entails. It is not uncommon for developers to be unaware of how attackers will try to exploit a piece of code.

With this in mind, static analysis is suitable for identifying problems for several reasons [5]:

Static analysis tools apply checks thoroughly and consistently, without any prejudice that the programmer may have about which parts of the code are "interesting" from a security perspective or which parts of the code can easily be performed by dynamic testing. The code review must be done without the subjective feeling of the programmer. Therefore, unbiased analysis can be valuable, which static code analysis tools provide.

By examining the code itself, static analysis tools can often point to the root cause of a security problem, not just one of its symptoms. This is especially important to ensure that defects are corrected. Static analysis can find defects in early development, even before the program is run for the first time. Finding a defect early reduces the cost of fixing the defect, but a cycle of quick feedback can help guide the programmer's work: The programmer has the opportunity to correct defects that he was not previously aware could even occur. Attack scenarios and code construct information used by the static analysis tool act to transmit knowledge.

When analysis detects a new type of attack, static analysis tools make it easy to recheck a large portion of the code to detect where a new attack might occur. Some security bugs have been in the software for years before they were discovered, making the assessment of outdated code of newly discovered types of defect invaluable.

However, despite all of the above, static code analysis tools are not used often. Software developers do not often decide to improve their source code's quality using these tools [1].

The standard code review cycle includes four main phases [5]:

1. defining the goal,
2. launching a tool for static code analysis,
3. review of the source code (based on the obtained result of static analysis),
4. source code reengineering.

In addition to the standard cycle, Figure 1 shows several potential feedbacks or minor iterations between standard cycle steps, making the cycle more complex than the standard one. The number of iterations of this cycle and the need for feedback within the cycle depends on applying the static code analysis report. This paper aims to identify strategies for applying static source analysis tools based on feedback within the standard cycle and iteration between steps. These strategies are described in Section 5 of this paper.

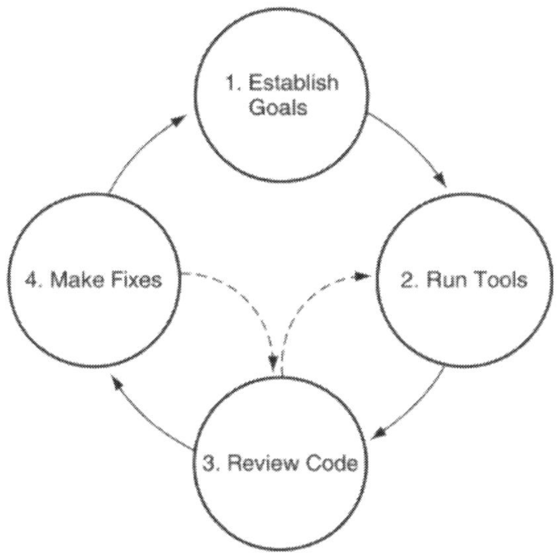

Figure 1. The code review cycle [5].

2.2. GPL and DSL

The GPL is a computer language that is widely applicable in application domains and does not have a special function for a specific domain. Static analysis tools and quality control tools, in general, are ready for GPL source code analysis.

In contrast, DSL specializes in a specific application domain. The difference is not always clear because the programming language may have specialized characteristics for a particular domain but be applicable more widely or capable of wide application, but in practice is used primarily for a particular domain [6]. Professionals in many different fields can develop and describe solutions to problems that take the form of source code artifacts using DSLs that are specifically defined for their field [7]. Static code analysis is a big challenge when it comes to DSL.

The experiment described in this paper refers to a single GPL, and the strategies identified relate to the GPL. Additional analyzes are needed to conclude the implementation of strategies over DSL.

3. Related work

Systematic literature review in the field of static analysis of code and its tools was conducted. A paper describing the process of the systematic literature review was published in [4]. The studies included in this review are published in the previous ten years. The paper review was performed based on guidelines for a systematic review of software engineering literature by Barbara Kitchenham [8], focusing on presenting the most commonly used tools divided by programming languages for which they have provided support and detecting defects.

To review the literature, two databases were searched:

1. SCOPUS and
2. Google Scholar.

The search terms for the mentioned databases are as follows: "Static code analysis" AND "tools" AND "detection" and pubyear> = 2010. The search resulted in 346 initial papers.

Based on the set inclusion criteria, by reviewing abstracts and titles, 22 primary studies were included in the literature review. Some of the representative examples of the included studies are [9], [10], [11],

[12], [13]. A detailed procedure of this systematic review of the literature step by step is presented in the research [4].

The results obtained show that many more papers represent results on tools that have secured support for the GPL than papers representing tools that have secured DSL support. The percentage is shown in Table 1.

Table 1. Percentage of primary studies by type of programming language [4].

Programming language	Number of studies	%
DSL	3	14%
GPL	19	86%

As part of this review, the tools for static code analysis that are most often used in research on different programming languages and detecting various defects were also analyzed.

Many of the static code analysis tools presented in different studies support several GPLs, while other tools have support for only one programming language. Most static code analysis tools support those programming languages that are most commonly used in the industry.

This systematic review presents the static code analysis tools most commonly used in research and which tools are analyzed based on the type of defects they detect and the programming languages they support.

Six tools were presented, which were most often analyzed and presented in primary studies. These are tools that have been presented three or more times in different studies. In comparison, other tools presented in the studies covered by this review were analyzed in two or one studies. Graphically presented results are shown in Figure 2.

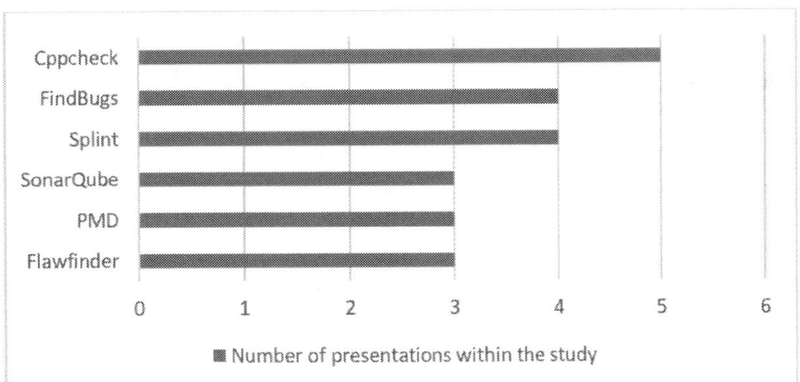

Figure 2. The most common static code analysis tools presented in primary studies [4].

Static code analysis tools are increasingly finding their application and use in the work of programmers in various positions, so it is important that these tools are constantly improved, but also to develop new tools that will offer their customers a better service by building confidence in the results of static analysis and allowing them to avoid certain defects in the source code and improve their software solution.

However, static code analysis tools only allow the analysis of a certain number of programming languages. The range of programming languages for which they have support is constantly expanding. In comparison, it is expected that work will be done on creating and improving tools that will have support for DSL in the coming period. For now, only a few tools have DSL support with a specification of the rules built into the tool using the plugin.

The results show that certain tools are used more often in studies than others. Based on the results obtained from research [4], it is concluded that research mainly analyzes the results obtained using a certain tool for static code analysis, under certain conditions, for certain types of defects, and over certain programming languages.

No paper attempts or proposes specific strategies for conducting static code analysis using some of the presented tools.

4. Methodology

To successfully set up the research methodology, it is first necessary to identify the tools of static code analysis that will be used for the realization of the research. An experiment conducted on tools was used as a research model to identify strategies for conducting static code analysis. In this regard, the tasks of this research are:

1. identify the tools that will be used to conduct the experiment,
2. identify strategies for conducting static code analysis.

The tools used to conduct this research have been identified based on a literature review conducted within [4], already described in Section 2 of this paper. The literature review identified the most commonly used tools for static code analysis in the primary studies covered by this literature review. Also, the selected tools must have java programming language support and a plug-in for the Eclipse framework.

Once the tools that will be used to conduct the experiment have been identified. An analysis was conducted on the selected tools, their characteristics and the features it offers. The experiment was conducted using selected tools on different software products and analysis of the implementation in relation to the standard cycle of code review.

As the first phase of conducting the experiment, it is necessary to establish the initial configuration of the selected tools on the device from which the experiment was conducted. Next, it is necessary to analyze the features of each tool and the ways in which it is possible to apply the tool over the source code of the software product. After the analysis, it is necessary to conduct an experiment, which means conducting a code review procedure on each selected tool and identifying different strategies. Above the identified strategies, it is necessary to detect standard steps for their implementation. The experiment is considered complete when all ways to perform static code analysis in relation to the standard code review cycle are detected and described.

A description of the strategy identification process is described in Section 5 of this paper.

4.1. Research limitations

In the research, the authors limit themselves to the possibility of additional strategies that can be identified in the process of conducting static analysis over other tools not covered by this research or over tools that have provided support for other programming languages not used in this research. The experiment was performed on a single GPL, and it is not possible to draw conclusions about DSL based on this experiment. Also, the experiment was conducted on one project, and there is a possibility of identifying other strategies on different types of projects that are not the subject of this research.

5. Results

This section describes the selected tools, identified strategies, sub-strategies of the static code analysis tools, and the steps of implementing each strategy concerning the standard code review cycle.

The selected tools that met the set requirements for conducting the experiment are::

1. SonarQube,
2. FindBugs,
3. PWD,
4. Checkstyle.

5.1. Strategies over static code analysis tools

The basic division into two major classes of strategies for using static code analysis tools is [14]:

1. Running a static analysis tool over a completed software project – **S1**.
2. Running a static analysis tool during software project development – **S2**.

In this paper, a completed software product will be considered a software product in which development has been completed and which should be tested before deployment. While the application during development is considered to be the use of tools for static analysis during development, ie the inclusion of tools from the beginning of software product development.

If a static analysis startup strategy is applied during software product development, it is essential that the static code analysis tool be included in the integrated development environment (IDE) being developed by the program or external startup at specified time intervals (S2). In this way, the development environment, with the help of a plugin, independently generates static code analysis reports, after which the software developer can check the detected defects and correct them before continuing with further development. Many static code analysis tools provide the ability to integrate into integrated development environments. However, static code analysis is often used only after development is complete, and it is necessary to apply a strategy to run analysis over the completed software product (S1).

Strategies for using static analysis tools will be described in detail, and their sub-strategies and steps within these strategies.

5.1.1 Running a static analysis tool over a completed software project – S1

For this research, a static code analysis was performed on a completed project done in an Eclipse development environment. The source code of the project is written in the Java programming language. The project is a backend part of a multi-layered application for monitoring students at faculties, divided by departments. Over the completed software project, a static analysis of the code was started using tools, and a report was generated which marked the defects in the project. Static code analysis tools independently classify defects that they detect according to the adopted classification scheme.

The tool's initial report detected a total of 18 defects at different levels in the project. After correcting all defects detected by the tool within the initial report, restarting the tool is expected to result in a total of 0 defects. However, after reengineering the source code and resolving all defects, restarting the tool over the source code generates a report that detects 8 new defects in the project's source code.

Based on the obtained report, it can be concluded that the source code corrections caused new defects marked within the new report, generated by restarting the tool on the same project.

After correcting newly created defects, another run of the tool over the source code, the number of all defects is equal to 0.

By conducting this experiment, which involves correcting all defects observed in the initial report and then restarting the tool to check with the second report, it is possible to identify the first subtype of the strategy, listed in section 5.1. of this paper, which is entitled "Starting the static analysis tool after correcting all defects "– S1.1.

If the identified strategy is compared with the presented standard code review cycle based on static analysis, described in section 2.1. it would be seen that the identified strategy fully follows the standard procedure without frequent feedback or iterations between standard steps within the cycle.

Finally, the standard steps for implementing this sub-strategy, relative to the standard code review cycle, are:

1. defining the goal,
2. launching a tool for static code analysis,
3. review of source code,
4. source code reengineering,
5. launching a tool for static code analysis,
6. review of source code,

IOP Conf. Series: Materials Science and Engineering 1163 (2021) 012012 doi:10.1088/1757-899X/1163/1/012012

7. source code reengineering.

However, if the possibility of more frequent use of feedback and smaller iterations within the standard procedure is set, another sub-strategy can be identified.

If the software developer starts correcting defects in the program's source code and, after eliminating one defect, restarts the static analysis over the source code, the report is expected to identify one error less than the initial report.

In this way, the software developer, every time he corrects one of the detected defects, checks whether that defect has been eliminated. On the other hand, there is a possibility that the correction of one defect will cause additional defects in the source code of the project. Frequent starting of the tool, i.e. starting after each correction, enables earlier detection of newly formed defects. In this way, even if new defects occur, the software developer can react and eliminate new defects promptly or otherwise correct the source code not to cause new defects.

Setting up this way of applying static analysis, by running the tool after correcting each defect, another subtype of the strategy listed in section 5.1 of this paper is identified, entitled "Running the static analysis tool after correcting any defect" – S1.2.

If the identified strategy is compared with the presented standard cycle of code review based on static analysis, described in section 2.1., It can be concluded that this sub-strategy requires more feedback, i.e. more iterations between standard steps within the cycle.

Finally, the steps of the second sub-strategy are:

1. defining the goal,
2. launching a static analysis tool,
3. review of the source code (based on the obtained result of static analysis),
4. correction of a defect.

Steps two, three, and four are repeated as many times as there are total defects in the project.

5.1.2 Running a static analysis tool during software project development – S2

If a different static analysis strategy is applied to the static code analysis process during software development, in this case, two sub-strategies can be separated. These sub-strategies differ depending on whether the static code analysis tool is included in the development environment in which the software solution is being developed or is used externally during development.

The first sub-strategy, "Starting static analysis tools by external tool calling during software project "– S2.1 involves the external use of static code analysis tools and the occasional running of tools over a project during development. Occasional running represents running the tool at different levels of the scope of written code, there are a number of levels such as: running after a single class is written, a specific block of code, a logical part, or after each line of code in a project during development.

It is also possible to start based on a time determinant. In defining this sub-strategy and its steps, there will be no more detailed sub-strategies depending on the written code's volume levels.

The steps of this sub-strategy concerning the standard code review cycle are:

1. defining the goal,
2. development of a part of the software project,
3. launching a static analysis tool,
4. review of the source code (based on the report of the obtained result of static analysis),
5. correction of detected defects,
6. continued development of the software project.

Steps two, three, four, five, and six are repeated until the software solution development is complete.

The second sub-strategy, "Running static analysis tools by including tools in the development environment in which the software project is developed," – S2.2 involves the integrated use of static

IOP Conf. Series: Materials Science and Engineering 1163 (2021) 012012 doi:10.1088/1757-899X/1163/1/012012

code analysis tools. Integrated use is possible only for those static analysis tools, which have provided support for the appropriate environment, ie, it is possible to place them as an adjunct in the development environment in which the project is developed. In this way, during the development of the project, a tool for static analysis is launched, which, while the project is being developed, performs analysis on the source code and detects the observed defects.

The advantage of this method is that the software developer finds out the cause that caused the defect and eliminates it at the moment of the defect's occurrence.

The steps of this sub-strategy concerning the standard code review cycle are:

1. defining the goal,
2. inclusion of tools for static analysis in the development environment of the project,
3. software project development,
4. the static analysis tool detects a defect,
5. review of the source code (based on the detected defect),
6. correction of detected defect,
7. continued development of the software project.

Steps three, four, five, six, and seven are repeated until the software solution's development is completed.

6. Conclusion

Finally, strategies for applying static analysis tools can be classified as presented in table 2.

Table 2. Identified strategies over tools for static code analysis

Strategy label	Strategy	Substrategy label	Substrategy
S1	Running a tool for static analysis over a completed software project	S1.1	Starting the static analysis tool after correcting all defects
		S1.2	Running the static analysis tool after correcting any defect
S2	Running a static analysis tool during software project development	S2.1	Starting static analysis tools by external tool calling during software project development
		S2.2	Running static analysis tools by including tools in the development environment in which the software project is developed

The strategies identified by this research relate to the type of project used to conduct the experiment, and the tools used in the experiment. The authors limit themselves to the possibility of identifying other strategies over other tools and projects.

Further research on identified strategies should include a comparative analysis of strategies across different tools and projects. It is necessary to conduct different experiments by applying the identified strategies over different static code analysis tools. During the analysis, it is necessary to measure the speed of strategy implementation to compare the strategies concerning the time required for their implementation. It is also necessary to compare the "accuracy" of the strategy, i.e., how many erroneous results the tool generates using the identified strategies.

IOP Conf. Series: Materials Science and Engineering 1163 (2021) 012012 doi:10.1088/1757-899X/1163/1/012012

The development and application of tools for static code analysis enable better software solutions and represent a step forward in software development.

References

[1] E. Penttila, "Improving C++ Software Quality with Static Code Analysis," Aalto, Aalto University, 2014.

[2] A. Moller and I. Schwartzbach, "Static program analysis," 2019.

[3] M. Beller, R. Bholanath and M. S., "Analyzing the state of static analysis: A large-scale evaluation in open source software," 2016.

[4] D. Stefanović, D. Nikolić, D. Dakić, I. Spasojević and S. Ristić, "Static code analysis tools: A systematic Literature Review," in Proceeding of of 31st DAAAM International Symposium, Vienna, 2020.

[5] J. West and B. Chess, "Secure programming with static code analysis," Pearson education, 2007.

[6] M. Fowler, "Domain-Specific Languages," in Pearson Education, London, 2010.

[7] I. Ruiz-Rube, T. Person, J. M. Dodero, J. M. Mota and J. M. Sánchez-Jara, "Applying static code analysis for domain-specific languages," Software & Systems Modeling, vol. 19, no. 1, pp. 95-110, 2020.

[8] B. Kitchenham, "Procedure for Undertaking Systematic Reviews," Keele University, Computer Science Department, 2004.

[9] A. Kaur and R. Nayyar, "A Comparative Study of Static Code Analysis tools for Vulnerability Detection in C/C++ and JAVA Source Code," Procedia Computer Science, vol. 171, no. 2019, pp. 2023-2029, 2020.

[10] D. Marcilio, C. Furia, R. Bonifacio and P. Gustavo, "SpongeBugs: Automatically generating fix suggestions in response to static code analysis warnings," The Journal of Systems & Software, vol. 168, 2020.

[11] P. Nunes, I. Medeiros, J. Fonseca, N. Neves, M. Correia and M. Vieira, "An empirical study on combining diverse static analysis tools for web security vulnerabilities based on development scenarios," Computing, vol. 101, no. 2, pp. 161-185, 2019.

[12] N. Meghanathan, "Identification and removal of software security vulnerabilities using source code analysis: a case study on a java file writer program with password validation features," Journal of Software, vol. 8, no. 10, 2013.

[13] K. Goseva-Popstojanova and A. Perhinschi, "On the capability of static code analysis to detect security vulnerabilities," Information and Software Technology, pp. 18-33, 2015.

[14] J. Xie, H. R. Lipford and B. Chu, "Evaluating interactive support for secure programming," Conference on Human Factors in Computing Systems, pp. 2707-2716, 2012.

Analysis of the Tools to Support Systematic Literature Review in Software Engineering

Darko Stefanovic[1], Sara Havzi[2*], Danilo Nikolic[3] Dusanka Dakic[4] and Teodora Lolic[5]

[1,2,3,4,5]Department of Industrial Engineering and Engineering Management, Faculty of Techincal Sciences, University of Novi Sad, 21 000, Serbia

E-mail: havzisara@uns.ac.rs

Abstract. A thorough and accurate review of the literature is a necessary condition for relevant research. However, the literature review is a laborious, error-prone process. For this reason, various software tools are being developed to support researchers in conducting literature reviews. In this paper, three tools to support a systematic literature review in software engineering were analysed and evaluated using the DESMET methodology. Tools were found by conducting mapping study and were selected since they were most promising to support an overall systematic literature review process. Although the tools do not yet support the whole systematic review process on a high level, they provide a reasonable basis for further development. It is needed to set features that need to be supported by the tools.

1. Introduction

A review of the literature is the basis for establishing comprehension of the state in any scientific area. Evidence-based research initially started in medicine, because research has shown that medical advice and conclusions based on expert opinion are not as reliable as advice based on the accumulation of the results of scientific experiments [1].

In order for the literature review to be correct and relevant, specific rules, as well as strategies and methods, need to be developed in the literature review process. For the reasons given, the literature review presents a complicated process, with a high possibility of error and subjectivity. All of this leads to a large amount of time required to develop a transparent literature review.

In software engineering, the most used approach for the review of the literature is a systematic literature review (SLR) developed by Barbara Kitchenham [2].

Approaches to reviewing the literature, especially the SLR, have reduced these problems to some extent. However, many authors point out that there are still problems with the amount of time spent reviewing the literature, the occurrence of errors and the subjectivity of researchers in this process [3]. In 2012. the research conducted by Hernandez and Craft showed that there are many activities of the SLR that need to be reduced in the difficulty level or time needed to perform an activity. The review of the literature is a manual process that requires a large amount of work [4]–[6] and is, therefore, the primary candidate to benefit aided by technology.

In recent years, software tools are being developed in order to solve the current problems that researchers are facing when conducting a literature review. In order to examine the usefulness of the developed tools in solving existing problems, it is necessary to evaluate these tools. Marshall [7]

conducted a feature analysis using DESMET methodology [8] on the set of candidate tools: SluRp [9], StArt [10], SLR-Tool [11] and SLRTOOL.

The study reported here aims to evaluate three other tools using the same method as Marshall. This way, there will be more tools evaluated by the same method, and researchers will have a broader scope of tools to consider.

The remainder of the paper is organised as follows. In Section 2, authors presented SLR as well as the literature review in the field software tools for literature reviews. In Section 3, the methodology for performing this study was presented. Results and discussion are presented in Section 4. Section 6 concludes the study, its limitations and suggests future research.

2. Tools to Support Systematic Literature Reviews in Software Engineering

SLR is one of the most used approaches in performing the review process of the literature [12]. Fink defined systematic literature review as "systematic, explicit, and reproducible method for identifying, evaluating, and synthesising the existing body of completed and recorded work produced by researchers, scholars, and practitioners" [13].

Therefore, researchers who conduct an SLR use transparent and a systematic process to define the research question, search the publication, assess their quality and qualitative or quantitative synthesis of research results.

SLR guidelines that are given by Kitchenham [2] consist of three main steps:

- planning,
- conducting and
- reporting the review.

Planning the review is the first phase, and it contains activities such as identification of the need for a review, development of protocol and validation of protocol. Conducting the review consists of the identification of research, study selection, quality assessment, data extraction, as well as data synthesis. Reporting means documenting the review, which is writing the review and validating the report. For the SLR to be correct, all of the steps above need to be well performed.

In order to support SLR process in:

- reducing the time required to complete the literature review process,
- increasing quality of assessment of the relevance of papers and review of literature,
- reducing the subjectivity of researchers in the process of reviewing the literature, as well as
- making it easier for researchers to review the literature,

software tools for supporting the SLR process where developed.

In order to map as many tools as possible that are developed to support the SLR, we conducted a mapping study and SLR study and found more than one hundred tools, that could, to some extent help in conducting the SLR. However, there are only a few tools that support the whole SLR process in a way, and some of them are not applicable in the field of software engineering.

3. Method

This section describes the method used to perform the evaluation of the three candidate tools, as well as the description of the chosen feature analysis approach. In addition, it describes how Marshall [14] applied this method evaluating three tools for SLR support, which is used in this paper.

3.1. DESMET method

DESMET is a methodology for evaluating tools or methods. It was developed by Barbara Kitchenham [8] and consisted of nine evaluation methods suitable for different tool analyses. There are quantitive and qualitative methods in DESMET methodology [15]. In the guidelines, the first step to perform DESMET methodology is selecting an appropriate evaluation method.

DESMET method is not flawless, so it has limits when trying to mix and match methods and tools as well as if there is no controllable development process in the organisation [15].

Kitchenham defined evaluation criteria on the basis of which the DESMET methods are chosen. Marshall [7] chose the feature analysis method for this type of evaluation.

3.2. Feature analysis

Feature analysis is a short type of evaluation method and is considered to have a low risk. The objective of the feature analysis provides an input for a decision whether to use or not use a tool in an organisation. Evaluation should consider the following areas [15]:

1. Suitability for purpose
2. Economic issues
3. Drawbacks
4. Other advantages

"A feature analysis type of evaluation is an attempt to put rationale and structure behind a "gut feeling" for the right product" [15]. This evaluation method is also considered as subjective.

Some of the criteria that should be considered when deciding whether to perform a feature analysis are [15]:

- a large number of methods/tools to assess and
- short timescales for evaluation exercise.

3.3. Candidates, Features and Scoring

This chapter presents candidate tools for evaluation as well as a brief overview of the features, feature analysis and scoring used in Marshalls evaluation. The detailed method can be found in [7]. The tool was evaluated by the authors based on a review of the tool as well as publicly available publications and tools documentation. Candidates were chosen from the unpublished SLR of software tools that support literature review in software engineering mentioned above. Candidate tools are chosen as their developers claim that they support each stage of the SLR process. Tools that were already evaluated by Marshal weren't taken into consideration of this evaluation.

The three candidate tools are:

1. SESRA - a web application to support systematic literature review and is a completely free tool. The tool is available in Portuguese and English.
2. ReLis – a web application to support the iterative systematic literature review process.
3. Parsifal - a web application that supports researchers in performing a systematic review of the literature.

A detailed explanation of the features and scoring can be seen in [7].

id	*Feature Set*	*id*	*Subfeature*	*Subfeature Level of Importance*	*Interpretation of Judgement Scale*	*Feature set Importance Weighting*
F1	Economic	F1-SF01	The tool does not require financial payment to use.	HD	JI1	0.1
		F1-SF02	Maintenance	HD	JI1	
F2	Ease of introduction and setup	F2-SF01	The tool has reasonable system requirements.	M	JI1	0.2
		F2-SF02	Simple installation and setup.	HD	JI2	
		F2-SF03	There is an installation guide.	HD	JI1	
		F2-SF04	There is a tutorial.	HD	JI1	
		F2-SF05	The tool is self-contained.	HD	JI1	
F3	SR activity support	F3-SF01	Protocol development	D	JI3	0.4
		F3-SF02	Protocol validation	D	JI3	
		F3-SF03	Supports automated searches	HD	JI3	
		F3-SF04	Study selection and validation	HD	JI3	
		F3-SF05	Quality assessment and validation	HD	JI3	
		F3-SF06	Data extraction and validation	HD	JI3	
		F3-SF07	Automated analysis	HD	JI3	
		F3-SF08	Text analysis	N	JI1	
		F3-SF09	Meta-analysis	N	JI1	
		F3-SF10	Report write up	N	JI3	
		F3-SF11	Report validation	N	JI3	
F4	Process Management	F4-SF01	Support for multiple users	M	JI1	0.3
		F4-SF02	Document management	M	JI1	
		F4-SF03	Security	D	JI1	
		F4-SF04	Management of roles	HD	JI1	
		F4-SF05	Support for multiple projects	M	JI1	

Figure 1. - Features used in the analysis [14]

Subfeature levels of importance are [7]:
1. Mandatory feature – M – 4 points
2. Highly Desirable – HD – 3 points
3. Desirable – D – 2 points
4. Nice to have – N – 1 point

Interpretations of the Judgement Scales are [7]:
1. J11 – Is the feature present?
2. J12 – Is the tool simple to install and setup?
3. J13 – Is the activity supported?

4. Results
In this section, the authors give an overview of the evaluation results for every candidate tool. Feature sets for every tool are briefly described, and a table with grades and the overall score is presented for every tool. Summarisation of the results of all the papers is presented in Table 4.

4.1. SESRA
Table 1 presents the scores for SESRA.

4.1.1. Feature Set 1
In conducting this study, the authors noticed that there are parts of the tool that have not been translated from Portuguese into English, in the English version. The tool is not maintained often and the last changes, according to the author, happened two years ago. SESRA score 3 out of 6 for this feature set.

4.1.2. Feature Set 2

SESRA is based on guidelines for conducting a systematic literature review by Kitchenham. As for web applications, SESRA does not require any special installation. In order for the user to use the tool, it is necessary to register by entering data such as name, surname and email. The developers have made the tutorial available for SESRA users. However, the tutorial is only in Portuguese, and there is no English version. SESRA scored 11.5 out of 13 for this feature set.

4.1.3. Feature Set 3

SESRA makes it possible to identify the need for a systematic review of the literature. By filling out the form within the tool, the user has the opportunity to see if there are reviews of the literature in the area under study. A query made this way gets results from the SESRA public knowledge repository. After that, users are provided with a checklist based on which the user can see whether the currently existing systematic reviews of the literature in the area are sufficient and whether there is really a need for a review of the literature.

SESRA supports the development of research questions according to PICOC components.

Adheres to the study selection stage, at which point studies should be examined and consequently included or excluded from further assessment. Although it is possible to undertake the selection process based on the title and abstract, researchers often opt to obtain a copy of the paper to assess the full text. If multiple researchers are involved, differences among their selection may occur. Hence, SESRA tools help in reaching an agreement through an advisor that reviews the selection, or by applying an inter-rater reliability test to evaluate a random sample of the selected studies.

The tool does not have advanced techniques for automating SLR activities, except when it comes to graph extraction when writing reports. This tool scored 14.5 out of 23 for this feature set.

4.1.4. Feature Set 4

Within the tool, the user can create multiple projects. The disadvantage of SESRA is that it does not have the ability to store the entire text of the publication, and researchers can see the title, authors, abstract and keywords within the tool. In addition, there are different roles that can be assigned to users. For this feature set, SESRA scored 15 out of 17.

Table 1. Scores for SESRA

ID	FEATURE SET	ID	WEIGHTED SCORE	FEATURE SET SCORE	MAX SCORE	% FEATURE SET SCORE
F1	Economic	F1SF01	3	3	6	50%
		F1SF02	0			
F2	Ease of itroduction and setup	F2SF01	4	10	13	77%
		F2SF02	3			
		F2SF03	0			
		F2SF04	3			
F3	SRL activity support	F3SF01	2	14.5	23	63%
		F3SF02	2			
		F3SF03	1.5			
		F3SF04	3			
		F3SF05	3			
		F3SF06	1.5			
		F3SF07	0			
		F3SF08	0			
		F3SF09	0			
		F3SF10	0.5			
		F3SF11	1			
F4	Process management	F4SF01	4	15	17	88%
		F4SF02	2			
		F4SF03	2			
		F4SF04	3			
		F4SF05	4			
				42.5	59	72%

4.2. ReLis

Table 2 presents the scores for ReLis.
Beside authors experience on using ReLis, the publication that presents [16] tool was also taken into consideration.

4.2.1. Feature Set 1

ReLis is entirely free and is maintained regularly. Users can register by filling an easy form. This tool scores maximum 6 points for this feature set.

4.2.2. Feature Set 2

It is easy to register users, as well as ReLis does not require any special system characteristics. There is a detailed tutorial on how to use ReLis, and the tool is self-contained. Regarding all this, ReLis scores 13 out of 13 for this feature set.

4.2.3. Feature Set 3

ReLis does not support automated searches, and it is one of its biggest disadvantages. Also, ReLis does not have report write up and report validation features, comparing to two other tools.

4.2.4. Feature Set 4

Within this tool, users can create multiple projects as well as to have multiple roles in multiple projects. Role management is possible in ReLis, and users can participate in projects in different ways. There is half-support when document management is considered.

Table 2. Scores for ReLis

ID	FEATURE SET	ID	WEIGHTED SCORE	FEATURE SET SCORE	MAX SCORE	% FEATURE SET SCORE
F1	Economic	F1SF01	3	6	6	100%
		F1SF02	3			
F2	Ease of itroduction and setup	F2SF01	4	13	13	100%
		F2SF02	3			
		F2SF03	3			
		F2SF04	3			
F3	SRL activity support	F3SF01	2	10	23	43%
		F3SF02	2			
		F3SF03	0			
		F3SF04	3			
		F3SF05	0			
		F3SF06	1.5			
		F3SF07	0			
		F3SF08	0			
		F3SF09	1.5			
		F3SF10	0			
		F3SF11	0			
F4	Process management	F4SF01	4	15	17	88%
		F4SF02	2			
		F4SF03	2			
		F4SF04	3			
		F4SF05	4			
				44	59	74.6%

4.3. Parsifal

Table 3 presents the scores for Parsifal.

4.3.1. Feature Set 1

This tool is completely free and can be found in [17]. Parsifal provides the organisation of a systematic review of high-level literature. Parsifal is very similar to SESRA in the possibilities it offers to users.

It also works on the principle, filling out various forms and generating documents. Unlike the SESRA tool, Parsifal is in English and provides detailed monitoring of a systematic review of the literature. The tool is maintained regularly and the last changes, according to the evaluation, happened nine months ago.

4.3.2. Feature Set 2

Parsifal does not require any system requirements, and since it is a web application, installation, and setup consist of registering with email. There is a video tutorial on how to use this tool as well as Help and Blog pages on the website.

4.3.3. Feature Set 3

Parsifal is a tool that, through a series of steps, provides support to researchers in performing a systematic review of the literature.
It enables researchers to easily plan, execute and document a systematic review of the literature through forms.

What is one of the biggest advantages of the Parsifal tool in relation to the SESRA tool, which enables almost the same functionalities, is that with the help of Parsifal it is possible to search the databases of Scopus and Science Direct publications directly from the tool.
Finally, the tool allows reporting, but according to the evaluation performed for the purposes of this paper, the tool only allows the creation of a technical report and tables that would help researchers more easily classify publications in the review.

4.3.4. Feature Set 4

Parsifal offers the possibility to work on multiple projects. However, at the time that this study was taken, the invitation link for collaboration on the project led to "Error 404" page, and there was no way to add other people to project, even that tool offers to add collaborators. There is no possibility to manage roles.

Table 3. Scores for Parsifal

ID	FEATURE SET	ID	WEIGHTED SCORE	FEATURE SET SCORE	MAX SCORE	% FEATURE SET SCORE
F1	Economic	F1SF01	3	4.5	6	100%
		F1SF02	1.5			
F2	Ease of itroduction and setup	F2SF01	4	13	13	100%
		F2SF02	3			
		F2SF03	3			
		F2SF04	3			
F3	SRL activity support	F3SF01	2	15	23	65%
		F3SF02	2			
		F3SF03	3			
		F3SF04	3			
		F3SF05	3			
		F3SF06	1.5			
		F3SF07	0			
		F3SF08	0			
		F3SF09	0			
		F3SF10	0.5			
		F3SF11	0			
F4	Process management	F4SF01	4	8	17	47%
		F4SF02	2			
		F4SF03	2			
		F4SF04	3			
		F4SF05	4			
				40.5	59	68%

Table 4. Feature Set Scores

TOOL	F1	F2	F3	F4	TOTAL
SESRA	50%	77%	63%	88%	72%
RELIS	100%	100%	43%	88%	74.6%
PARSIFAL	100%	100%	65%	47%	68%

4.4. Discussion
This section presents the discussion of the results, as well as the limitations of the study. ReLis achieved the best results in this analysis, i.e. 74.6% of the total score. As the differences in the percentages of other tools are not overly different, what ReLis pointed out as better than other tools is the regular maintenance of the tools and the management of roles within the tools. The tools show similar results, indicating that the authors chose tools with relatively similar capabilities and options for researchers. Comparing ReLis to two other tools that are evaluated in this study, its one of the biggest disadvantages is in feature set 3. ReLis does not support automated search, while Parsifal and SESRA do support it. Researchers that conduct SLR pointed out this feature as important [18].

The disadvantages of all tools are reflected when looking at the F3 feature set, which is the most important part of the evaluation. All tools support SLRs in organising reviews, but none of these tools supports researchers in increasing the quality of literature reviews or automating certain processes.

5. Conclusion
The paper presents an evaluation of three identified tools to support the entire SLR process. The methodology used within the paper is the DESMET, i.e. Feature analysis approach, based on the evaluation performed by Marshall in 2014.

There are a sufficient number of tools that support the whole SLR process. However, they do not support it by automating, but by managing the organisation of the process.

In the further development of the tool, developers should focus on automation based on machine learning, extracting knowledge from text and similar, to help reduce time and increase the quality of the SLR.To be able to fully conclude if these tools are useful to researchers, more researchers should use their tools for a longer period of time and conduct a qualitative analysis of their experience. The question that arises for further research is which tool combination gives the best results.

References
[1] B. Kitchenham, O. Pearl Brereton, D. Budgen, M. Turner, J. Bailey, and S. Linkman, "Systematic literature reviews in software engineering - A systematic literature review," *Inf. Softw. Technol.*, vol. 51, no. 1, pp. 7–15, 2009.

[2] B. Kitchenham and S. Charter, *Guidelines for performing Systematic Literature Reviews in Software Engineering*. 2007.

[3] M. J. Grant and A. Booth, "A typology of reviews: An analysis of 14 review types and associated methodologies," *Health Info. Libr. J.*, vol. 26, no. 2, pp. 91–108, 2009.

[4] P. Brereton, B. A. Kitchenham, D. Budgen, M. Turner, and M. Khalil, "Lessons from applying the systematic literature review process within the software engineering domain," *J. Syst. Softw.*, vol. 80, no. 4, pp. 571–583, 2007.

[5] M. A. Babar and H. Zhang, "Systematic literature reviews in software engineering: Preliminary results from interviews with researchers," *2009 3rd Int. Symp. Empir. Softw. Eng. Meas. ESEM 2009*, pp. 346–355, 2009.

[6] M. Riaz, M. Sulayman, N. Salleh, and E. Mendes, "Experiences Conducting Systematic Reviews from Novices' Perspective," no. May 2014, 2010.

[7] C. Marshall, P. Brereton, and B. Kitchenham, "Tools to support systematic reviews in software

engineering: A feature analysis," *ACM Int. Conf. Proceeding Ser.*, no. May, 2014.

[8] B. Kitchenham, S. Linkman, and D. Law, "A methodology for evaluating software engineering methods and tools," *Lect. Notes Comput. Sci. (including Subser. Lect. Notes Artif. Intell. Lect. Notes Bioinformatics)*, vol. 706 LNCS, no. June, pp. 121–124, 1993.

[9] D. Bowes, T. Hall, and S. Beecham, "SLuRp - A tool to help large complex systematic literature reviews deliver calid and rigorous results," *EAST'12 - Proc. 2nd Int. Work. Evidential Assess. Softw. Technol.*, pp. 33–36, 2012.

[10] E. Hernandes, A. Zamboni, S. Fabbri, and A. Di Thommazo, "Using GQM and TAM to evaluate StArt – a tool that supports Systematic Review," *CLEI Electron. J.*, vol. 15, no. 1, 2012.

[11] A. M. Fernández-Sáez, M. G. Bocco, and F. P. Romero, "SLR-Tool a tool for performing systematic literature reviews," *ICSOFT 2010 - Proc. 5th Int. Conf. Softw. Data Technol.*, vol. 2, no. January, pp. 157–166, 2010.

[12] D. Moher, A. Liberati, J. Tetzlaff, and D. G. Altman, "Preferred reporting items for systematic reviews and meta-analyses: The PRISMA statement," *Int. J. Surg.*, vol. 8, no. 5, pp. 336–341, 2010.

[13] F. Arlene, *Conducting research literature reviews : from the internet to the paper*. Thousand Oaks, CA, USA: Sage Publications, 2005.

[14] C. Marshall, P. Brereton, and B. Kitchenham, "Tools to support systematic reviews in software engineering: A feature analysis," *ACM Int. Conf. Proceeding Ser.*, no. September, 2014.

[15] B. A. Kitchenham and S. G. Linkman, *DESMET : A method for evaluating Software Engineering methods and tools*. 2000.

[16] B. Bigendako and E. Syriani, "Modeling a tool for conducting systematic reviews iteratively," *Model. 2018 - Proc. 6th Int. Conf. Model. Eng. Softw. Dev.*, vol. 2018-Janua, no. Modelsward, pp. 552–559, 2018.

[17] "Parsifal Tool." [Online]. Available: https://parsif.al/.

[18] J. C. Carver, E. Hassler, E. Hernandes, and N. A. Kraft, "Identifying barriers to the systematic literature review process," *Int. Symp. Empir. Softw. Eng. Meas.*, pp. 203–213, 2013.

Preparing Fluorescent Conjugated Polymers through One Pot Enzymatic Polymerization for Sensing Applications

W Kiratitanavit[1]

Department of Mining and Materials Engineering, Prince of Songkla University, Hatyai, Songkhla, Thailand

E-mail: kweeradech@eng.psu.ac.th

Abstract. Fluorescence based polymeric sensors are simple, extremely sensitive, versatile and adaptable for detection of different types of analysts. However, the multi-step synthesis using toxic chemicals followed by purification is often required for obtaining these fluorophores. Here the possibility of utilizing enzymes as the catalyst, for the one step polymerization of naturally occurring monomers to yield fluorescent conjugated polymers is presented. Four monomers, 4-Hydroxyphenylacetic acid (HPA), Hydroxytyrosol (HDT), Chlorogenic acid (CGA) and Serotonin (Sero) were polymerized using Horseradish peroxidase as the biocatalyst. The broad peaks of IR spectrum were obtained from all polymers comparing with the sharp peaks of monomers. The decrease of UV-Vis intensity at significant wavelength of each monomer was found when the polymer was formed. These polymers exhibit fluorescence with significant stokes shift around 100 nm rendering them useful in fluorescence quenching-based sensors. In the detection of lead ion, DNT and TNT in solution through fluorescence-quenching are studied. Both nitro aromatic compounds were detected from electron interaction between phenol and nitro-aromatic compound structures that poly(HPA) and poly(Sero) detected at 1 ppm level. From having carboxylic group, poly(HPA) detected lead at 50 ppb at neutral pH.

1. Introduction

Fluorescence-based sensors have been used to detect various specific analytes such as explosive materials, metal ions and biological materials.[1,2] These sensors can be interacted with the analytes (receptors) depending on their chemical structure and mechanism. The fluorescent characteristic changing is relied on the interaction and also the contribution of the analytes concentration. From many reported sensor mechanisms, the turn off or quenching based sensors, reduction in fluorescent intensity or shifting the particular wavelength range, are simplest fluorescent techniques. However, many fluorescent polymers used in sensors, for example polythiophenes and poly(*p*-phenylene ethynylene)s, require multi-step synthesis and purification by using more toxic chemicals.[3,4] To solve these problems, enzyme catalysis has been introduced. This catalyst has been introduced in many chemical reactions including polymerization from the non-toxic nature of enzymes. Enzymes have a unique ability to simplify transformations with less step like one pot method. The enzymatic chemical reactions can be carried out under mild reaction conditions including pressure, temperature and pH and also in greener solvents (aqueous based or mixed solvent systems).[5]

In this research, four naturally occurring phenol monomers were chosen as starting materials to polymerize fluorescent materials. These are including 4-hydroxyphenylacetic acid (HPA), hydroxytyrosol (HDT), chlorogenic acid (CGA), and serotonin (Sero).[6-9] These monomers were selected from having different functional groups that have a potential to detect different analytes through different mechanisms. The oxidoreductase enzyme namely Horseradish peroxidase (HRP) had been selected as catalyst to polymerize these four monomers. The biocatalytic synthesis in this work is one-step reaction. Enzymatically synthesized polyphenols were then characterized by using spectroscopic techniques including UV-Vis absorption, FTIR, NMR and fluorescence spectroscopy. To determine the sensing abilities of these four polyphenols, the quenching studies were carried in the presence of nitroaromatic compounds including dinitrotoluene (DNT) and trinitrotoluene (TNT) in an organic solvent of acetonitrile and Lead (II) nitrate in water.

2. Materials

Hydroxytyrosol (TCI America), Serotonin hydrochloride (Alfa Aesar), 4-Hydroxyphenylacetic acid, Chlorogenic acid and Horseradish Peroxidase (HRP-Type II, RZ = 1.9 from Sigma Aldrich) were purchased and used without further purification. 3%(v/v) of Hydrogen peroxide (or H_2O_2 from Alfa Aesar) stock solution was prepared by diluted with de-ionized water. All analytes, 2,4-dinitrotoluene (DNT), 2,4,6-trinitrotoluene (TNT) and lead (II) nitrate, were purchased from Sigma Aldrich and used as received. Other remaining organic solvents (such as Acetonitrile and Ethanol, etc.) were reagent grade.

3. Methods

3.1. Enzymatic Synthesis

10 milligrams of each monomer was firstly dissolved in the mixtures of deionized water and ethanol (9:1 v/v) to prepare the starting monomer solution. The pH of the solution was then adjusted to 7.5 by using acidic or alkaline solution depending on starting pH. Then, 4 mg of HRP catalyst was added into the adjusted monomer solution following with slow dropping of H_2O_2 (1 ml). The solution was continuing stirred at room temperature for 24 hours and removed of unreacted monomer by dialyzing against deionized water at least 48 hours.

3.2. Characterization

3.2.1. UV-Vis Absorption Spectroscopy. To monitor the enzymatic polymerization, the reactions were investigated by using a UV-Vis spectrometer (Model - Agilent 8453). The sample solution (about 0.1 ml) was added into 2.9 ml of a mixture solution (water and ethanol at 9:1 v/v) in a quartz cuvette. The UV spectrum (in the wavelength range of 190-1100 nm) was then recorded.

3.2.2. Spectroscopy Studied. The spectroscopic spectra of monomers and polymers were characterized with two techniques. Thermo Scientific Nicolet 4700 FTIR instrument with a Smart Orbit Attenuated Total Reflectance (ATR) accessory in the frequency range of 4000 - 400 cm^{-1} with a resolution of 4 cm^{-1} and 32 scans was used to obtain FTIR spectrum. The ^1H-NMR spectra were recorded using a Bruker & Spectrospin Advance 500 spectroscopy (500 MHz) with using d6-DMSO as solvent at room temperature (~20 °C).

3.2.3. Fluorescence Analysis Studies of Polyphenols. To compare the fluorescent characteristic, the absorption of each monomer and polymer was measured by using a fluorescence spectrophotometer (Cary Eclipse). The sample solution was diluted with deionized water to optimize the fluorescent intensity. The excitation maxima for each of the solutions was firstly determined in the wavelength of 200 to 700 nm. The fluorescence emission of each sample solution was then measured by using that maximum excitation wavelength.

3.2.4. Fluorescence Quenching Studied in Analysts Detection. Fluorescent quenching behaviour or reduction in the fluorescence intensity of each polyphenol in the presence of DNT, TNT, and lead (II) nitrate, was measured by using the same spectrophotometer in 3.2.3. To eliminate the solubility limit of nitroaromatic compounds (DNT and TNT), the polymer solution was diluted with acetonitrile instead of water while the water was used as solvent for lead (II) nitrate). Moreover, to optimize lead detection condition, the pH of polymer solution was varied in the range of 4 to 9. Fluorescence quenching sensitivity was calculated based on Stern-Volmer constant (K_{SV}), provided in the following equation.[10,11]

$$\frac{I_0}{I} = 1 + K_{SV}(M_c)$$

Where I_0 and I are the fluorescent intensity before and after adding the analyte, respectively. The M_c represents the molar analyte concentration. The K_{SV} can be determined from a slope of the graph of analyte concentration versus I_0/I in this equation.

4. Results and Discussion

At the starting condition before the enzymatic reaction, all monomer solutions were clear and colorless (except in a yellow tinge for CGA). However, the color of all monomer mixture solution changed rapidly to orange-yellow or dark brown (depending on monomer type) after addition of the HPA and H_2O_2 to start polymerization reaction.

4.1. UV-Vis Analysis of Polyphenols

As mention earlier, the UV-Vis spectrophotometer was used to monitor the enzymatic polymerization reaction. The phenol monomers have well-known exhibit high UV absorbance at less than 350 nm).[12] The examples of UV-Vis spectrum of phenol monomer and polymer are shown in figure 1 and 2 while the spectra peaks of all monomers and polymers are summarized in table 1. Normally, all phenol monomers have a dominant peak represented the $\pi \rightarrow \pi^*$ transition of the phenyl ring (at 224 nm for HPA as shown in figure 1). The remaining small peaks are from the substituted groups such as carbonyl group in HPA monomer structure (from $n \rightarrow \pi^*$ at 276 nm) or the double bond between phenyl ring and ester linkage (figure 2) in CGA chemical structure at 324 nm.[13,14] After polymerization, most of polymer peaks exhibited a redshift (or disappeared) with longer tail absorption to 400-600 nm depended on monomer type. This redshift, shift to negative, is from electron delocalization of polymer formation, having more conjugated structure, and with the lower energy required for the $\pi \rightarrow \pi^*$ transition.

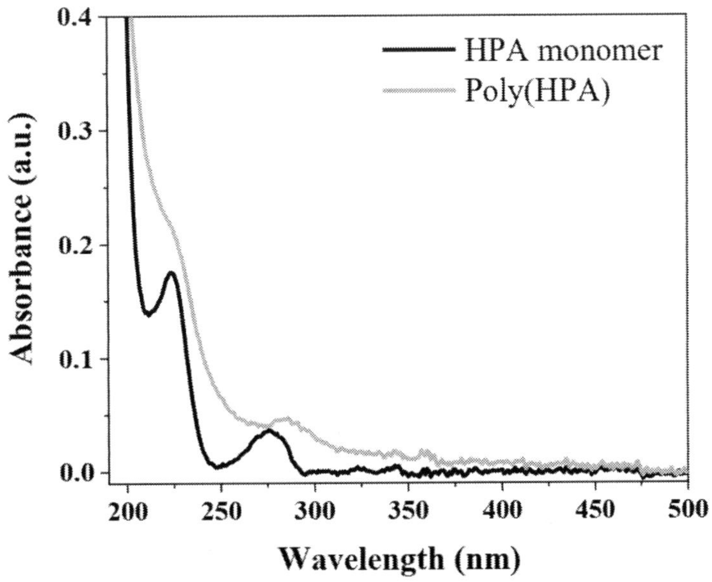

Figure 1. UV-Visible absorption of HPA monomer and polymer.

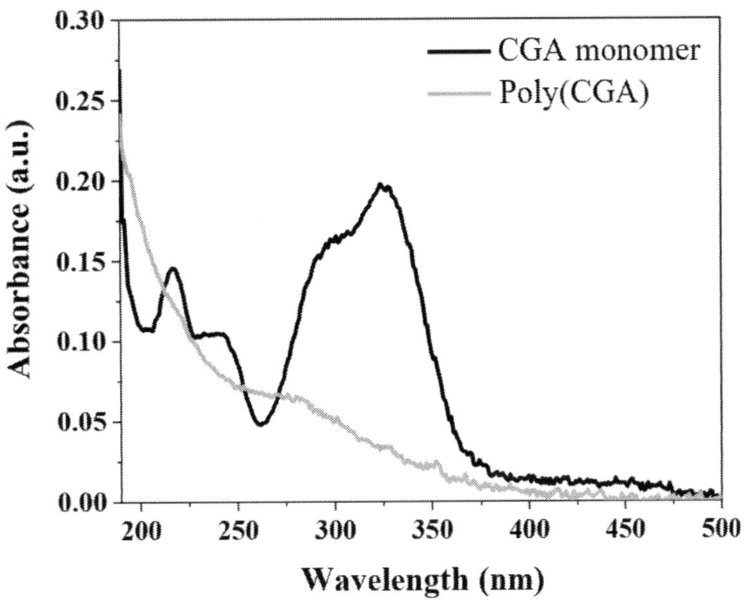

Figure 2. UV-Visible absorption of CGA monomer and polymer.

IOP Conf. Series: Materials Science and Engineering 1163 (2021) 012014 doi:10.1088/1757-899X/1163/1/012014

Table 1. UV-Visible Wavelength Peaks of Phenol Monomers and Polymers.

Materials	Wavelength peak (nm)
HPA monomer	224 and 276
Poly(HPA)	228 and 285 with long tail till 450 nm
HDT monomer	225 and 283
Poly(HDT)	250 and 350 with long tail till 465 nm
CGA monomer	218, 240 and 324
Poly(CGA)	275 with long tail till 500 nm
Sero monomer	260, 270 and 300
Poly(Sero)	Long tail till 600 nm

4.2. Chemical Structure Characterization by FTIR and ^{1}H-NMR

Normally, polyphenols from enzymatic polymerization have highly articulated structures with a combination of monomer units coupled through C-C and C-O-C linkages making it more complicated for clear structure elucidation using NMR spectroscopy (as shown in figure 3 for ^{1}H-NMR spectra of HPA and poly(HPA)). Fourier transform infrared spectroscopic technique is usually for suitable for confirming the polymer structure for most of polyphenols synthesized from the enzymatic polymerization reaction.[15] Figure 4 show the example of FTIR spectra of HPA monomer and poly(HPA) polymerized through enzymatic reaction. After purification by removing the unreacted monomer and HRP catalyst from the polymer solution by dialysis, the characteristic peaks especially amide group of HRP are not recognized in the FTIR spectra of all polyphenols. Most polymers have broader peaks as compared with monomers from having more conjugated structure from a combining of phenyl rings with C-C and/or C-O-C linkages.

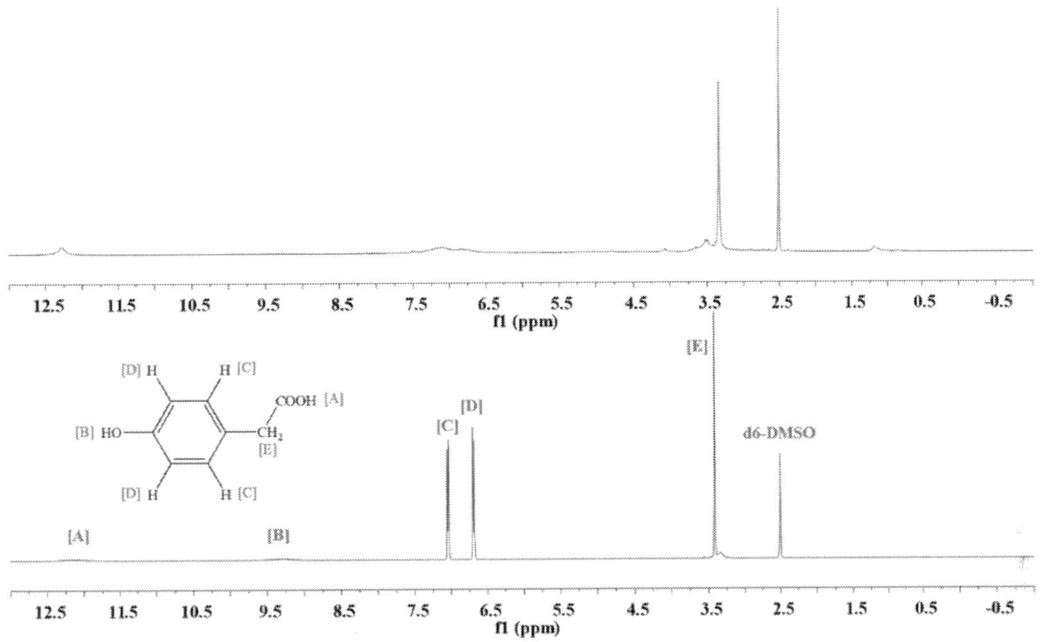

Figure 3. ^{1}H-NMR spectra of HPA monomer and polymer.

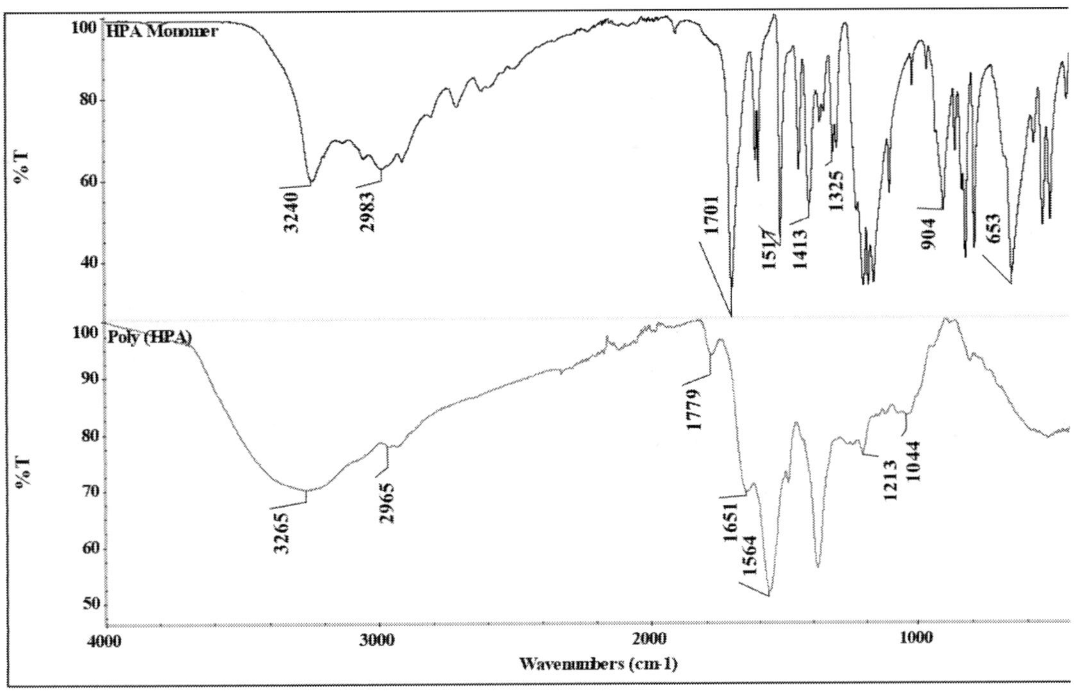

Figure 4. FTIR spectra of HPA monomer and polymer.

4.3. Fluorescence Characteristic Properties of Polyphenol

All enzymatic polymerization reactions were carried out in the water/ethanol mixtures instead of buffer solution as reported from the literature. This is from avoiding of buffer salts effect, the interaction between polyelectrolytes (salts) and conjugated polymers leading to quench the fluorescence emission.[16]

Figure 5 and 6 show the example of fluorescence characteristic strum of monomer and polymer of HPA. The fluorescent results; excitation, emission and Stokes shift, of all materials are summarized in table 2. The excitation wavelength in each sample solution was determined at the maximum absorbance. The Stokes shift, the difference between the excitation and emission maxima of a particular fluorophore, is an important property for utilization of these fluorescent materials in sensing applications. It allows photons emitted by the fluorophore to be detected against the background, without interference from the photons used for excitation. Therefore, the shift varies with the molecular structure of fluorescent samples (or fluorophores). From having higher conjugated structure, all polymers (except poly(CGA)) emit at a higher wavelength as compared to their monomers. These results are the evidence to confirm that the chemical conformation differences between monomer and polymer in the solution have an influence on the shift of excitation and emission.[17]

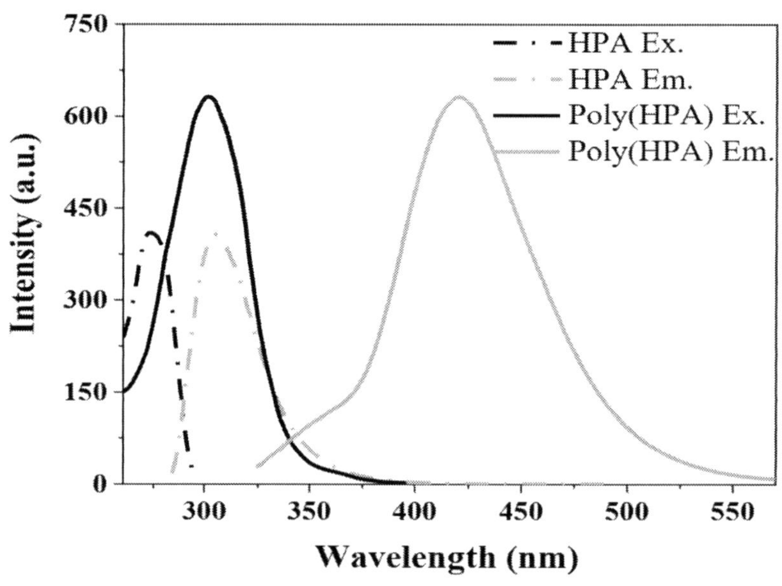

Figure 5. Fluorescence spectra of excitation (EX.) and emission (Em.) of HPA monomer and polymer.

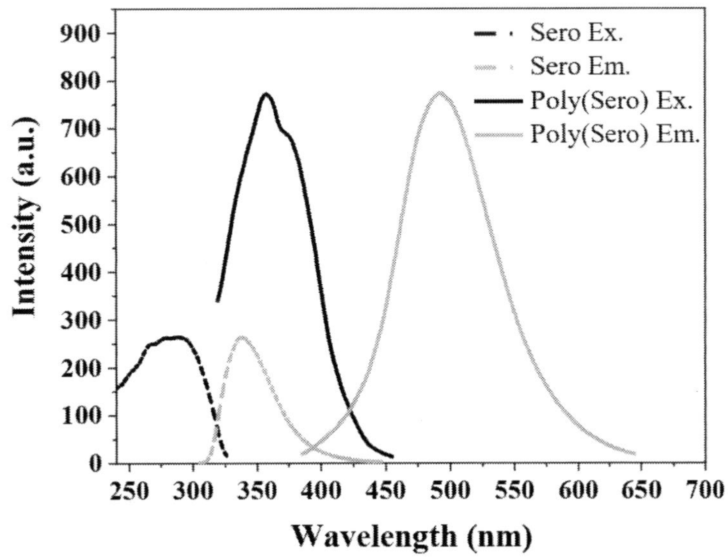

Figure 6. Fluorescence spectra of excitation (EX.) and emission (Em.) of Sero monomer and polymer.

Table 2. Excitation, Emission and Stokes Shift of Phenol Monomers and Polymers.

Materials	Excitation Wavelength ($\lambda_{excitation}$, nm)	Emission Wavelength ($\lambda_{emission}$, nm)	Stokes shift ($\lambda_{emission}$ -$\lambda_{excitation}$, nm)
HPA monomer	275	305	30
Poly(HPA)	301	421	120
HDT monomer	277	319	42
Poly(HDT)	330	417	87
CGA monomer	340	455	115
Poly(CGA)	329	460	131
Sero monomer	280	339	59
Poly(Sero)	360	490	130

4.4. Fluorescence Quenching Studies of Polyphenols in a Presence of DNT and TNT

The quenching mechanism and detective sensitivity of fluorescence sensor are relied on the interaction between analytes and fluorophore resulting in fluorescence intensity. Since nitroaromatic compounds as DNT and TNT are electron deficient, they will quench the fluorescence of electron-rich as conjugated polyphenols after the interaction between them.[11] All polyphenols have good ability to detect TNT at slightly better than DNT. However, the analytical detection ability limit for all conjugated polyphenols are in the range of parts per hundreds of analyte concentration. The fluorescence quenching sensitivity can be calculated based on Stern-Volmer constant (K_{SV}). The K_{SV} is the slope of the graph of the ratio of fluorescence intensity in the absence of the analyte to intensity in the presence of the analyte ($\frac{I_0}{I}$) versus the analyte concentration (as shown in figure 7). A higher K_{SV} value of represents the high sensitivity of the fluorophore towards the analyte.[17] All enzymatic synthesized polyphenols showed the K_{SV} values in the thousands as shown the results in table 3. The limited sensitivity is attributed to extensive aggregation of the polymers in the solutions.

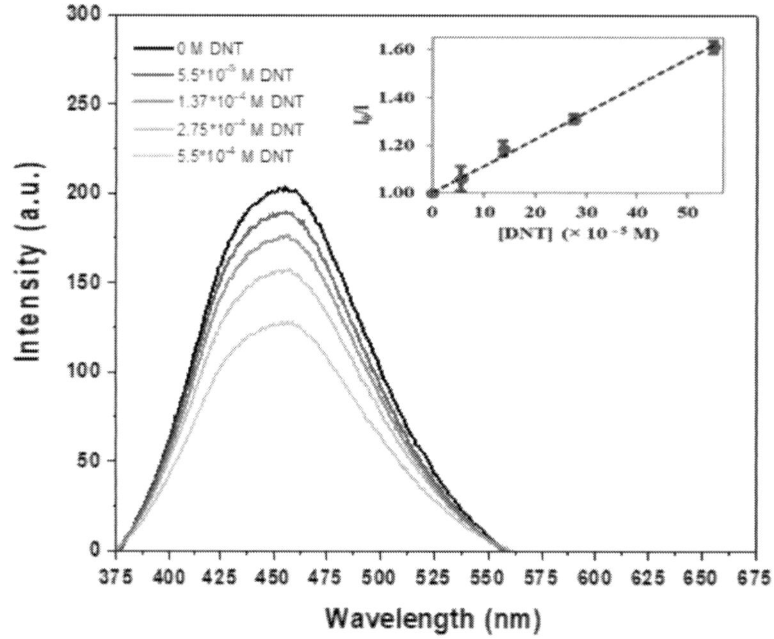

Figure 7. Fluorescence spectra with Stern-Volmer plot of poly(CGA) at various DNT concentrations.

Table 3. Summary of Sensitivity of DNT and TNT Detection by Polyphenols through Fluorescence Quenching.

	Minimum DNT concentration detection			Minimum TNT concentration detection		
Polyphenol	$(\times 10^{-6})$ M	ppm	K_{SV} (M^{-1})	$(\times 10^{-6})$ M	ppm	K_{SV} (M^{-1})
Poly(HPA)	5.5	1.0	3460	4	0.9	3410
Poly(HDT)	55	10	1120	20	4.6	2170
Poly(CGA)	55	10	1150	20	4.6	2280
Poly(Sero)	5.5	1.0	2690	4	0.9	2620

4.5. Fluorescence Quenching Studies of Polyphenols in a Presence of Lead

The preliminary results showed that only poly(HPA) and poly(CGA) had a potential to detect lead ion in water while the fluorescence quenching ability to detect lead ion of poly(HDT) and poly(Sero) required at the high amount of lead ions (more than 50 ppm). This is from having a carboxylic group (-COOH) in their chemical structure. The carboxylic group has a potential to absorb and interact (form a complex) with lead ion (and also other metal ions as described in the literatures).[18,19] The effect of pH on the detection ability of poly(HPA) and poly(CGA) was investigated as shown the results in table 4. It was found that both polyphenols have a high ability to detect lead ions at lower concentration during pH 6 to 8. The pKa reported values of carboxyl position of HPA and CGA monomer are 4.25 and 3.5, respectively.[20,21] At lower pH closed to pKa or lower than, the carboxylic groups are protonated leading to the reduction of the adsorptive activity.[18] The deprotonation of carboxyl groups to carboxylate ions (-COO^-) starts at high pH resulting in highly complex forming between polyphenols and lead ions. The detection ability decreased at high pH from lower solubility of lead ions in water (precipitation).[19]

Table 4. Summary of Sensitivity of Lead (II) Nitrate Detection by Polyphenols through Fluorescence Quenching.

	Minimum Pb^{2+} concentration detection by Poly(HPA)			Minimum Pb^{2+} concentration detection by Poly(CGA)		
pH	(×10^{-6}) M	ppm	K_{SV} (M^{-1})	(×10^{-6}) M	ppm	K_{SV} (M^{-1})
4	55	10.0	620	5.5	1.0	5400
5	28	5.0	900	5.5	1.0	4010
6	0.55	0.1	5300	2.8	0.5	6720
7	0.28	0.05	8900	2.8	0.5	4870
8	0.55	0.1	12400	2.8	0.5	5400
9	2.8	0.5	3670	2.8	0.5	3920

Conclusion

The synthesis of four naturally occurring phenols; HPA, PDT, CGA and Sero, have been successfully prepared by using HRP as catalyst under mind condition in this research. The enzymatic reaction is one pot reaction method with less purification step. All phenol monomers and polymers were characterized using UV-Vis, FTIR and Fluorescence spectroscopy to determine and confirm the formation of conjugated polyphenolic structure. By using fluorescence spectroscopic studies, all fluorescent polyphenols exhibit significant Stokes shift up to 100 nm making them suitable for being utilized in quenching based sensors. The nitroaromatic compounds and metal ion detection ability of these polyphenols showed in fifty to hundred parts per billion with a reasonable Stern-Volmer constant number.

Acknowledgement

The author thanks US Army Natick Soldier RDEC (Grant number: W911QY-14-2-0001) for financial support for this project. The author also thanks the Core Research Facilities, Harnessing Emerging Research Opportunities to Empower Soldiers (HEROES) lab facilities, Prof. Ramaswamy Nagarajan's research lab and Prof. Jayant Kumar's research lab at University of Massachusetts Lowell for the characterization facilities.

References

[1] Guo Z, Wang X, Wei P, Gao Y, Li Q 2019 *J. Anal. Methods. Chem.* **8130767** 1
[2] McNamara K, Nguyen T, Dumitrascu G, Ji J, Rosenzweig N and Rosenzweig Z 2001 *Anal. Chem.* **73** 3240

[3] McQuade D, Pullen A amd Swager T 2000 *Chem. Rev.* **100** 2537
[4] Yang J and Swager T 1998 *J. Am. Chem. Soc.* **120** 11854
[5] Kobayashi S, Uyama H and Kimura S 2001 *Chem. Rev.* **101** 3793
[6] Boss A, Bishop K, Marlow G, Barnett M and Ferguson L 2016 *Nutrients* **8** 513
[7] Liang N, Lu X, Hu Yand Kitts D 2016 *J. Arg. Food Chem.* **64** 681
[8] Xue C, Wang X, Zhu W, Han Q, Zhu C, Hong J, Zhoa X and Jiang H 2014 *Sensor. Actuat. B-Chem.* **196** 57
[9] Ramakrishma A, Giridhar P and Ravishankar G 2011 *Plant Signal. Behav.* **6** 800
[10] Satapathi S, Bheemaraju A, Surampudi S, Venkataraman D and Kumar J 2014 *IEEE Sensor. J.* **14** 4334
[11] Sun X, Wang Y and Lei Y 2015 *Chem. Soc. Rev.* **44** 8019
[12] Bruno F, Akkara J, Samuelson L, Kaplan D, Mandal B, Marx K, Kumar J and Tripathy S 1995 *Langmuir* **11** 889
[13] Belay A, Kim H and Hwang Y 2016 *Lumin.* **31** 565
[14] Antosiewicz J and Shugar D 2016 *Biophys. Rev.* **8** 151
[15] Kiratitanavit W, Ravichandran S, Xia, Z, Kumar J and Nagarajan R 2013 *J. Renew. Mater.* **1** 289
[16] Bouldin R, Singh A, Magaletta M, Connor S, Kumar J and Nagarajan R 2014 *Bioeng.* **1** 246
[17] Thiessen A, Würsch D, Jester S, Aggarwal A, Idelso A, Bange S, Vogelsang, J, Höger S and Lupton J 2015 *J. Phys. Chem. B* **119** 9949
[18] Balakrishnan C, Theetharappan M, Natarajan S, Thalamuthu S and Neelakantan M 2015 *RSC Adv.* **5** 105453
[19] Atieh M, Bakather O, Altawbini B, Bukhari A, Abuilaiwei F and Fettouhi M 2010 *Bioinor. Chem. Appl.* **603978** 1
[20] Manoel H and Moya H 2015 *Molecules* **20** 22411
[21] Šeruga M and Tomac I 2014 *Int. J. Electrochem. Sci.* **9** 6134

An Assessment of the Implications of Biodiesel Consumption Promotion in the Transportation Sector for Thailand

T Purathanang[1] and S Wattana[2]

[1] Faculty of Engineering, Mahasarakham University, Mahasarakham, Thailand
[2] Solar Energy and Energy Resources Research Unit, Faculty of Engineering, Mahasarakham University, Mahasarakham, Thailand

Email: supannika.w@msu.ac.th

Abstract. This paper assesses the implications of the promotion of biodiesel consumption in the Thai transportation sector in terms of energy, environment and agriculture. In order to analyse the implications, scenarios are modelled quantitatively to assess their impacts for the period 2019-2037. Scenarios developed in this paper represent various percentages of biodiesel blending mandates including REF (7% biodiesel blend), B10 (10% biodiesel blend) and B20 (20% biodiesel blend). The analytical tool employed to assess the scenario impacts is the Low Emissions Analysis Platform (LEAP) model. The analyses suggest that an increase in percentage share of biodiesel blend would contribute to a reduction in diesel consumption and hence crude oil requirement. Consequently, it would generate less CO_2 emissions and especially less Fine Particulate Matter (PM2.5) - an issue of contemporary environmental health concerns. It is, further, revealed that higher shares of biodiesel in diesel would result in an increase demand for oil palm production in order to meet a rising growth in biodiesel production. Such an increase in demand for oil palm feedstock would result in higher commodity prices and therefore higher income for farmers. However, a high oil palm production demand would require a substantial crop cultivation area. The effort to increase palm planting and productivity to meet the growing demand is a challenging task. In order to overcome this challenge, this paper suggests the implementation of agricultural zoning and the advancement of crop species and biodiesel conversion technology.

1. Introduction

For Thailand, oil has played an important role in the country's economic and social development. In 2019, petroleum products consumption in Thailand accounted for 49% of final energy consumption [1]. Of the total petroleum products, diesel consumption accounted for 47%. Moreover, Thailand has heavily relied on imported oil to meet the rising demand because of its limited indigenous energy resources. Given concerns about the energy security and environmental health, the Thai government has implemented policies on the promotion of biodiesel consumption on the transportation sector over the last twenty years. In 2005, the government began a campaign to promote the production and consumption of biodiesel. B5 (a blend of 5% biodiesel and 95% fossil diesel) has been on sale since 2005. In 2014, the government has increased the proportion of biodiesel blending to 7% (B7). In 2020, the government has made further efforts to push to increase the proportion of biodiesel blending to B10 and B20 with a view to balance the entire palm oil system in the country, stabilize palm oil prices and help reduce pollutions including carbon dioxide emissions and Fine Particulate Matters (PM 2.5). An increase proportion of biodiesel in diesel would certainly result in an increase of biodiesel production. A rising demand for biodiesel would further require a substantial amount of energy crops especially palm oil and hence have an impact on land for growing crops. With this background, this paper aims to assess the implications of the promotion of biodiesel consumption in the Thai transportation sector in terms of energy, environment and agriculture.

2. Research Methodology

The methodology employed in this paper is the Low Emissions Analysis Platform (LEAP). LEAP is a widely-used software tool for energy policy analysis and climate change mitigation assessment, developed by the Stockholm Environment Institute (SEI). It is an integrated, scenario-based modelling tool originally developed to track energy consumption, production and resource extraction in all sectors of an economy. It can account for both energy sector and non-energy sector greenhouse gas (GHG) emission sources and sinks [2]. LEAP has been employed by several studies to assess the energy and environmental impacts of bioenergy including biofuels [3-7].

In this paper, the impacts of an increase share of biodiesel in diesel are assessed in terms of energy, environment and agriculture. The impacts on energy are assessed in terms of diesel consumption and crude oil requirements. The impacts on environment are assessed in terms of CO_2 emissions and Fine Particulate Matter (PM2.5). And, the impacts on agriculture are assessed in terms of production demand for oil palm and future land requirements.

In order to analyze the implications, scenarios are developed to assess their implications for the period 2019-2037. In this paper, three scenarios (namely REF, B10 and B20) represent various percentages of biodiesel blending mandates. The REF scenario reflects a situation in which the proportion of the biodiesel blending for standard diesel is B7 and B10 is optional. The B10 scenario is developed to represent the case that biodiesel blending for standard diesel is B10. In the B20 scenario, 20% biodiesel blending (B20) is set to be standard diesel. For more details of each scenario, Table 1 provides an overview of the key scenario descriptions. In addition, the refinery capacity throughout the entire studied period is assumed to be the same as the capacity in 2019. The capacity for crude oil refinement in 2019 was 1.2 million barrel per day [1].

Table 1. Key scenario descriptions

Scenario theme	Key scenario descriptions
REF scenario	• The demand growth for diesel consumption in the transportation sector is assumed to be 3.3% annually [8]. • The proportion of the biodiesel blending for standard diesel is 7% (B7) and B10 is optional. • The consumption shares of B7 and B10 are 95% and 5% respectively.
B10 scenario	• The demand growth for diesel consumption in the transportation sector is assumed to be 3.3% annually [8].

Scenario theme	Key scenario descriptions
	• Percentage share of biodiesel blending for standard diesel is 10% (B10) and B20 is optional. • The consumption shares of B10 and B20 are 70% and 30% respectively.
B20 scenario	• The demand growth for diesel consumption in the transportation sector is assumed to be 3.3% annually [8]. • 20% biodiesel blending (B20) is set to be standard diesel which is 100% consumption share.

3. Data Consideration
In accordance with the multidisciplinary nature of the research, this paper requires extensive data including energy, environment and agriculture. The data is in the form of biofuel policies, the Alternative Energy Development Plan (AEDP), statistical data of energy supply and consumption. The information on energy (for example, consumption of crude oil, diesel and biodiesel) is available from various Thailand Energy Balance reports and Thailand Alternative Energy Situation reports, annually published by the Department of Alternative Energy Development and Efficiency (DEDE) [1], [9]. The growth for final energy demand can be obtained from the Ministry of Energy (MOE) and the DEDE [10]. Oil palm yield and productivity of oil palm can be taken from the Office of Agricultural Economics (OAE) and the DEDE [11-12].

4. Empirical Results
This paper provides an assessment of the scenario impacts on agriculture, energy and environment. The assessment is accordingly divided into three sub-sections, namely, energy, environment and agriculture.

4.1. Energy
In this paper, the scenario impacts on agriculture are assessed in terms of a decrease in projected diesel consumption and crude oil requirements.

4.1.1. Projected diesel demand
Figure 1 reveals that projected diesel demand in the transportation sector under the REF scenario would increase continuously, from 20,960 Million litres in 2019, to 39,058 Million litres in 2037. In 2037, the B10 scenario would contribute to a reduction of 4,673 Million litres in comparison with the REF scenario. The demand for diesel in the case of the B20 scenario in 2037 would be 10,064 Million litres less than diesel demand under the REF scenario. Therefore, the demand for diesel in the case of B10 and B20 scenarios would be 12% and 26%, respectively, lower than in the REF scenario. This is a result of an increasing share of biodiesel in diesel. It is evident that higher shares of biodiesel in diesel would undoubtedly cause less demand for diesel.

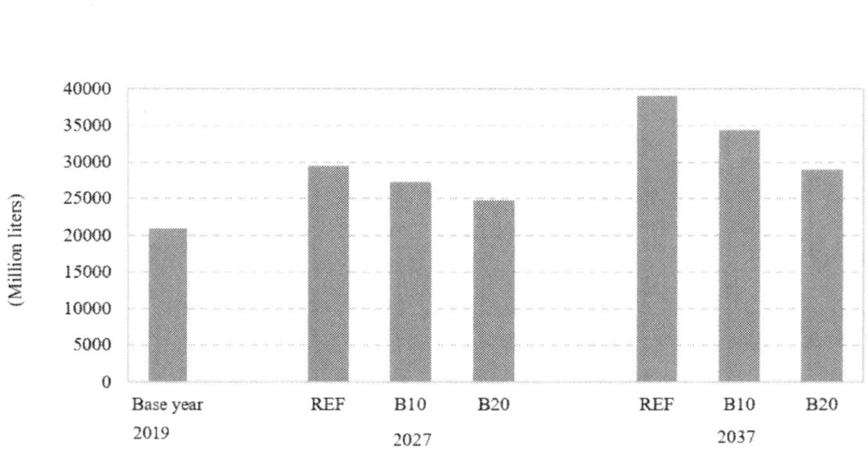

Figure 1. Projected diesel demand for the period 2019-2037

4.1.2. Crude oil requirements

In order to meet an increasing demand for diesel, crude oil requirements in the REF scenario is expected to grow considerably. Table 2 shows crude oil requirement for the period 2019-2037. The results from Table 2 reveals that crude oil requirements would increase from 19,383 KTOE in 2019 to 36,119 KTOE in 2037 - a double increase over the period 2019-2037. In 2037, crude oil requirements in the case of B10 and B20 scenarios would be 10% and 22%, respectively, lower than in the case of the REF scenario. Table 2 further shows that crude oil requirements in the B10 and B20 scenarios is expected to decrease continuously as compared with the REF scenario. For example, a decrease in crude oil requirements, under the B10 scenario, in 2027 would be 1,775 KTOE, in 2037 would be higher - 3,747 KTOE in comparison with the REF scenario. And, in the B20 scenario, a reduction in crude oil requirements in 2027 would reach 3,728 KTOE, in 2037 would be higher - 7,870 KTOE, as compared with the REF scenario. This signifies that increased share of biodiesel in diesel would contribute to higher crude oil savings.

Table 2. Crude oil requirements for the period 2019-2037

Year	REF scenario	B10 scenario		B20 scenario	
	Crude oil requirement	Crude oil requirement	Changes from REF scenario	Crude oil requirement	Changes from REF scenario
	(KTOE)	(KTOE)	(KTOE)	(KTOE)	(KTOE)
2019	19,383	-	-		-
2027	27,310	25,535	-1,775 (-6%)	23,582	-3,728 (-14%)
2037	36,119	32,372	-3,747 (-10%)	28,249	-7,870 (-22%)

Notes: Number in brackets show percentage change from the REF scenario.

4.2. Environment

In this section, the implications of biodiesel consumption promotion on the Thai transportation sector are analyzed. This analysis incorporates three main attributes, namely, CO_2 emissions and emissions of Fine Particulate Matters (PM2.5).

4.2.1. CO_2 emissions

The results from Table 3 reveal that CO_2 emissions under the REF scenario is estimated to reach from 56 million tonnes in 2019, to 105 million tonnes in 2037, an increase of 49 million tonnes over the 2019 emission level. CO_2 emissions in the B10 scenario is expected to grow to 92 million tonnes in 2037 – a decrease of 13% in CO_2 emissions as compared with the REF scenario. In the case of B20 scenario, CO_2 emissions in 2037 would be 27 million tonnes (29%) lower as compared to the CO_2 emissions in the REF scenario. Furthermore, Table3 shows that the B10 and B20 scenarios would contribute to a slowdown in a rise of CO_2 emissions in comparison with the REF scenario. In the B10 scenario, a reduction of CO_2 emissions in 2027 would reach 6 million tonnes, in 2037 would be higher – 13 million tonnes, as compared with the REF scenario (as shown in Table 3). It is further noticed that the B20 scenario would result in highest CO_2 savings – 27 million tonnes in 2037, as compared with the REF scenarios. Such a reduction in CO_2 emissions accords with the decrease in diesel consumption. It is evident that an increase in percentage share of biodiesel blend would contribute to a reduction in diesel consumption and hence lower CO_2 emissions.

Table 3. CO_2 emission in the case of various scenarios over the period 2019-2037

Year	REF scenario	B10 scenario		B20 scenario	
	CO_2 emission (million tonnes)	CO_2 emission (million tonnes)	Changes from REF scenario (million tonnes)	CO_2 emission (million tonnes)	Changes from REF scenario (million tonnes)
2019	56	-	-		-
2027	79	73	-6 (-8%)	92	-13 (-17%)
2037	105	92	-13 (-14%)	78	-27 (-29%)

Notes: Number in brackets show percentage change from the REF scenario.

4.2.2. Fine Particulate Matters (PM2.5)

Figure 2 reveals that annual emissions of PM2.5 under the REF scenario would increase from 18 thousand tonnes in 2019, to 34 thousand tonnes in 2037, an increase of 16 thousand tonnes over the 2019 emission level. In 2037, annual emissions of PM2.5 in the case of B10 scenario would be 12% lower than the emissions under the REF scenario. Moreover, the B20 scenario would contribute to a decrease of 24% in comparison with the REF scenario. Figure 3 further show that annual emissions of PM2.5 in the B10 and B20 scenarios would decrease continuously in comparison with the REF scenario. For example, a decline in emissions of PM2.5, in the case of B10 scenario, in 2027 would be 2 thousand tonnes, in 2037 would be higher – 4 thousand tonnes in comparison with the REF scenario. And, in the B20 scenario, a decrease in emissions of PM2.5 in 2027 would reach 4 thousand tonnes, in 2037 would be higher – 8 thousand tonnes, as compared with the REF scenario. Such a reduction could be a result of a decrease consumption of diesel which is one major source of PM2.5 emissions. It is widely known that most PM2.5 derives from inefficient combustion in diesel engines, wood burning and dust from combining of pollutant gases and the industrial activities. The issue of PM 2.5 has currently been moved to the forefront of global health concern discussion. This is because PM2.5 could cause several health problems including respiratory and cardiovascular morbidity and lung cancer. The promotion of diesel substitution by biodiesel would contribute to a slowdown an increase of PM2.5 emissions and therefore provide benefit to the public health (as shown in figure 2).

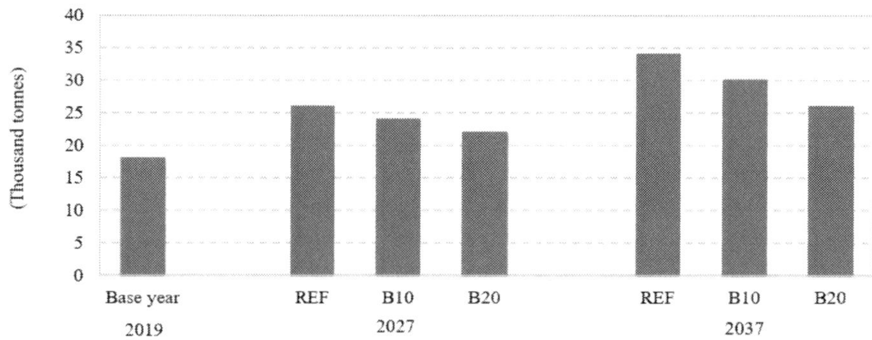

Figure 2. Fine Particular Matters (PM2.5) emissions over the period 2019-2037

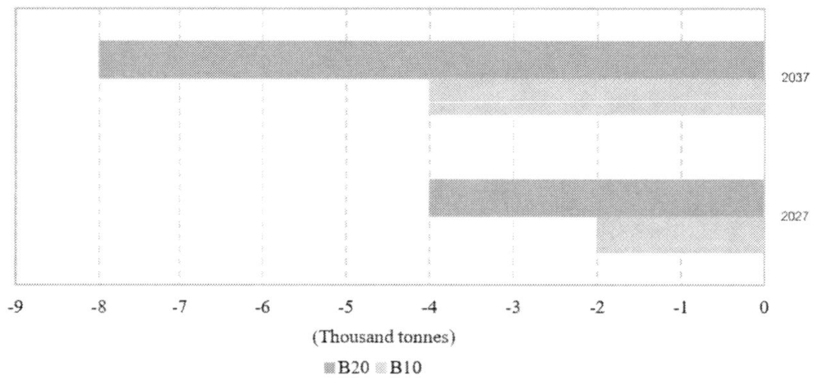

Note: This figure presents the changes in Fine Particular Matters (PM2.5) in the B10 and B20 scenarios as compared with the REF scenario.

Figure 3. Changes in Fine Particular Matters (PM2.5) emissions in the case of various scenarios

4.3. Agriculture

In this paper, the scenario impacts on agriculture are assessed in terms of production demand for oil palm and future land requirements for growing oil palm.

4.3.1. Production demand for oil palm

The policy to promote biodiesel consumption in the Thai transportation sector would result in a rising demand for oil palm. Figure 4 shows the projected oil palm production demand for the period 2019-2037. It should be noted that crop production for oil palm in this paper requires 4 kg of fresh fruit bunches (FFBs) for 1 litre of biodiesel [13]. The results from Figure 4 reveals that the demand for oil palm production in the case of REF scenario would increase by about 0.4 million tonnes, from 0.5 million tonnes in 2019 to 0.9 million tonnes in 2037. The demand for oil palm in the B10 scenario is expected to grow to 3.5 million tonnes in 2037 - more than four-fold increase in oil palm production demand in comparison with the REF scenario. Especially, the demand for oil palm, in the case of B20 scenarios, in 2037 would increase by about 8 times as compared with the demand under the REF scenario.

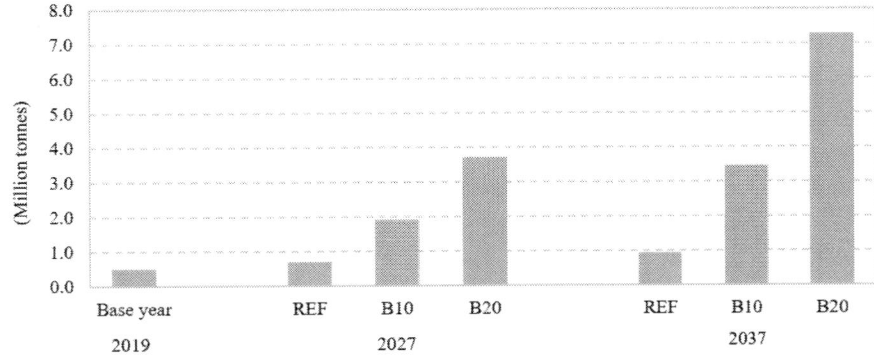

Figure 4. Projected production demand for oil palm over the period 2019-2037

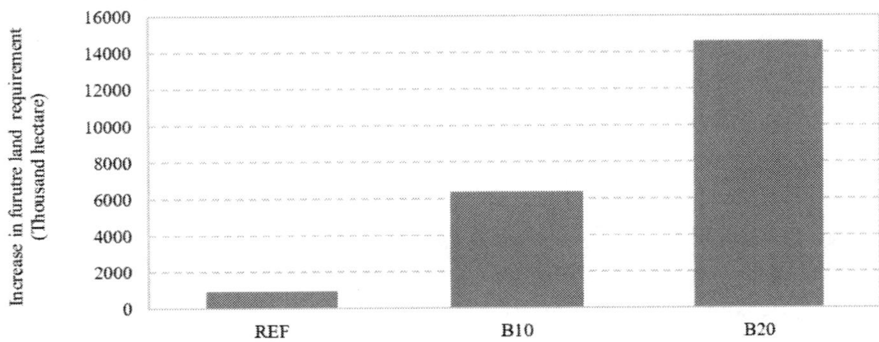

Figure 5. Projected future land extension for growing oil palm

4.3.2. future land requirements

To supply the increasing demand for oil palm production, the projected future land requirement is expected to rise considerably. The results from Figure 5 reveals that the projected future land extension for growing oil palm in the REF scenario is expected to rise by about 936 Thousand hectares. The requirement for land extension in the B10 scenario would be about 6,364 Thousand hectares - approximately seven-fold increase in comparison to the land extension in the case of REF scenario. In the case of B20 scenarios, the demand for future land extension would increase by 14,595 Thousand hectares - about 15 times higher than in the REF scenario. A substantial increase in crop production demand would directly result in a rise of future land requirements for growing oil palm.

The inference drawn from the above analyses is that an increase in percentage share of biodiesel blend would provide beneficial impacts on various aspects. In terms of energy, increased share of biodiesel in diesel would contribute to a reduction in diesel consumption and hence crude oil requirement. For example, in the case of B20 scenario, crude oil requirements in 2037 would be lower by 15% and 28%, respectively, as compared with the B10 and REF scenarios. For the environmental impacts, increased share of biodiesel in diesel would help slowdown an increase of CO_2 and PM2.5 emissions - an issue of contemporary environmental health concerns. For instance, CO_2 emissions in the case of B20 scenario would be 19% and 35%, respectively, lower than the emissions in B10 and REF scenarios. And, PM2.5 emissions in the case of B20 scenario would be 15% and 31%, respectively, lower than the emissions in B10 and REF scenarios. Even though the B20 scenario would provide a positive impact on several perspectives, it would require more energy crops for biodiesel production and future land requirements in comparison with other scenarios. For example, in the case of B20 scenario, demand for oil palm in 2037 would be higher by 52% and 87%, respectively, as compared with the B10 and REF scenarios. Additionally, future land requirements in the case of REF, B10 and B20 scenarios in 2037 would be 2 time, 7 times and 14 time, respectively, as compare with land requirement for growing oil palm in 2019. In order to reduce the demand for oil palm production, the efficiency improvement of the biodiesel conversion technology is essentially important. In fact, a rise in oil palm demand would help farmers generate extra income. This should be cautiously considered in the context of Thailand - the country that always face crop price volatility. In addition to the increased demand, a substantial increase in future land requirements would be one major challenge. This paper, therefore, recommends that the advancement of crop species and the agricultural zoning could be effective strategies in order to overcome this challenge. The advancement of crop species would help improve crop yield and hence help reduce crop cultivation area. In addition, the implementation of agricultural zoning is one of the key strategies to help balancing between food crops and energy crops production and increasing productivity of crops.

5. Conclusion

This paper assesses the implications of the promotion of biodiesel consumption in the Thai transportation sector in terms of energy, environment and agriculture. The assessment reveals that an increase in percentage share of biodiesel blend would provide beneficial impacts on several perspectives. For example, increased share of biodiesel in diesel would contribute to a reduction in diesel consumption and hence crude oil requirement. Increased biodiesel share would contribute to about 28% reduction in crude oil requirement. It would also help mitigating CO_2 and PM2.5 emissions. CO_2 emissions could be reduced by about 35% as a result of increased share of biodiesel. Also, PM 2.5 emissions could be reduced by 31% due to increased share of biodiesel. It would, however, require high crop production demand and a substantial increase in future land requirements. Increase in biodiesel share would require 87% more oil palm. This would result in a 14 times future land required for growing oil palm. The effort to increase palm planting and productivity to meet the growing demand is a challenging task. In order to overcome this challenge, this paper suggests the advancement of crop species and biodiesel conversion technology as well as the implementation of agricultural zoning would be effective strategies.

6. References

[1] Department of Alternative Energy Development and Efficiency 2019a. *Energy balance of Thailand: 2012-2019, annual reports*, Bangkok, Thailand.

[2] Stockholm Environment Institute 2020 Low Emissions Analysis Platform (LEAP): training

materials, Stockholm, Sweden.

[3] Sritong N, Promjiraprawat K and Limmeechokchai B 2014 CO_2 mitigation in the road transport sector in Thailand: Analysis of energy efficiency and bio-energy *Energy Procedia* 52 pp 131-141.

[4] Kemausuor F, Nygaard I and Mackenzie G 2015 Prospects for bioenergy use in Ghana using longrange energy alternatives planning model *Energy* 93 pp 672-682.

[5] Islas J, Manzini F and Masera O 2007 A prospective study of bioenergy use in Mexico *Energy* 32 pp 2306-2320.

[6] Suganthi L and Samuel A A 2012 Energy models for demand forecasting—A review *Renewable and Sustainable Energy Reviews* 16 pp. 1223-1240.

[7] Pagnarith K and Limmeechokchai B 2009 Biomass and Solar Energy for Rural Electrification and CO_2 Mitigation in Cambodia *International Journal of Renewable Energy* 4 pp 25-34.

[8] Department of Alternative Energy Development and Efficiency 2015 *Alternative Energy Development Plan (AEDP2015)*, Bangkok, Thailand.

[9] Department of Alternative Energy Development and Efficiency 2019b. *Thailand alternative energy situation: 2012-2019, annual reports*, Bangkok, Thailand.

[10] Ministry of Energy 2015. Thailand Energy Outlook 2015, Bangkok, Thailand.

[11] Office of Agricultural Economics 2018. Agricultural Statistics of Thailand 2018, Bangkok, Thailand.

[12] Department of Alternative Energy Development and Efficiency 2019c. Handbook for the Development and Investment in Renewable Energy Series 8 Biodiesel, Bangkok, Thailand.

[13] Nupueng S, Oosterveer P and Mol P J A 2018 Implementing a palm oil-based biodiesel policy: The case of Thailand *Energy Science & Engineering* 6 pp 643-657.

Acknowledgement
The authors would like to express special thanks to the Solar Energy and Energy Resources Research Unit, Faculty of Engineering, Mahasarakham University for providing research facilities.

IOP Conf. Series: Materials Science and Engineering 1163 (2021) 012016 doi:10.1088/1757-899X/1163/1/012016

An Analysis of the Impacts of Solar Electricity Generation on the Electricity Industry for Thailand

P Aungyut[1] and B Wattana[2]

[1] Faculty of Engineering, Mahasarakham University, Mahasarakham, Thailand
[2] Solar Energy and Energy Resources Research Unit, Faculty of Engineering, Mahasarakham University, Mahasarakham, Thailand

Email: buncha.w@msu.ac.th

Abstract. This paper provides an analysis of the impacts of solar electricity generation on the electricity industry for Thailand. In this paper, three scenarios (REF, Solar2015 and Solar2018) are developed to represent an increased levels of electricity produced from solar energy. A Low Emissions Analysis Platform (LEAP) model is employed in this paper to assess the impacts for the period 2019-2037. This paper assesses the scenario impacts in terms of diversification of electricity generation, generation technology mix, primary energy requirement and CO_2 emissions.The analysis reveals that greater role of solar energy in the electricity generation would have positive impacts on the Thai electricity generation from several aspects including improving the diversification of primary energy supply for electricity generation, reducing fossil fuel consumptions for power production, less reliance on fossil fuel sources and environmentally friendly electricity generation. For example, 20% share of solar in fuel mix for power generation would result in a 17% reduction in fossil fuel consumption. Furthermore, it would contribute to a 16% decrease in CO_2 emissions and 28% reduction in SO_2 emissions. Despite several benefits, there are a number of emerging barriers for promoting solar based electricity generation in Thailand. These include inefficient and unreliable nature of solar energy for baseload electricity demand, high-initial cost, unsupportable grid infrastructure and unfavourable regulatory framework. This paper, therefore, recommends that development of energy storage technologies, improvement of grid flexibility and the revision of the regulations to support solar energy business could be effective strategies in order to address the barriers facing the Thai electricity industry.

1. Introduction

Electricity is one major factor in meeting the basic needs of the country. It also plays a very important role in the country's economic and social development because electricity is one of the most important factors driving the development of the industrial, commercial and residential sectors. For Thailand, a country that has continued economic and social expansion which is widely distributed to all parts of the country according to government's policies. Therefore, a growth for electricity demand has consistently increased and tends to rise continuously. Between the years 2009 and 2019, electricity demand in Thailand increased annually by about 4.2%, from 135,209 GWh in 2009, to 203,714 GWh in 2019 [1-2]. In order to meet the increasing electricity demand, Thailand has been mainly dependent on fossil fuels for electricity generation. However, Thailand has limited energy resources, for example, crude oil, natural gas and coal. As a result, a large amount of energy was imported to be used as fuel in power generation. With a view to reduce dependency on energy import and to diversify the energy sources used for generating electricity, the government promotes the use of renewable energy as a matter of priority. Renewable energy, specifically solar energy, appears to be an attractive option for the government due to the fact that solar energy is clean and abundant renewable energy source available. In addition, Thailand has great solar potential. The average annual solar energy is relatively high compared to many countries, with most parts of the country exposed to the highest solar radiation between April and May in the range 20-23 $MJ/m^2/day$ [3]. In 2018, the government has developed the Alternative Energy Development Plan (AEDP2018) for the period 2018-2037. Under this plan, renewable energy in Thailand is projected to be 30% of the total energy production by 2037 [4]. According to the AEDP, the proportion of electricity from solar electricity production would increase to 20% in 2037 - more than three-fold increase as compare to 2018. The solar generating capacity is estimated to increase substantially, from 2,962 MW in 2018 to 15,574 MW in 2037. This paper, therefore, aims to analyze the impacts of solar electricity generation on the Thai electricity industry in terms of diversification of electricity generation, generation technology mix, primary energy requirement and CO_2 emissions.

2. Methodology

To assess the impacts of solar electricity generation, the Low Emissions Analysis Platform (LEAP) is selected. LEAP is a widely-used software tool for energy policy, climate change mitigation and air pollution abatement planning developed by the Stockholm Environment Institute (SEI) [5]. LEAP has been adopted by thousands of organizations in more than 190 countries worldwide. Its users include government agencies, academics, non-governmental organizations, consulting companies, and energy utilities, and it has been used at scales ranging from cities and states to national, regional and global applications. LEAP has been employed by several studies to assess the energy and environmental impacts of renewable energy [6-18].

In this paper, the development of scenario is mainly based on the Power Development Plan (PDP) and the Alternative Energy Development Plan (AEDP). The scenario analysis covers for the time period of 2019-2037. In this study, scenario development is mainly focused on solar electricity generation. For this purpose, this paper develops two alternative scenarios, namely, the Solar2015 and Solar2018 scenarios. In addition to the two alternative scenarios, this paper includes a REF scenario which represent a continuation of current trends in primary energy and technology mix for power generation. The two alternative scenarios (Solar2015 and Solar2018 scenarios) are primarily intended to increase shares of solar energy in electricity generation. In the Solar2015 scenario, the generating capacity of solar energy would increase to 6,000 MW in 2037. The solar2018 scenario is developed to represent greater role of solar energy in power production. The installed capacity of solar energy, under the solar2018 scenario, is expected to grow to 15,574 MW in 2037. Table 1 provides the details of key features of the three scenarios.

3. Data Consideration

The data required by this study include electricity consumption, power generation by fuel types, installed capacities by generating technologies, electricity losses, efficiencies of power plant technologies, electricity load curve and electricity demand growth. The electricity consumption data, power generation by fuel types, installed capacities and electricity losses are available from Energy Balance report and Alternative Energy Situation report, published by the Department of Alternative Energy Development and Efficiency (DEDE) [1], [19]. The information on the electricity demand growth can be achieved from the Power Development Plan (PDP2018) developed by the Energy Policy and Planning Office (EPPO), Ministry of Energy [20]. The shares of installed capacity by each plant type (e.g., steam turbine, combine-cycle, co-generation, gas turbine, hydro, solar, wind, biomass, biogas, municipal solid waste and geothermal) can be taken from the PDP2018 [20]. The information on the efficiencies of power plant technologies can be taken from external sources including International Energy Agency (IEA) and Energy Information Administration (EIA) [21-22]. The data on electricity load curve can be obtained from the Electricity Generating Authority of Thailand (EGAT) and the relevant literature [23-24].

Table 1. Key features of three scenarios

Scenario	Scenario features
REF scenario	• Continue with current primary energy and technology mix for electricity generation. • The share of solar energy in electricity production would be 5% until the year 2037. • The generating capacity of solar energy is expected to grow to 4,517 MW in 2037
Solar2015 scenario	• Slight increase of solar electricity installed capacity to 6,000 MW in 2037.
Solar2018 scenario	• Greater role of solar in electricity production, thus resulting in a considerable increase of solar electricity installed capacity to 15,574 MW in 2037.

4. Empirical Results and Discussion

This paper assesses the impacts of solar electricity generation on the Thai electricity industry in terms of diversification of electricity generation, fossil fuel consumption, CO_2 emissions and SO_2 emissions for the period 2019-2037.

4.1. Diversification of electricity generation

Figure 1 shows that electricity production under the REF, Solar2015 and Solar2018 scenarios would increase from 207 TWh in 2019, to 340 TWh in 2037. Over the entire studied period, the electricity produced from natural gas under the REF, Solar2015 and Solar2018 scenarios would increase, from 127 TWh, to 209 TWh, 203 TWh and 184 TWh respectively. Over the period 2019-2037, the generation of electricity from coal and lignite under the REF, Solar2015 and Solar2018 scenarios would increase, from 37 TWh, to 60 TWh, 56 TWh and 49 TWh respectively. It is noticed from Figure 1 that the electricity generation from solar energy would increase considerably. Electricity generated from solar energy in 2037 is expected to rise by 1.6 times in the REF scenario, 2.2 times in the Solar2015 scenario, and 5.8 times in the Solar2018 scenario as compared to 2019. Electricity produced from Solar in 2037 is expected to grow to 10 TWh in the Solar2015 and 26 TWh in the Solar2018 - more than double increase as compared to 2019.

In view of electricity generation share by fuel type, the percentage share of natural gas in total electricity generation in the case of Solar2015 and Solar2018 scenarios in 2037 would decrease to 59% and 54% respectively (as shown in Figure 2). Moreover, the share of coal and lignite would also decrease to 17% in the Solar2015 scenario and 15% in the Solar2018 scenario. These decreases would be contributed by the government's policy to diversify primary energy supply for electricity generation. The decreases would be substituted by an increased share of solar energy in power production. The share of solar energy in power production under the Solar2018 scenario is expected to grow considerably from 2% in 2019, to 8% in 2037.

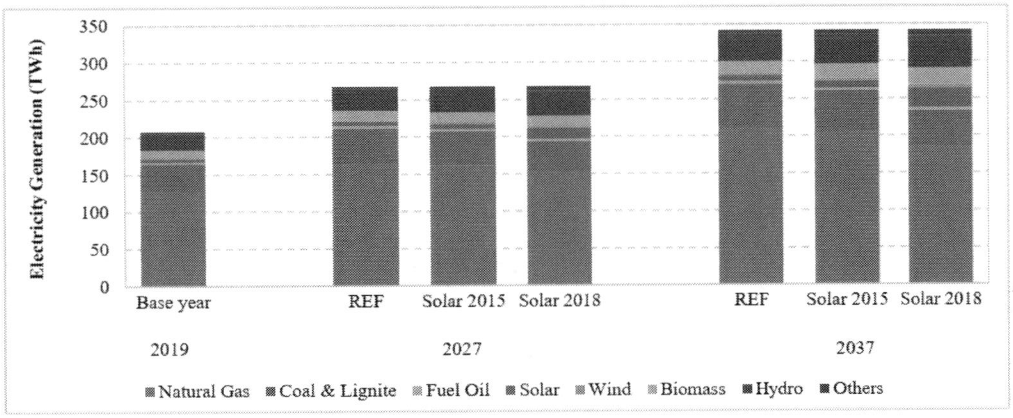

Figure 1. Electricity generation by fuel type for the period 2019-2037

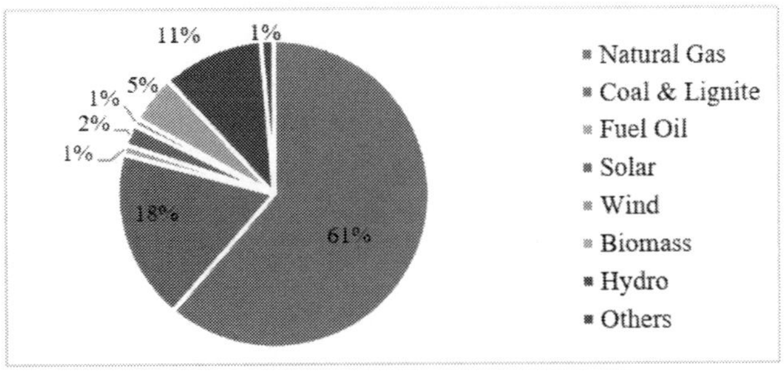

(a) in 2019

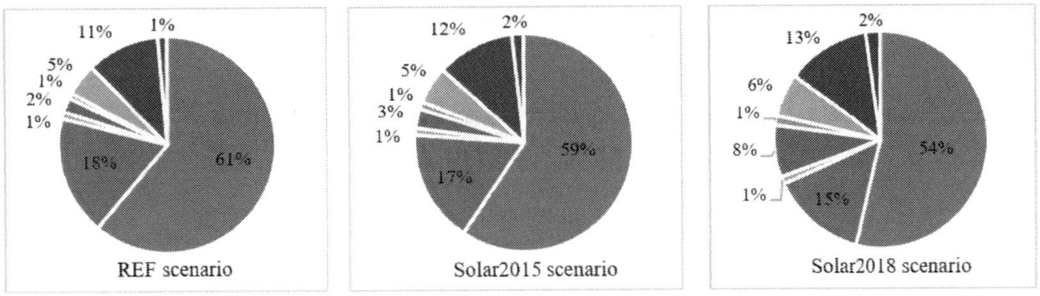

Figure 2. Electricity generation by fuel type

4.2. Fossil fuel consumption

This paper employs fossil fuels consumption as an indicator to assess a decline in fossil fuel consumption attributed to an increased electricity generation from solar. Table 2 shows the fossil fuels consumption for power generation in the case of the REF, Solar2015, and Solar2018 scenarios. And, Figure 3 provides changes in fossil fuel consumption by fuel type in 2037 for the Solar2015 and Solar2018 scenarios in comparison with the REF scenario.

It can be seen from Table 2 that over the period 2019-2037, fossil fuel consumption in the case of REF scenario increased considerably, from 33,143 KTOE in 2019, to 55,996 KTOE in 2037. In 2037, fossil fuel consumption in the Solar2015 and Solar2018 scenarios would be 9.9% and 17.5%, respectively, lower than in the case of the REF scenario. The fossil fuel inputs for electricity generation under the Solar2015 scenario is expected to grow to 50,408 KTOE in 2037 - a decrease of 5,588 KTOE in comparison with the REF scenario. In the Solar2018 scenario, fossil fuel consumption in 2037 is expected to be 46,204 KTOE - a reduction of 9,792 KTOE in fossil consumption for power production as compared with the REF scenario.

Table 2. Fossil fuels consumption for electricity generation

Year	REF scenario	Solar2015 scenario Changes from REF scenario		Solar2018 scenario Changes from REF scenario	
	(KTOE)	(KTOE)	(%)	(KTOE)	(%)
2019	33,143	-	-	-	-
2027	43,166	-2,208	-5.1	-3,965	-9.2
2037	55,996	-5,588	-9.9	-9,792	-17.5

Notes: Number in brackets show percentage change from the REF scenario.

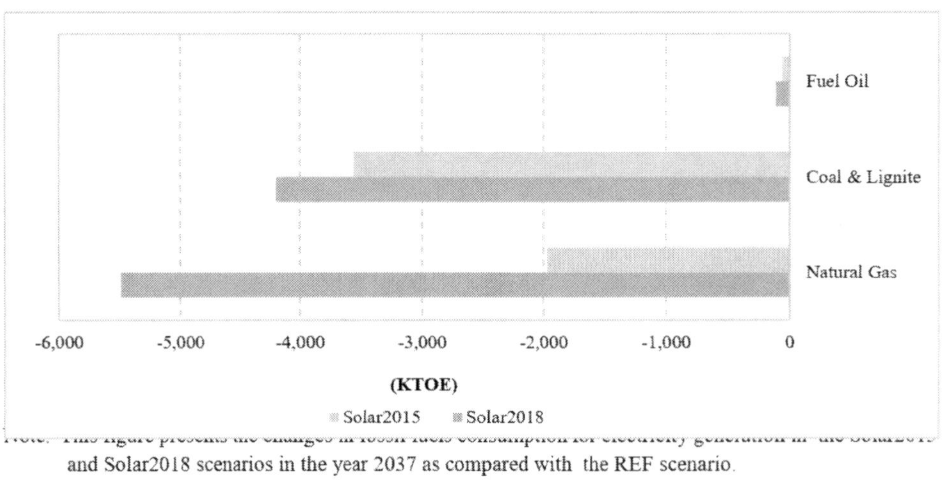

Note: This figure presents the changes in fossil fuels consumption for electricity generation in the Solar2015 and Solar2018 scenarios in the year 2037 as compared with the REF scenario.

Figure 3. Changes in fossil fuel consumption by fuel types in 2037

IOP Conf. Series: Materials Science and Engineering 1163 (2021) 012016 doi:10.1088/1757-899X/1163/1/012016

From Figure 3, it appears that major contribution to a decline in fossil fuels consumption for electricity generation in the case of Solar2015 and Solar2018 scenarios would be from natural gas and coal and lignite. The decline in the demand for both natural gas and coal and lignite would help reduce fossil fuels imports for electricity generation. This is because natural gas and coal imports have been consistently increasing since 2000s [1-2]. In addition, natural gas imports accounted for 30% of total natural gas supply in 2019 and more than 75% of total natural gas consumption has been employed as fuel inputs for power production [1]. For coal and lignite, coal consumption for power generation accounted for more than 30% of coal imports in 2019 [1]. An increased electricity generation from solar energy would, therefore, help reduce fossil fuel imports and hence help enhance security of energy supply.

4.3. CO_2 emissions

In view of CO_2 emissions, Table 3 shows that CO_2 emissions under the REF scenario is estimated to increase from 92 million tonnes in 2019, to 150 million tonnes in 2037, an increase of 58 million tonnes over the 2019 emission level. In the case of Solar2015 and Solar2018 scenarios, CO_2 emissions in 2037 would be, respectively, 9.4 million tonnes (6.3%) and 24.8 million tonnes (16.6%) lower than the emissions in the REF scenario. A reduction in CO_2 emissions is a result of a decrease in fossil fuels consumption for power generation.

Table 3. CO_2 emissions over the period 2019-2037

Year	REF scenario	Solar2015 scenario		Solar2018 scenario	
		Changes from REF scenario		Changes from REF scenario	
	(Million tonnes)	(Million tonnes)	(%)	(Million tonnes)	(%)
2019	92	-	-	-	-
2027	118	-3.81	-3.2	-10.10	-8.6
2037	150	-9.40	-6.3	-24.82	-16.6

4.4. SO_2 emissions

Table 4 reveals that SO_2 emissions, under the REF scenario, would increase from 351 thousand tonnes in 2019, to 573 thousand tonnes in 2037, an increase of 222 thousand tonnes over the 2019 emission level. The SO_2 emissions in the case of Solar2015 and Solar2018 scenarios in 2037 would be, respectively, 23.2% and 27.5% lower than the emissions in the REF scenario. A decline in SO_2 emissions would be mainly from a decrease in coal and lignite consumption for power generation (as previously discussed in Section 4.2).

Table 4. SO_2 emissions over the period 2019-2037

Year	REF scenario	Solar2015 scenario		Solar2018 scenario	
		Changes from REF scenario		Changes from REF scenario	
	(Thousand tonnes)	(Thousand tonnes)	(%)	(Thousand tonnes)	(%)
2019	351	-	-	-	-
2027	450	-53.2	-11.8	-63.0	-14.0
2037	573	-133.3	-23.2	-157.8	-27.5

The foregoing results suggests that greater role of solar energy in the electricity generation would help diversifying primary energy mix for power generation. Moreover, it would contribute to a reduction in fossil fuels consumption for electricity generation and hence result in a decrease in fossil fuel imports. An improvement in the diversification of primary energy supply supplemented by a decline in fossil fuels imports would, therefore, help enhance the energy security of the country. Furthermore, increased electricity generation from solar energy would have a positive impact on the environment. As previously discussed, increased contribution of solar energy to electricity generation would considerably reduce the use of fossil fuels for electricity generation and, therefore, help mitigating CO_2 and SO_2 emissions. The attractiveness of solar energy would considerably enhance when considering in the context of Thailand - country that has great solar potential.

Despite the fact that solar energy would provide several benefits to the Thai society, the promotion of solar energy could face numerous barriers. For example, the intermittency of solar energy would make it inefficient and unreliable for supplying electricity to the system. Another barrier would be the capital-intensive investment for solar energy. Due to its high initial capital cost but inefficient and unreliable nature would make solar energy get a low rate of return on investment. Unsupportable grid infrastructure would emerge as one major barrier. This is because the current structure of the Thai electricity sector is in the form of centralized electricity system which is not designed to support distributed generation characterizing by small-scale and intermittent generators like solar energy. The issue of regulatory framework would be also important. The regulatory arrangements for the Thai electricity industry for the last sixty years have been primarily designed to support centralized electricity system. The traditional regulatory framework would, therefore, hinder the penetrations of solar energy. To address the barriers facing the Thai electricity industry, this paper, therefore, recommends that development of energy storage technologies, improvement of grid flexibility and the revision of the regulations to support solar energy business could be effective strategies.

5. Conclusion

This paper analyses the impacts of solar electricity generation on the Thai electricity industry in terms of diversification of electricity generation, fossil fuel consumption, CO_2 emissions and SO_2 emissions for the period 2019-2037. The analyses reveal that greater role of solar energy in the electricity generation would have positive impacts on the Thai electricity generation from several aspects including improving the diversification of primary energy supply for electricity generation, reducing fossil fuel consumptions for power production, less reliance on fossil fuel sources and environmentally friendly electricity generation. For example, 20% share of solar in fuel mix for power generation would result in a 17% reduction in fossil fuel consumption. Furthermore, it would contribute to a 16% decrease in CO_2 emissions and 28% reduction in SO_2 emissions. Despite several benefits, the promotion of solar energy could face numerous barriers including inefficient and unreliable nature of solar energy for baseload electricity demand, high-initial cost, unsupportable grid infrastructure and unfavorable regulatory framework. This paper, therefore, recommends that development of energy storage technologies, improvement of grid flexibility and the revision of the regulations to support solar energy business could be effective strategies in order to address the barriers facing the Thai electricity industry.

References

[1] Department of Alternative Energy Development and Efficiency 2019a *Energy Balances Report 2012-2019*, annual reports, Bangkok, Thailand.

[2] Department of Alternative Energy Development and Efficiency 2011 *Thailand Energy Situation Report 2000-2011*, annual reports, Bangkok, Thailand.

[3] Department of Alternative Energy Development and Efficiency 2014 *Handbook for the Development and Investment in Renewable Energy Series 2 Solar Energy*, Bangkok, Thailand.

[4] Department of Alternative Energy Development and Efficiency 2019 *Summary of the public hearing on the draft of the Alternative Energy Development Plan 2018 (AEDP2018)*, Bangkok, Thailand.

[5] Stockholm Environment Institute 2020 *Low Emissions Analysis Platform (LEAP): training materials*, Stockholm, Sweden.

[6] Kusumadewi TV, Winyuchakrit P, Misila P and Limmeechokchai B 2017 GHG Mitigation in Power Sector: Analyzes of Renewable Energy Potential for Thailand's NDC Roadmap in 2030 *Energy Procedia* 138 pp 69-74.

[7] Hu G, Ma X and Ji J 2019 Scenarios and policies for sustainable urban energy development based on LEAP model - A case study of a postindustrial city: Shenzhen China *Applied Energy* 238 pp 876-886.

[8] Emodi N, Emodi C, Girish G P and Emodi A 2017 Energy Policy for Low Carbon development in Nigeria: A LEAP Model Application *Renewable and Sustainable Energy Reviews* 68 pp 247-261.

[9] Rivera-González L, Bolonio D, Mazadiego L F and Valencia-Chapi R 2019 Long-Term Electricity Supply and Demand Forecast (2018-2040): A LEAP Model Application towards a Sustainable Power Generation System in Ecuador *Sustainability* 11 pp 5316.

[10] Mirjat N H, Harijan K and Valasai G 2018 Long-Term Electricity Demand Forecast and Supply Side Scenarios for Pakistan (2015-2050): A LEAP Model Application for Policy Analysis *Energy* 163 pp 512-526.

[11] Wang J, Hua C and Li L 2018 Long-term Energy Sustainability Development Analysis to Reducing Carbon Emissions and Air Pollutions of China Based on LEAP Simulation Model *Ekoloji* 27(106) pp 173-179.

[12] Kresnawan M R, Safitri I A and Darmawan I 2018 Long Term Projection of Electricity Generation Sector in East Kalimantan Province: LEAP Model Application. 12th South East Asian Technical University Consortium (SEATUC) Yogyakarta, Indonesia, doi: 10.1109/SEATUC.2018.8788875, pp 1-5.

[13] Bhuvanesh A, Christa S T J and Kannan S 2017 Least Cost Electricity Generation Planning for China with Low GHG Emission Using LEAP and EnergyPLAN, 2017 IEEE International Conference on Computational Intelligence and Computing Research, ICCIC 2017, 8524458.

[14] Shahinzadeh H, Fathi S H and Hasanalizadeh-Khosroshahi A 2016 Long-term energy planning in IRAN using LEAP scenario: Using combined heat and power (CHP) 2016 Iranian Conference on Renewable Energy & Distributed Generation (ICREDG), Mashhad, doi: 10.1109/ICREDG.2016.7875915, pp 32-37.

[15] McPherson M and Karney B 2014 Long-term scenario alternatives and their implications: LEAP model application of Panama's electricity sector *Energy Policy* 68 pp 146-157.

[16] Park N, Yun S and Jeon E 2013 An analysis of long-term scenarios for the transition to renewable energy in the Korean electricity sector *Energy Policy* 52 pp 288-296.

[17] Afreen S and James L Wescoat Jr 2013 Energy use in large-scale irrigated agriculture in the Punjab province of Pakistan *Water International* 38:5 pp 571-586, DOI: 10.1080/02508060.2013.828671.

[18] Lin J, Cao B, Cui S, Wang W and Bai X 2010 Evaluating the effectiveness of urban energy conservation and GHG mitigation measures: The case of Xiamen city, China *Energy Policy* 38

pp 5123-5132.

[19] Department of Alternative Energy Development and Efficiency 2019b *Thailand alternative energy situation2019*, Bangkok, Thailand.

[20] Energy Policy and Planning Office 2018 *Power Development Plan (PDP2018)*, Bangkok, Thailand.

[21] International Energy Agency 2013 *Southeast Asia energy outlook: World energy outlook special report*, Paris, France.

[22] Sorapipatana C2013 *An Assessment of Energy Security in Thailand's Power Generation*, Final Report submitted to the National Science and Technology Development Agency (NSTDA). (In Thai).

[23] Siritiprussamee P, Suwannarat W and Dulyarittirong R 2014 Analysis of Thailand Electricity Demand Pattern *Journal of Environmental Management* 10(1) pp 1-18. (In Thai)

[24] Electricity Generating Authority of Thailand 2018 *Electricity Peak Demand of National Grid*, Nonthaburi, Thailand.

Acknowledgement

The authors would like to express special thanks to the Faculty of Engineering, Mahasarakham University for financial supports. The authors also wish to extend appreciation to the Solar Energy and Energy Research Unit, Faculty of Engineering, Mahasarakham University for providing research facilities.

Sustainability indicators for rubber plantations in Thailand: Environmental integrity dimension

Khamphe Phoungthong[1], Sutinee Sinutok[2], Oramas Suttinun[1], Suriya Palamae[1], Jitti Mungkalasiri[3], Prakaytham Suksatit[3], Charongpun Musikavong[4]

[1]Environmental Assessment and Technology for Hazardous Waste Management Research Center, Faculty of Environmental Management, Prince of Songkla University, Songkhla 90112, Thailand
[2]Coastal Oceanography and Climate Change Research Center, Faculty of Environmental Management, Prince of Songkla University, Hat Yai, 90112, Thailand
[3]National Metal and Materials Technology (MTEC), National Science and Technology Development Agency, Pathumthani 12120, Thailand
[4]Environmental Assessment and Technology for Hazardous Waste Management Research Center, Department of Civil and Environmental Engineering, Faculty of Engineering, Prince of Songkla University, Hatyai, Songkhla, 90112, Thailand

E-mail: mcharongpun@eng.psu.ac.th, charongpun@gmail.com

Abstract. The natural rubber industry has played a vital role in Thailand' s economic, environmental, and social contexts. Therefore, sustainable rubber production is a significant concern. At present, a wide variety of international tools has been developed to assess the agricultural sector' s sustainability performance. The Food and Agriculture Organization of the United Nations developed the Sustainability Assessment of Food and Agriculture systems (SAFA) in 2013. Environmental integrity is one of the dimensions of sustainability in SAFA. SAFA is applied to assess the impact of agriculture operations on the environment. This research aims to assess suitable environmental integrity indicators in compliance with SAFA for rubber plantations in Thailand. The system boundary is set to cover rubber plantations, harvesting of fresh latex, and rubber tree felling. This work conducted focus group discussions, using the SAFA, guidelines, indicators, and tools for rubber plantations, the review process by rubber plantation experts, representatives of rubber associations, SAFA researchers, and environmental scholars, and site visits to interview rubber farmers in the south, northeast, and north of Thailand. Environmental themes, including the atmosphere, water, land, biodiversity, and material and energy were included in assessing the environmental integrity of a rubber plantation. Animal welfare was not included.

1. Introduction

Thailand has an area of rubber plantations of 3.27 million ha and produces natural rubber of about 4.84 million tonnes yearly. This amount accounts for 33 percent of the total natural rubber in the world [1]. The rubber industry's life cycle chain consists of rubber plantations, the production of intermediate rubber products, and the production of final products. The intermediate rubber products are ribbed smoked sheets (RSS), block rubber (Standard Thai Rubber, STR), and concentrated latex. The final rubber products are gloves, condoms, belts, soles, vehicle tires, and industrial rubber parts [2] Therefore, the natural rubber industry has played a vital role in Thailand's economic, environmental, and social contexts.

A rubber plantation starts from site preparation followed by rubber seedling cultivation, rubber cultivation, rubber tapping, and rubber tree felling, which takes approximately 25 years. Fresh latex and cup lumps are the products from a plantation. The fresh latex is the raw material of RSS and concentrated latex. The cup lump is used for producing block rubber. Hevea wood and branches are the products from the rubber tree felling and are used as the manufacturing of lumber AB [3]. The rubber plantation is the upstream phase of the life cycle of the rubber industry; therefore, the sustainable production of a rubber plantation is a significant concern.

A wide variety of international tools have been developed to assess the agricultural sector's sustainability performance. The Food and Agriculture Organization of the United Nations developed the Sustainability Assessment of Food and Agriculture systems (SAFA) in 2013. SAFA is a universal framework for sustainability assessment. According to SAFA, the publication consists of SAFA guidelines, SAFA indicators, the SAFA tool, and SAFA smallholders App. [4] SAFA includes four sustainability dimensions: good governance, environmental integrity, economic resilience, and social well-being dimensions. Each dimension consists of themes, sub-themes, and indicators for assessing the sustainability of the target food or agriculture supply chains [5].

For assessing the sustainability of rubber plantations in Thailand by SAFA, the reviews of SAFA guidelines, SAFA indicators, the SAFA tool, analysis, and discussion among stakeholders have been conducted to obtain suitable themes, sub-themes, and indicators for each dimension. We aim to assess suitable indicators of environmental integrity in compliance with SAFA for rubber plantations in Thailand for this work.

2. Material and methods

2.1. System boundary and study area

The system boundary of the study is set to cover rubber seedling cultivation, rubber cultivation, rubber tapping, and rubber tree felling. Fresh latex, cup lump, and hevea wood and branches are the main products of rubber plantations. Southern Thailand has the largest rubber plantation area of about 2.17 million ha (60% of the total area of rubber plantations), followed by northeastern (0.84 million ha, 23%), central (0.39 million ha, 11%), and northern (0.22, million ha, 6%) Thailand, respectively [6]. The study areas for site visits, interviewing, and collecting data to develop and test the questionnaire of the environmental integrity dimensions were Songkhla province in southern Thailand, Bueng Kan province in northeastern Thailand, and Phitsanulok province in northern Thailand.

2.2 Methodology

According to SAFA, this work started by reviewing literature data and other sustainable assessment methods for agricultural products, especially rubber products. The goal, system boundary, and studied products were set. The secondary data of the study areas were collected. Then, the questionnaire for assessing the environmental integrity of a rubber plantation was developed by a working group concerned with the rubber plantation practices in all regions in Thailand. The questionnaire structures were mainly divided into five themes: atmosphere, water, land, biodiversity, and material and energy. Each theme is comprised of sub-themes and performance, practice-based, and target indicators compliant with the SAFA guidelines [5].The technical committees, which included rubber plantation experts, representatives of rubber associations, SAFA researchers, and environmental scholars were set. The technical committees reviewed each indicator's questionnaire and rating score to assess the environmental integrity of rubber plantations in Thailand. The working groups visited the study area and collected data to build and test the questionnaire by interviewing 20 rubber farmers in Hat Yai, Songkhla, in November 2019, 23 rubber farmers in Seka, Bueng Kan in February 2020, and 30 rubber farmers in Wang Thong and Nakhon Thai, Phitsanulok in February 2020. In the final stage, the technical committee finalized the questionnaire and rating score to assess the environmental integrity of rubber plantations in Thailand based on SAFA.

3. Results and Discussion

3.1 Atmosphere

The environmental integrity dimensions have six themes: E1 Atmosphere, E2 Water, E3, Land, E4 Biodiversity, E5 Material and energy, and E6 Animal welfare. The theme E1 Atmosphere had two sub-themes E1.1 Greenhouse gasses (GHG) and E1.2 Air quality. Each sub-theme consisted of indicators. SAFA divided the indicators of environmental integrity into three types, consisting of performance (P), practice-based (R), and target (T) indicators [5]. All of the indicators in theme E1 (Table 1) are included to assess the environmental integrity of rubber plantations. The GHG reduction target was not directly related to rubber farmers. This must be evaluated by the national policy of GHG reduction from an agricultural section. For the GHG mitigation practices, the farmer can employ such practices as optimizing chemical fertilizer use, use organic fertilizer, and not conduct open burning. The sub-theme E1.2 (Table 1) included three indicators. The air pollution reduction target must be assessed by national policy. The Pollution Control Department (PCD) monitors the ambient air quality in all parts of Thailand. This monitoring data can be used to assess the ambient concentration of air pollutants in the plantation area. The farmers could support the air pollution prevention practices by using biofuels such as ethanol and biodiesel and solar cells in the activities related to their rubber plantation.

3.2 Water

Theme E2 Water consisted of two sub-themes, which were water withdrawal and water quality and seven indicators (Table 1) [5]. For the water conservation target and clean water target, these should be proposed by the policymaker. The secondary data on the rain intensity (mm/year) in the rubber plantation area can be used to determine the suitable areas for rubber plantations. A plantation in a low precipitation area is not recommended due to the potential of water scarcity problems and the low production yield. The water stress index value that is evaluated from the ratio of water withdrawal to water availability [7] can represent the water scarcity level in the rubber plantation area. A farmer could support water sustainability by recording the amount of water as the baseline and reducing the amount of water use from the baseline. A storage pond and other structures for storing water in a plantation area should be constructed. The planting of ground cover plants in a rubber plantation area could help to prevent water pollution.

Table 1. Selected sub-themes and indicators for themes E1 Atmosphere and E2 Water for rubber plantations

Theme	Sub-Themes	Selected Indicators
E1 Atmosphere	E1.1 Greenhouse Gases	E1.1.1 GHG Reduction Target
		E1.1.2 GHG Mitigation Practices
		E1.1.3 GHG Balance
	E1.2 Air Quality	E1.2.1 Air Pollution Reduction Target
		E1.2.2 Air Pollution Prevention Practices
		E1.2.3 Ambient Concentration of Air Pollutants
E2 Water	E2.1 Water Withdrawal	E2.1.1 Water Conservation Target
		E2.1.2 Water Conservation Practices
		E2.1.3 Ground and Surface Water Withdrawals
	E2.2 Water Quality	E2.2.1 Clean Water Target
		E2.2.2 Water Pollution Prevention Practices
		E2.2.3 Concentration of Water Pollutants
		E2.2.4 Wastewater Quality

Remark: The themes, sub-themes, and indicators were obtained from SAFA guideline [5]

3.3 Land

The theme E3 Land had two sub-themes E3.1 Soil quality and E 3.2 Land degradation [5]. All of the indicators in theme E3 (Table 2) were used to assess environmental integrity. For the soil improvement practices, the farmer must carry out the activities such as using organic fertilizers, applying a soil conditioner, and growing ground cover plants as well as multiple cropping. Soil sampling and analysis are required to assess the physical structure of the soil, soil chemical quality, soil biological quality, and soil organic matter [5]. When the larger area in Thailand was considered, the Land Development Department, Thailand studied the soil properties for all regions in Thailand. This data should be used to determine the sustainability of the soil quality of rubber plantations. In addition, the indirect indicators such as diseases of rubber trees and the number of ants, termites, earthworms can represent the soil quality. The land conservation and rehabilitation plan should be established by the policymaker. The rubber farmer can undertake land conservation and rehabilitation practices such as constructing or maintaining a good irrigation system in rubber plantations and agroforestry.

Table 2. Selected sub-themes and indicators for themes E3 Land and E4 Biodiversity for rubber plantations

Theme	Sub-Themes	Selected Indicators
E3 Land	E3.1 Soil Quality	E3.1.1 Soil Improvement Practices
		E3.1.2 Soil Physical Structure
		E3.1.3 Soil Chemical Quality
		E3.1.4 Soil Biological Quality
		E3.1.5 Soil Organic Matter
	E3.2 Land Degradation	E3.2.1 Land Conservation and Rehabilitation Plan
		E3.2.2 Land Conservation and Rehabilitation Practices
		E3.2.3 Net Loss/Gain of Productive Land
E4 Biodiversity	E4.1 Ecosystem Diversity	E4.1.1 Land scape/Marine Habitat Conservation Plan
		E4.1.2 Ecosystem Enhancing Practices
		E4.1.3 Structural Diversity of Ecosystems
		E4.1.4 Ecosystem Connectivity
		E4.1.5 Land Use and Land Cover Change
	E4.2 Species Diversity	E4.2.1 Species Conservation Target
		E4.2.2 Species Conservation Practices
		E4.2.3 Diversity and Abundance of Key Species
		E4.2.4 Diversity of Production
	E4.3 Genetic Diversity	E4.3.1 Wild Genetic Diversity Enhancing Practices
		E4.3.2 Agro-biodiversity in-situ Conservation
		E4.3.3 Locally Adapted Varieties and Breeds
		E4.3.4 Genetic Diversity in Wild Species
		E4.3.5 Saving of Seeds and Breeds

Remark: The themes, sub-themes, and indicators were obtained from SAFA guideline [5]

3.4 Biodiversity

Theme E4 Biodiversity consisted of sub-themes: E4.1 Ecosystem diversity, E4.2 Species diversity, and E4.3 Genetic diversity (Table 2) [5]. The farmers can be involved in proposing the landscape and habitat conservation plants and species conservation target in their rubber plantation area. The rubber plantation practices such as single crops, agroforestry, and growing ground cover plants, and multiple cropping directly affects the biodiversity indicators. The land-use change from the past 20 years and types of land use around the rubber plantation area must be determined in the assessment. The number of rare and endemic species must be determined. The data of rubber verities and the collection of seeds for seedlings are required.

3.5 Material and energy
Table 3 presents the selected sub-themes and indicators for themes E5 Material and energy. The theme E5 Land had three sub-themes E5.1 Material use, E 5.2 Energy use, and E 5.3 Waste reduction and disposal [5]. All of the indicators in theme E5 (Table 3) are undertaken to access environmental integrity. Theme E5 is directly related to rubber farmer practices. Farmers must plan for the use of alternative energy, recycle and reuse the raw material in the plantation, and reduce the amount of waste. Farmers must keep records of the amount of material, energy, alternative energy use, and waste generated. The generated waste must be managed and disposed of by the proper methods. For the theme E6 Animal welfare, the rubber plantations were not involved in animal products; therefore, animal welfare was not included in the assessment of the environmental integrity of rubber plantations.

Table 3. Selected sub-themes and indicators for themes E5 Material and energy for rubber plantations

Theme	Sub-Themes	Selected Indicators
E5 Materials and Energy	E5.1 Material Use	E5.1.1 Material Consumption Practices
		E5.1.2 Nutrient Balance
		E5.1.3 Renewable and Recycled Materials
		E5.1.4 Intensity of Material Use
	E5.2 Energy Use	E5.2.1 Renewable Energy Use Target
		E5.2.2 Energy Saving Practices
		E5.2.3 Energy Consumption
		E5.2.4 Renewable Energy
	E5.3 Waste Reduction and Disposal	E5.3.1 Waste Reduction Target
		E5.3.2 Waste Reduction Practices
		E5.3.3 Waste Disposal
		E 5.3.4 Waste Reduction

Remark: The themes, sub-themes, and indicators were obtained from SAFA guideline [5]

4 .Conclusion
This work determined the suitable indicators for assessing the environmental integrity of rubber plantations in Thailand. The system boundary is set to cover rubber plantations, the harvesting of fresh latex, and rubber tree felling. A focus group discussion, using SAFA, guidelines, indicators, and tools for rubber plantations, the review process by rubber plantation experts, representatives of rubber associations, SAFA researchers, and environmental scholars, and the site visit for interviewing rubber farmers in the south, northeast, and north of Thailand were carried out. Environmental themes, including the atmosphere, water, land, biodiversity, and material and energy were included in assessing the environmental integrity of a rubber plantation. Animal welfare was not included. The policymakers must promulgate the plan with the target according to SAFA indicators for supporting the environmental integrity of rubber plantations. According to SAFA indicators, rubber farmers should carry out the recommended activities with the technical and financial support from the governance to get close to sustainable rubber plantations in Thailand.

References

[1] Food and Agriculture Organization (FAO), 2020. UN Food and Agriculture Organization, Corporate Statistical Database (FAOSTAT, 2018). Available at : http://www.fao.org/faostat/en/#data/QC (accessed 25.12.20).

[2] Jawjit W, Kroeze C and Rattanapan S 2010 Greenhouse gas emissions from rubber industry in Thailand. Journal of Cleaner Production. 18 403-411

[3] Pyay S, Thanungkano W, Mungkalasiri J and Musikavong C 2019 A life cycle assessment of intermediate rubber products in Thailand from the product environmental footprint perspective. Journal of Cleaner Production. 237 117632

[4] FAO 2013.Sustainability assessment of food and agriculture systems (SAFA) Available at: http://www.fao.org/nr/sustainability/sustainability-assessments-safa/en/ (assessed 25.12.20)

[5] FAO 2014. SAFA Guidelines version 3.0. Available at: http://www.fao.org/nr/sustainability/sustainability-assessments-safa/en/ (assessed 25.12.20)

[6] Office of Agricultural Economic (OAE), 2018. Information. Available at: http://www.oae.go.th/view/1/Information/EN-US (assessed 25.12.20)

[7] Pfister S, Koehler A and Hellweg S 2009 Assessing the Environmental Impacts of Freshwater Consumption in LCA. Environmental Science & Technology. 43 (11) 4098-4104

Acknowledgements

This research was funded by the National Science and Technology Development Agency, under contract number FDA-CO-2562-8681-TH. The authors would like to thank the Rubber Authority of Thailand, the Thai Rubber Association, the Thai Latex Association, and the Thai Hevea Wood Association.

Progressive Web App for Crop Field Data Collection

Phunon Koysawat[1] , Chayanon Boonprakob[1] , Khwantri Saengprachatanarug[2] , Arnut Chaosakul[1] , Panupong Wanjantuk[1] , Mahisorn Wongphati[3] , Santawat Santiteerakul[3] , Adulwit Chinapas[1] , Kanda Runapongsa Saikaew[1]

[1]Department of Computer Engineering, Faculty of Engineering, Khon Kaen University, 123 Mittraparp Rd., A. Muang, Khon Kaen, 40002
[2]Department of Agricultural Engineering, Faculty of Engineering, Khon Kaen University, 123 Mittraparp Rd., A. Muang, Khon Kaen, 40002
[3]HG Robotics Co., Ltd., 1407-1410, 14th Floor Chulalongkorn Research Building, 254 Chulalongkorn University, Phayathai Rd. Wang Mai, Pathum Wan District, Bangkok, 10330

Email: krunapon@kku.ac.th (Kanda Runapongsa Saikaew)

Abstract. General farming needs to have a matter of data collection. For efficient cultivation planning and harvesting, information must be collected for evaluation and analysis to make each harvest decision. For example, we need to assess plants' ages, changes in size, color, and other senses (sweetness, smell, etc.) to plan for different crop field growth phases. Also, data collection problems occur due to the need to collect data in multiple locations simultaneously. Collecting information manually in the form of documents makes the process challenging to analyze. There is a need to collect field data more efficiently. Thus, this article proposes developing a web application to replace manual note-taking to reduce errors in recording and gathering data and missing data. Users can locate and modify crop field locations using latitude and longitude and work seamlessly with the Google Earth application. Furthermore, to accommodate practical use in real-life, the web application developed using Progressive Web Application (PWA) can perform without the Internet. Besides, PWA web applications are as fast as using an application on a smartphone without downloading and saving space. Based on the experimental result, compared with the process without using the application, it was found that the proposed application could reduce time in collecting sugarcane sampling data by about 45.28%

1. Introduction

Nowadays, various technologies are involved in people's daily lives to support and solve work or activities problems more efficiently and conveniently. Smart farming is a new form of agriculture by bringing technology in farming more productively in several phases, such as plant selection and harvesting. To perform such tasks, we need a method to collect agricultural data effectively. The traditional approach in collecting data tends to cause errors and time-consuming. Thus, this article proposes and develops a progressive web application to assist farmers or merchants in collecting crop field data for making a more appropriate decision in crop growing or crop sale.

2. Related work

This section describes the basic knowledge needed to develop an application and related applications.

2.1. Progressive web application

Web applications generally require the Internet all the time. However, sometimes web application users may be in an environment that the Internet is unavailable or unstable. Several mobile applications can function without the Internet, but that comes with the extra space usage on mobile devices and continuous application updates. Nevertheless, there is a type of web application that can work without the Internet and does not need users to update the application continuously. That type of web application is called a progressive web application. A progressive web application can have customized notifications and support an offline mode. It also helps to increase the efficiency of the use of web applications even faster than before. The development will not look much different from general web application development. The development language would be HTML, CSS, JavaScript, which are typical web application development languages.

2.1.1. Operation of the service worker

To develop a progressive web application, we need to have a system called the service worker, which runs in the background of web applications. JavaScript controls the behavior of applications and caches (a type of memory) to increase web applications' performance. Working with a transmission network as the active API when running the API system is a network request Web server directly. Then it will store the data in the cache. And when the API is rerun instead of the network request, it is changed to the previously held cache instead. In addition to making web applications faster, caching can also define parts of the web application to be available offline. It also can set a notification. All of these functions can be controlled through the system. Service worker keeps web applications on the home screen of your device, just like regular applications. The app manifest customizes colors, themes, and icons.

2.2. Local storage and indexed DB

In general, browsers have a variety of methods and formats, local storage, and indexedDB. The two will be different from each other, which can be used according to the web application's operation's advantages and disadvantages. Local storage is suitable for the type of data storing that uses cookies. It is one of the HTML5 web storage APIs. On the other hand, IndexDB collects data that can be stored as a small database on the device that stores data as a key like web storage but will use an index to find information faster. It is asynchronous to avoid conflicts in the application system.

2.3. Related applications

Examples of applications that are close to use

2.3.1. Chaona Thai. It is an application developed by the Internal Trade Department for Thailand's rice farmers. The system will estimate cost profit to make decisions before planting and creating an agricultural calendar that guides farmers on what to do and when to apply weed fertilizers with cost-saving and ready to be sold on the market. This application's advantage is that its recommendation system helps users make decisions, but it does not have functions that support crop field data collection.

2.3.2. Insect Shot. It is an application that automatically counts the number of Brown Planthoppers in the rice field. Images will be sent to the server and return the result, the estimated number of Brown Planthoppers. Unlike Insect Shot, the proposed application was developed to collect general data in the crop field and not focus on the automatic counting of images in the fields.

2.3.3. Chao Kaset Agriculture. It is an application developed by the National Electronics and Computer Technology Center (NECTEC) to collect plant types and areas from users to assist them in planning to

IOP Conf. Series: Materials Science and Engineering 1163 (2021) 012018 doi:10.1088/1757-899X/1163/1/012018

plant a crop field. However, unlike Chao Kaset Agriculture, the proposed application focus on collecting the crop field data products, such as their location and quality.

3. System Implementation

This section will discuss functional specification, system architecture, and system implementation of the proposed system. The proposed app can collect time information, GPS location, support offline mode, and support various types of users, including general user, guest, user admin, and system admin. It also supports scanning and recognizing text by using Google Vision API. Figure 1 depicts the system architecture of the proposed web application.

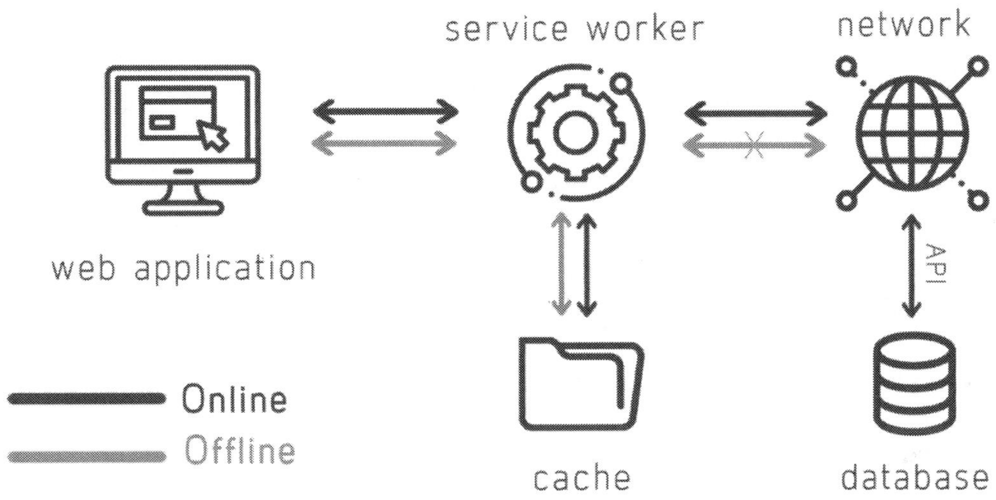

Figure 1. Progressive web application

Figure 1 shows the system is operated through a tool that controls the server's communication, called the service worker. There are two types of work through the cache and the work through the network. Working through caching is a function that will enable faster work on areas of previously used data such as images, text, and web applications when there is no Internet connection (offline). Working over a network is controlled by the service worker. The system will set the usage priority on network usage first. When there is an internet connection, it will be able to access the database's information.

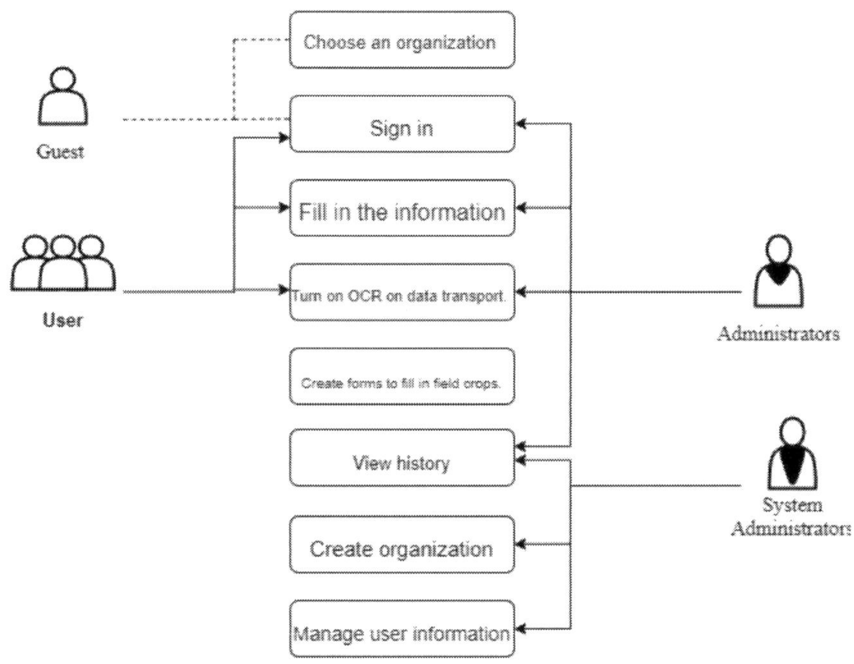

Figure 2. Role user diagram

There are four types of user access classification. The first category is a guest user who has not yet registered. Such users cannot access the main functions of the system. Because accessing the system requires an organization to exist, this user type can only select the organization to request the organization login to the system to collect information. The second category is a general user who can use the system login section, fill in the field crops, and view the history of filling in the database. The last two types are administrators. A user administrator can have access to all functions; on the other hand, a system administrator can create an organization and assign every user role.

The system has three main functions into three parts: 1) plant selection for data recording, 2) hold data for offline usage, and 3) show recordings in the database. The system authentication uses Firebase Authentication. Furthermore, the system has an OCR part to allow users to take photos and use Google Cloud Vision API to automatically scan and save data to fields according to each form data topic. The system used MongoDB as a NoSQL database to define the limits of the data (the Constraint) or relationship information (Entity Relationship) and saved it as JSON (JavaScript Object Notation).

4. Implementation Result

This section describes the screenshots of the proposed and developed web application.

IOP Conf. Series: Materials Science and Engineering 1163 (2021) 012018 doi:10.1088/1757-899X/1163/1/012018

Figure 3. Login page

Figure 4. Plant type

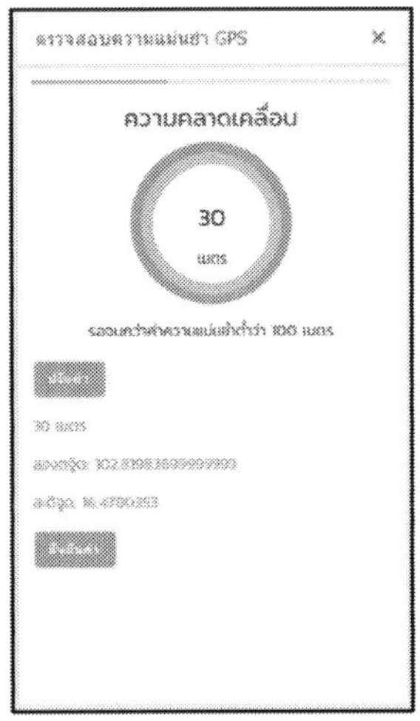

Figure 5. GPS accuracy measurement

Figure 3 shows the login page to which users can log in via Google mail. Figure 4 illustrates the working section of this page has tabs at the top to redirect the pages and display the affiliated organizations' information. Figure 5 shows the current user's GPS location with some errors within the threshold, adjustable by the administrator.

Figure 6. Upload page

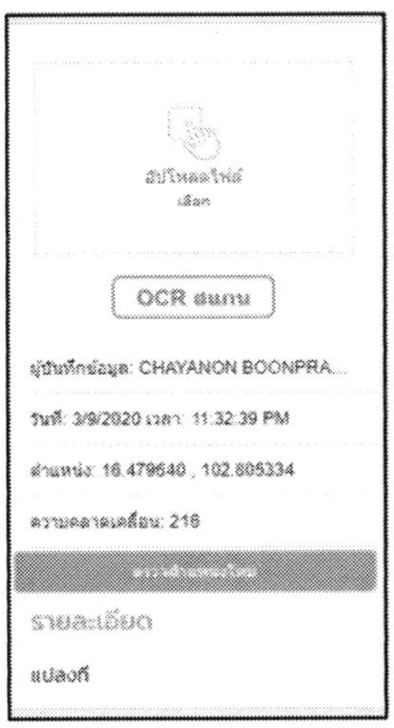

Figure 7. OCR scan page

Figure 8. Report page

Figure 6 shows the page that holds data for working offline and managing data in the card form. When the Internet is available, users can update data to the system. Figure 7 illustrates the page where the user can record the data, including GPS location and a button to scan the paper with some text. Figure 8 shows the logging reports tab. This page is divided into sections that display user information with a logout button at the bottom, and the next section shows the details of the crop field data.

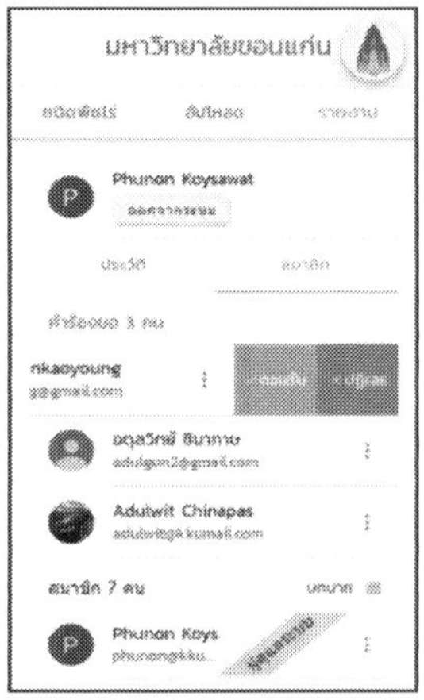

Figure 9. Organization page

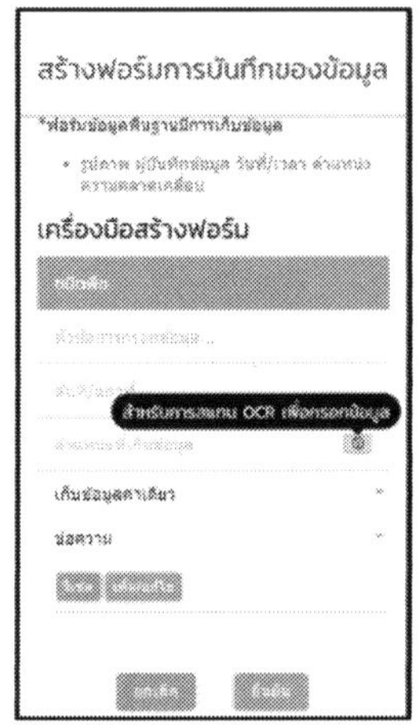

Figure 10. Form creation

Figure 9 illustrates all organization members whose administrator can manage all users and can accept/reject access requests under the organization. Figure 10 shows the creation form of recording farm crops within the organization. Creating a form requires an internet connection.

5. Experimental Result Analysis

We also performed experiments to compare the data collection processes' execution time without using the app and using the app, as shown in Figure 11 and Figure 12.

Flow Process Chart					
SUBJECT CHARTED: Sugarcane Sampling Process				**DATE:** February 2, 2021	
				CHART BY: BSFR LAB	
DEPARTMENT: FMM				**CHART NO :** 00-0-001	
Method: Without using the app				**SHEET NO:** 1/1	
Step	**Process name**	**Units (Bat)**	**Distance (Metre)**	**Time (Second)**	**Symbol**
1	Set the cone against the cane to take a picture	1	15	15	●
2	Taking a picture of cane	1	0	10	●
3	Check the GPS location from the phone and take notes	1	0	60	●
4	Count the number of cane trunks and take notes	2	20	150	●
5	Cut two cane samples	1	0	60	●
6	Take the sugarcane sampel to the working point	1	20	20	➡
7	Brake cane leaves from the trunk	1	0	60	●
8	Weigh trunks and cane leaves and take notes	1	0	100	●
9	Measure cane sweetness and take notes	1	0	60	●
10	Write a cane sample bag	1	0	40	●
11	Two soil samples were collected	1	10	90	●
12	Chop the cane into pieces to fill the bag	1	0	40	●
13	Check the cane sample	1	0	30	■
14	Collect the cane sample into the bag	1	0	30	▼
15	Check and validate information	1	0	60	■
16	**Fill data and pictures into an Excel file**	1	0	200	●
17	**Import Excel data and pin GPS to Google Earth app**	1	0	300	●
Total		2	65	1,325	13 1 0 2 1

Figure 11. Flow process without using the app

Flow Process Chart					
SUBJECT CHARTED: Sugarcane Sampling Process			**DATE:** February 2, 2021		
			CHART BY: BSFR LAB		
DEPARTMENT: FMM			**CHART NO :** 00-0-002		
Method: With using the app			**SHEET NO:** 1/1		
Step	**Process name**	**Units (Bat)**	**Distance (Metre)**	**Time (Second)**	**Symbol**
1	Set the cone against the cane to take a picture	1	15	15	
2	Taking a picture of cane	1	0	10	
3	Check the GPS location from the phone and take notes	1	0	10	
4	Count the number of cane trunks and take notes	2	20	120	
5	Cut two cane samples	1	0	60	
6	Take the sugarcane sampel to the working point	1	20	20	
7	Brake cane leaves from the trunk	1	0	60	
8	Weigh trunks and cane leaves and take notes	1	0	80	
9	Measure cane sweetness and take notes	1	0	50	
10	Write a cane sample bag	1	0	40	
11	Two soil samples were collected	1	10	90	
12	Chop the cane into pieces to fill the bag	1	0	40	
13	Check the cane sample	1	0	30	
14	Collect the cane sample into the bag	1	0	30	
15	Check and validate information	1	0	20	
16	**Update data**	1	0	10	
17	**Export data from the app to an Excel file**	1	0	10	
18	**Import Excel data to Google Earth app**	1	0	30	
Total		2	65	725	13 1 0 2 1

Figure 12. Flow process with using the app

Figures 11 and 12 notice that steps 1-15 of the flow processes both without and using the app are the same. However, in several steps, it took less time when using the app than when not using the app. Moreover, it took about 500 seconds to fill in data in excel and export the data to Google Earth app for the process without using the app in the last few steps. On the other hand, for the process of using the app, it took about 40 seconds to export the data from the developed app to Google Earth app.

Using the app dramatically reduces the time when it comes to storing data in Excel and Google Earth because people do not need to sit and fill in the information before, thus reducing the time up to 460 seconds. The process with using the app had a total time less than that without using the app, about 45.28%

6. Conclusions

We proposed and developed a web application that can store field crop data in a database. Such data includes photos, locations, and details of crop fields in different organizations. The objective of the developed app is to shorten the record-keeping time and minimizes human-incurred errors. The developed progressive web application can run as native mobile applications and use cache to speed up data access and work even without an internet signal. Our experimental result showed that the proposed application could save the processing time in collecting sugarcane sample data by about 45.28%. In the future, we can use the web application data for further analysis of agronomic crops such as cassava, which is another critical economic crop.

References

[1] *Progressive Web Apps | Web.* (n.d.). Google Developers. Retrieved September 15, 2019, from https://developers.google.com/web/progressive-web-apps/

[2] *Build a PWA With Vue & Ionic 4.* (n.d.). Retrieved September 15, 2019, from https://www.youtube.com/watch?v=6H1wftPS0oo

[3] Hartington, M. (2019, April 3). *Announcing the Ionic Vue Beta*. The Ionic Blog. https://ionicframework.com/blog/announcing-the-ionic-vue-beta/
[4] *Introduction—Vue.js*. (n.d.). Retrieved September 15, 2019, from https://vuejs.org/v2/guide/
[5] Ionic. (n.d.). *Ionic Framework—Ionic Documentation*. Ionic Docs. Retrieved September 15, 2019, from https://ionicframework.com/docs/

IOP Conf. Series: Materials Science and Engineering 1163 (2021) 012019 doi:10.1088/1757-899X/1163/1/012019

Synthesis and characterization of activated carbon derived from rubberwood sawdust by using KOH/KMnO4 as multiple function activation agents

S Phainuphong[1], J Taweekun[2,3], T Theppaya[2] and K Maliwan[2]

[1]Energy Technology Program, Faculty of Engineering, Prince of Songkla University, Hatyai, Songkhla, 90112, Thailand
[2]Department of Mechanical and Mechatronics Engineering, Faculty of Engineering, Prince of Songkla University, Hatyai, Songkhla, 90112, Thailand.

E-mail: juntakan.t@psu.ac.th, juntakan2016@gmail.com

Abstract. This research is focused on the effects of impregnation ratios using $KOH/KMnO_4$ as multiple function activation agents on microporous of activated carbon. Rubberwood sawdust, the waste from wood processing industry, was used as a raw material in synthesis of activated carbon by chemical activation. The rubberwood sawdust was carbonized at 400 °C for 1 hour under an inert atmosphere to produce char. Chars were impregnated with different proportions of KOH and $KMnO_4$. Impregnation time for each mixture proportion (Char: KOH: $KMnO_4$) was 24 hours before activation for 3 hours at 800 °C. The surface area, pore volume, micropore volume, pore size distribution, porous structure, and adsorption isotherm were examined and considered as the properties of activated carbon. The results showed that the activated carbon with an impregnation ratio of 10:10:5, 10:30:0, and 10:30:5 have the surface area 750.89 m^2/g, 1574.39 m^2/g, and 1346.10 m^2/g, respectively. The pore volume and micropore volume were in the range between 0.3289-0.6405 cm^3/g and 0.1920-0.3533 cm^3/g. The average pore sizes were 1.75 nm, 1.63 nm, and 1.65 nm, respectively. These synthesized activated carbons were characterized as type I isotherm and microporous solids.

1. Introduction

Activated carbon is a carbonaceous material containing a pore structure with a high specific surface area and pore size distribution [1]. According to it has large surface area and high pore volume, it can be used in many applications such as an outstanding adsorbent [2] and an electrode material for supercapacitor [3]. To prepare the activated carbon, the raw material is carbonized at a temperature between 200°C to 400°C to eliminate volatile compounds and tar and produce char [4]. After that, the activation process is applied. This activation can be divided into two methods. The first method is physical activation, carried out at a high temperature range of 700-1100 °C by using CO_2, steam, or a mixture of these gases. However, activated carbon obtained from this method has a low surface area. The second method is chemical activation involved the use of chemical agent to impregnate and carbonize raw material at a temperature range of 300°C to 800°C under an inert atmosphere. Therefore, chemical activation has benefits over physical activation, such as lower activation temperature, higher yield of activated carbon, and higher surface area. A widely chemical agent including the acidic group; sulfuric acid (H_2SO_4), phosphoric acid (H_3PO_4) [5,6], the alkaline group; potassium hydroxide (KOH), sodium hydroxide (NaOH) [7,8], and the metallic salt; zinc chloride ($ZnCl_2$) [9] are used as a dehydrating agent in chemical activation. Many researches shown that the alkaline group generate more micropore surface area than acidic group [2,10]. Therefore, among the alkaline group, KOH is the most effectively used for chemical activation since It can produce micropores, it can easily use for various applications.

Biomass is an attractive raw material to produce activated carbon because it contains high cellulose, hemicellulose, and lignin [11]. The elemental compositions of biomass are carbon (C), hydrogen (H), oxygen (O), nitrogen (N), and slightly sulphur (S) [12]. Many studied related to the chemical activation of selected biomass are summarized in table 1. All activated carbons using KOH as a chemical activation agent are carried out at a temperature of 800 °C. Because the activation temperature greater than 800 °C has not shown a significant increase in the surface area [13]. Impregnation ratio is an essential parameter for chemical activation. It also affects active carbon porosity. The activated carbon impregnated at the ratio of 1:3 has a higher surface area. In Thailand, Rubberwood is a high potential biomass. However, a little number of researches using rubberwood sawdust as a raw material for physical activation [14] or chemical activation with $ZnCl_2$ presented in table 1, without studied the effect of KOH to the characteristic of activated carbon derived from this raw material.

Activated carbon can be used as an electrode material for electrostatic double layer capacitors (EDLCs). To improve the electrochemical properties of activated carbon, surface functional groups or metal oxide can be used. Manganese monoxide (MnO) has higher specific capacitance (1,350 F/g) compared with manganese oxides [15,16]. However, It has low conductivity and cycle stability. Therefore, MnO was used as an additive to improve the specific capacitance of electrode materials. Potassium permanganate ($KMnO_4$) and KOH also have similar activation mechanisms that generate micropores [17]. $KMnO_4$ was decomposed into K_2O, MnO, and O_2 at an activation temperature of 750 °C, and MnO was added to activated carbon [18]. Moreover, $KMnO_4$ is low cost, environment friendly, and water-soluble. Thus, $KMnO_4$ is useful to improve the specific capacitance of activated carbon.

In this research, the activated carbon derived from rubberwood sawdust by using KOH and $KMnO_4$ as an chemical activation agent at the activation temperature of 800 °C were characterized. Surface area (S_A), average pore diameter (D_P), total pore volume (V_{Total}), pore size distribution, micropore fraction (M_F), and surface morphology were determined. In addition, the effect of $KMnO_4$ on porosity textural properties were discussed.

Table 1. Characteristics of activated carbons by chemical activation.

Materials	Activation condition			Surface Characteristic				Ref
	Agent	Ratio	Temp (°C)	S_{BET} [a] (m^2/g)	V_{total} [b] (cm^3/g)	V_{Mic} [c] (cm^3/g)	D_p [d] (nm)	
Polyalthia longifolia seed	KOH	1:2	600-900	226.26-664.45	0.129-0.276	0.046-0.234	1.03-3.09	[13]
Bambusa vulgaris striata	KOH	1:3	800	980.00	0.559	-	2.28	[19]
Oleaster fruit	KOH	1:3	800	1621.00	0.924	0.506	-	[20]
Corncob	KOH	1:3	800	800.00	0.340	-	-	[21]
Rubberwood sawdust	$ZnCl_2$	1:1.5	500	1,301.00	-	0.540	1.77	[22]

[a] BET surface area.

IOP Conf. Series: Materials Science and Engineering 1163 (2021) 012019 doi:10.1088/1757-899X/1163/1/012019

[b] Total pore volume.
[c] Micropore volume.
[d] Average pore diameter.

2. Experimental Methods

2.1. The precursor and characteristics
Rubberwood sawdust collected from Southern Biomass Fuel Company, Songkhla province, Thailand, was used as the precursor. The particle sizes of sawdust are around 0.2-1 mm. The rubberwood sawdust was cleaned and dried at 105 °C for 24 hours. Then, it was collected into a sample bag and kept in a dehumidifying box. The thermogravimetric analyzer, model TGA 701, LECO, USA (based on ASTM D7582-15) from the office of Scientific Instrument and Testing, Prince of Songkla University, Thailand was used to examined the proximate analysis of the raw material, carbonized char, and activated carbons. The characteristics are presented in Table 3.

2.2. Activated carbon preparation
The activated carbon was prepared from the Rubberwood sawdust (RS) by a two-step process. The first step was carbonization process. Rubberwood sawdust was dried and placed in a 4-inch diameter tube to produce char. This reactor was loaded into the tube furnace and heated up to 400 °C at a heating rate of 6 °C/min and maintained for 1 hour, and cooled down in the furnace for 12 hours. The second step, the chemical activation process. Using KOH and $KMnO_4$ as the activating agents. The mixture of char (10 g), KOH (10 g, 30 g), and $KMnO_4$ (0g, 5g) was stirred manually. The well-mixed samples were left for 24 hours and then loaded into the electric tube furnace and heated up to 800 °C for 3 hours. Nitrogen gas was introduced at 500 cm^3/min to create inert atmosphere during the carbonization and activation process. The activated carbon derived from rubberwood sawdust was soaked with 0.1 Molar Hydrochloric acid (HCL) for 3 hours in room temperature and washed with distilled water until the pH 7 was reached to remove impurities in activated carbon. Finally, the sample was dried in an oven at 105 °C for 24 hours, collected into a sample zip-lock bag, and kept in a silica gel box to prevent moisture absorption. The activated carbon samples were labeled as 10:10:5 (Char: KOH: $KMnO_4$), 10:30:0, and 10:30:5. The schematic preparation of activated carbon was shown in figure 1.

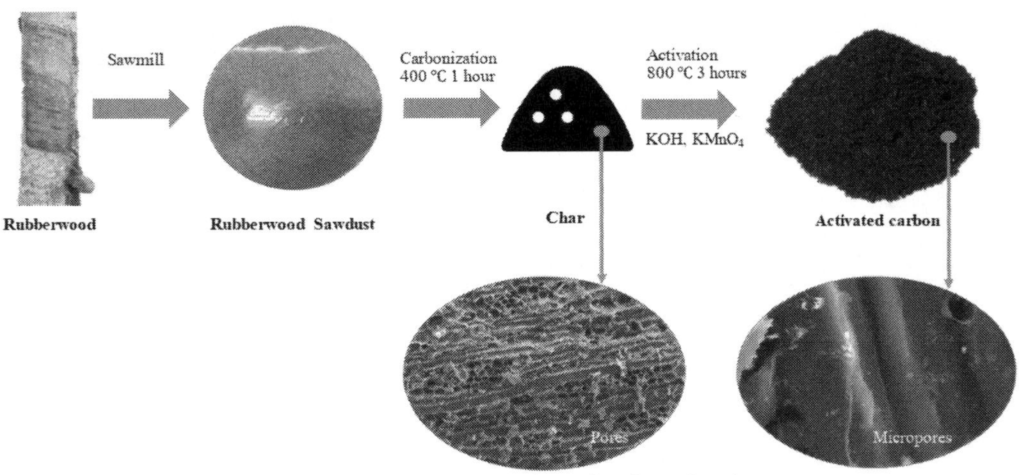

Figure 1. The schematic preparation of activated carbon.

2.3. *Textural characterization*

The textural properties of activated carbons such as surface area (S_A), total pore volume (V_{total}), micropore volume (V_{Mic}), average pore diameter (D_p), pore size distribution, micropore fraction (M_F), and adsorption isotherm were analyzed by using the high throughput surface area and porosity analyzer, Micromeritics (ASAP2460) from the office of Scientific Instrument and Testing, Prince of Songkla University, Thailand. Field Emission Scanning Electron Microscope (FESEM) images were used to examine the morphological properties of activated carbon by Merlin compact, ZEISS Microscopy, Jena, Germany from the Center for Scientific and Technological Equipment, Walailak University, Thailand.

3. Results and Discussions

3.1. *The yield of rubberwood salwdust char and activated carbon*

The rubberwood sawdust char and activated carbon yield and activated carbon were presented in figure 2. The yield of char was considered as equation (1) and equation (2), respectively. The yield of rubberwood sawdust char was around 32.85% by weight, and the yield of activated carbons was ranged between 15.10-18.96% by weight. The lowest yield of activated carbon occurred at proportion 10:30:0 (Char: KOH: $KMnO_4$) chemically reaction at 800 °C. At 10:30:5 proportion obtained the highest yield by 18.96 %. The yield of 10:10:5 was 17.37%. The trend of the mixtures with $KMnO_4$ have higher yield than the mixture without $KMnO_4$. Therefore, the trend of activated carbon yield was similar in the case of activation by using KOH. Nuilerd et al. [23] revealed that activated carbon yield decreased by increasing of impregnation ratio.

$$\text{The yield of char } (\%wt) = \frac{\text{Weight of char}}{\text{Weight of Raw material}} \times 100 \tag{1}$$

$$\text{The yield of activated carbon } (\%wt) = \frac{\text{Weight of Activated carbon}}{\text{Weight of Raw material}} \times 100 \tag{2}$$

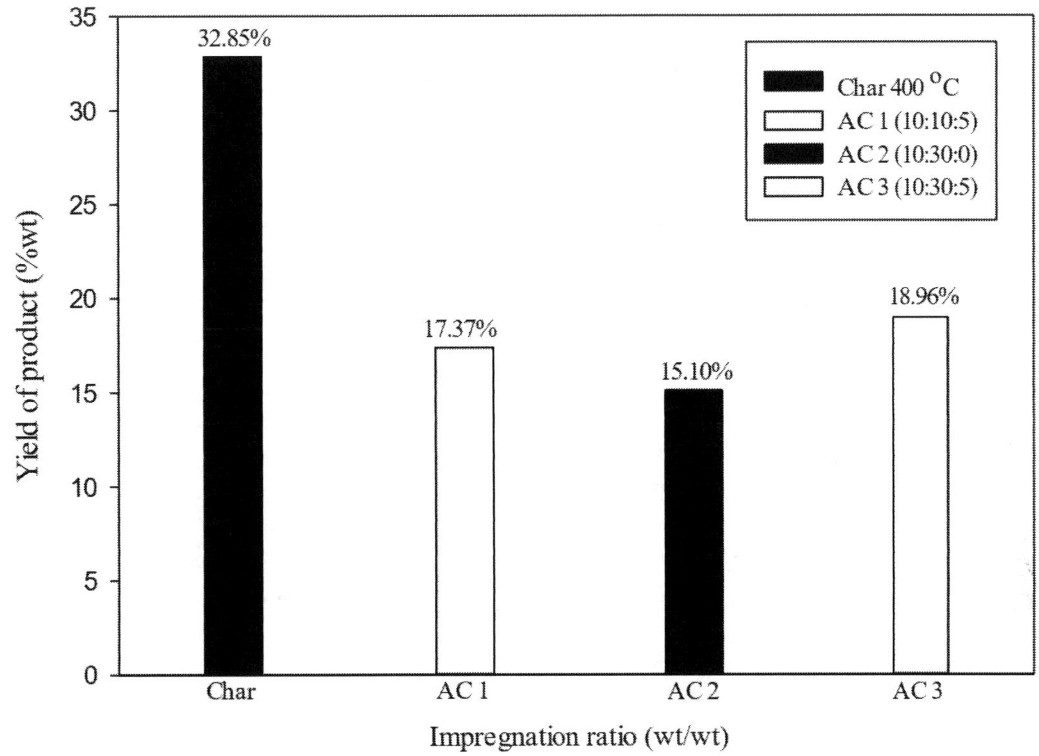

Figure 2. The yield of rubberwood sawdust char and activated carbon.

Table 2. Proximate analysis of rubberwood sawdust, char, and activated carbon.

Samples	Proximate contents (%wt)			
	Fixed carbon	Moisture	Volatile	Ash
Rubberwood sawdust	15.67	7.47	73.93	2.93
Char	59.26	5.26	30.13	5.35
AC (10:30:0)	76.91	4.95	12.95	5.19

3.2. Proximate analysis

Proximate analysis of rubberwood sawdust, char, and activated carbon are showed in Table 2. It is shown that the fixed carbon increased significantly after the carbonization and activation process. Fixed carbon increased from 15.67% to 59.26% and 76.91%, respectively. The raw material contains high volatile, high moisture, and ash content as 73.93%, 7.47%, and 2.93% by weight. The result showed that after the carbonization process, char has lower volatile and moisture but higher ash content as 30.13%, 5.26%, and 5.35%, respectively. The activated carbon with mixing proportion of 10:30:0 (Char: KOH: $KMnO_4$) and at activation temperatures of 800 °C has the lowest volatile, the lowest moisture, and ash content as 12.95%, 4.95%, and 5.19%.

3.3. Textural properties

Adsorption isotherm of activated carbons was shown in figure 3. The activated carbon with ratios 10:10:5, 10:30:0, and 10:30:5 have similar adsorption isotherm trends. In this study, the adsorption isotherm was obtained at a relative pressure between 0.01 to 0.99 P/P_o. All samples were characterized as Type I isotherm. According to the IUPAC classification of adsorption isotherm, the type I isotherm implies the activated carbon as microporous solids and gas molecules adsorb as a monolayer [24]. The surface area (S_A), total pore volume (V_{Total}), micropore volume (V_{Mic}), and average pore diameter (D_P) of activated carbons were shown in Table 3. The surface area of activated carbons with ratios 10:10:5, 10:30:0, and 10:30:5 were 750.89, 1,574.39, and 1,346.10 m^2/g, respectively. Kim et al. [18] reported that using $KOH/KMnO_4$ as the activation agent decreased the surface area (S_A) because $KMnO_4$ decomposed to K_2O, MnO, and O_2, then MnO was formed into activated carbon and blocked the pores of activated carbon. The addition, $KMnO_4$ decreased surface area around 14.48% by comparing between ratios 10:30:0 and 10:30:5. The decreases in surface area of 10:30:5 can be traced to detecting Mn (4.0 wt%) composition by EDX analysis in figure 7 (b). The total pore volume and micropore volume were in the range of 0.3289-0.6405 cm^3/g, and 0.1920-0.3533 cm^3/g. The ratio 10:30:0 without $KMnO_4$ also has higher total pore volume and micropore volume than ratio 10:30:5. The micropore fraction of ratios 10:10:5, ratio 10:30:0, and ratio 10:30:5 were 58.37%, 55.15%, 59.19%, respectively. Thus, $KMnO_4$ did not show significant effect in micropore fractions [17,18]. The average pore diameters of ratios 10:10:5, ratio 10:30:0, and ratio 10:30:5 were 1.75 nm, 1.63 nm, and 1.65 nm. These results indicated that the pore sizes of all activated carbons are micropores. The activated carbon pores were presented on the FESEM image, as shown in figure 6 (a)-(d).

Table 3. Characteristic properties of activated carbon

Samples	S_{BET} [a] (m^2/g)	V_{total} [b] (cm^3/g)	V_{Mic} [c] (cm^3/g)	D_p [d] (nm)	M_F [e] (%)
Char	0.56	0.0004	0.0003	187.63	-
10:10:5	750.89	0.3289	0.1920	1.75	58.37
10:30:0	1,574.39	0.6405	0.3533	1.63	55.15
10:30:5	1,346.10	0.5543	0.3281	1.65	59.19

[a] BET surface area.
[b] Total pore volume.
[c] Micropore volume.
[d] Average pore diameter.
[e] Micropore fraction.

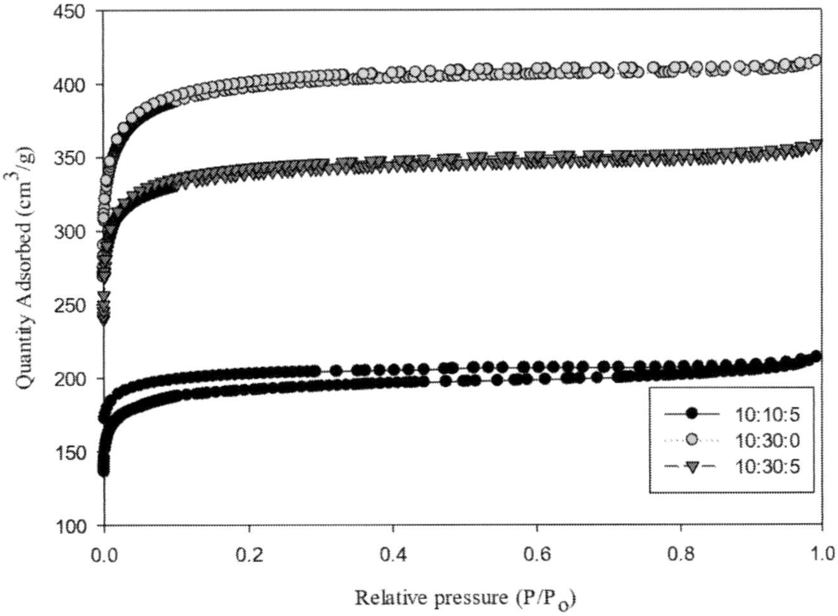

Figure 3. Adsorption and desorption isotherm of activated carbons.

3.4. Pore size distribution

Figure 4 presents the activated carbon pore size distribution range between 0-10 nm at ratios 10:10:5, 10:30:0, and 10:30:5. As the result of pore size distribution curves, it can be observed in two different pore size regions; micropores region (pores diameter less than 2 nm) and mesopores region (pores diameter between 2-50 nm). Ratios 10:10:5, 10:30:0, and 10:30:5 were comprised pores with the highest diameter around 1.26 nm with a maximum pore volume of 0.1792 cm^3/g/nm, 1.26 nm with a maximum pore volume of 0.3461 cm^3/g/nm, and 1.26 nm with a maximum pore volume of 0.2679 cm^3/g/nm respectively. The pore size distribution curves, activated carbon ratio 10:10:5 has distribution between 1.26-3.69 nm, ratio 10:30:0 has distribution between 1.18-2.73 nm, and ratio 10:30:0 has distribution between 1.18-4.00 nm, respectively.

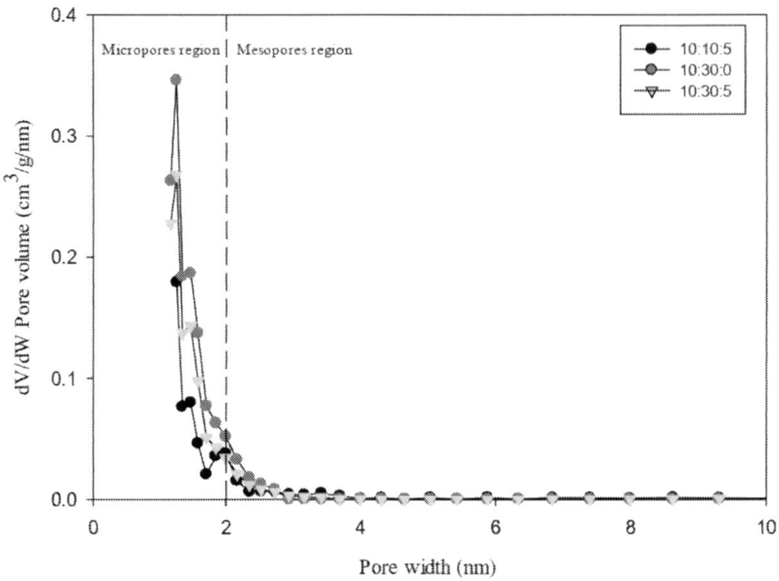

Figure 4. Pore size distributions of activated carbons.

3.5. *Cumulative pore size distribution*

The cumulative pore volume of activated carbons is presented in figure 5. The micropores volume at less than 2 nm of pore width indicates the same sharply increases pattern for ratios 10:10:5, 10:30:0, and 10:30:5 of activated carbons. Micropore volume between a range of mesopore regions (2-50 nm) remains constant for all samples. As the result shown activated carbon ratios 10:10:5, 10:30:0, and 10:30:5 have micropore volume 0.2364 cm^3/g, 0.4980 cm^3/g, and 0.4246 cm^3/g at pores width range less than 2 nm.

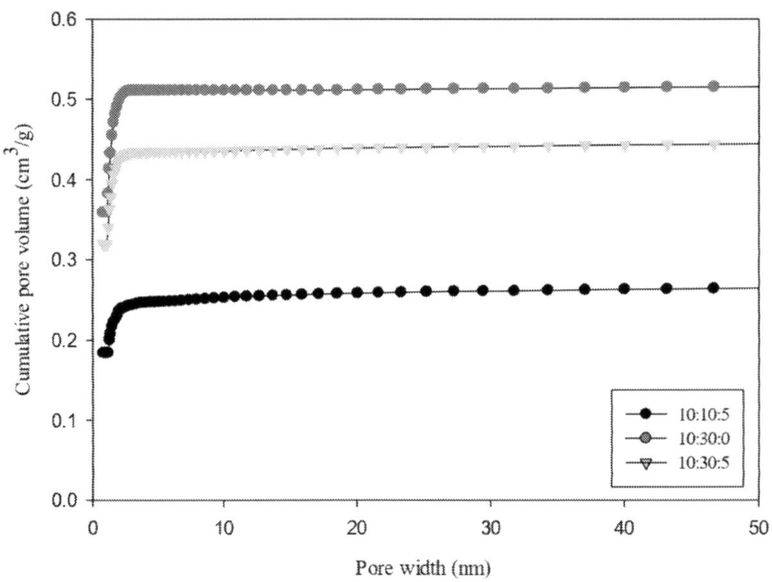

Figure 5. Cumulative pore size distributions of activated carbons.

3.6. Field Emission Scanning Electron Microscope analysis (FESEM) and EDX analysis

The images of rubberwood sawdust char, ratio 10:10:5, ratio 10:30:0, and ratio 10:30:5 are shown in figure 6, (a)-(d), respectively. The Field Emission Scanning Electron Microscope (FESEM) was used to observe the textural surface of activated carbons with 500x magnification. Char are macroporous (>50 nm) structures undeveloped into mesopore or micropore range. Therefore macropores are unnecessary for the process of adsorption compared with mesopore and micropore structure. Ratio 10:10:5 has less porous structure than 10:30:0, and 10:30:5, but all of those were micropores (< 2 nm), which is related to average pore diameter, as explained in table 3. Thus, chemical activation by using $KOH/KMnO_4$ as a multiple function activation agent can produce activated carbon with a well-formed porous structure from rubberwood sawdust char to increase its adsorption capability. Figure 7 (a) and (b) represented the Energy-dispersive X-ray spectroscopy (EDX) of activated carbon ratio 10:30:0 and ratio 10:30:5, which verified the presence of elemental composition of C, O, Ca, and Mg. However, in the ratio 10:30:5 detected, another significant elemental composition was Mn (4.0 wt%).

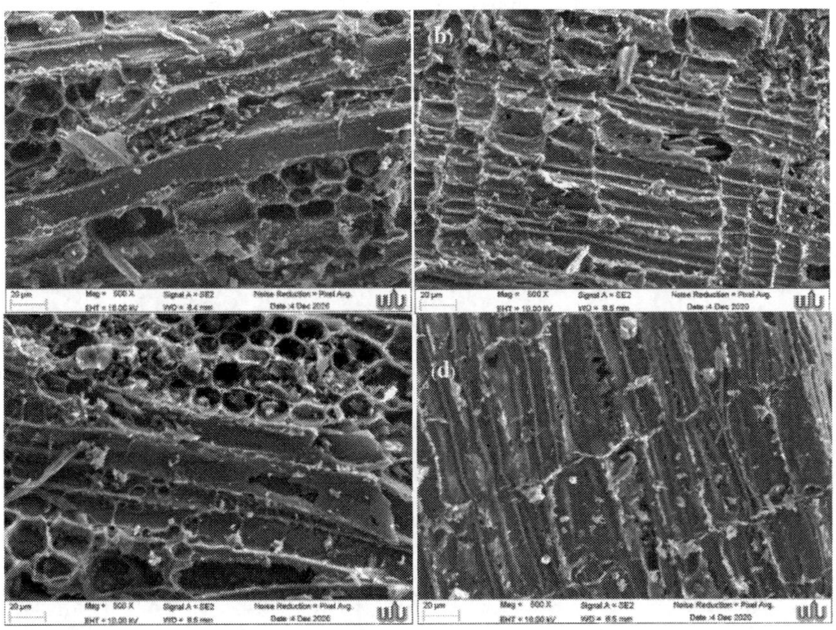

Figure 6. FESEM images with 500x magnificent of the samples, (a) Char, (b) ratio 10:10:5, (c) ratio 10:30:0, and (d) ratio 10:30:5.

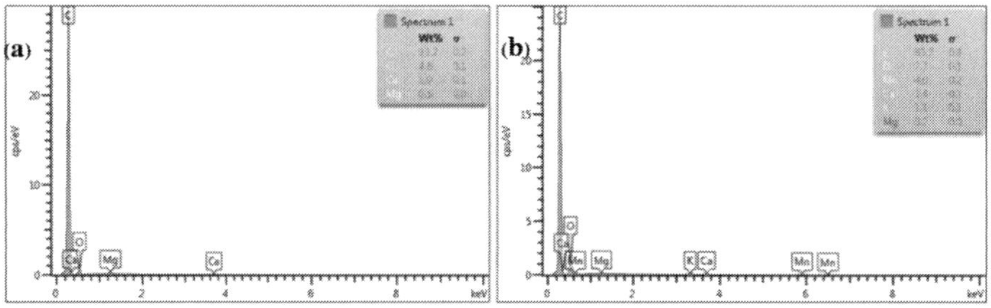

Figure 7. EDX analysis of activated carbons, (a) ratio 10:10:5, and (b) ratio 10:30:5.

4. Conclusions

In this study, the rubberwood sawdust char with low surface area, pore volume, and large pore size can improve its properties after the chemical activation process. The yield of product with a ratio of 10:30:5 has the highest value of 18.96% by weight. Activated carbon without $KMnO_4$ (ratio 10:30:0) contains the highest surface area, total pore volume, and micropore volume by 1,574.39 m^2/g, 0.6405 cm^3/g, and 0.3533 cm^3/g, respectively. The ratio 10:30:5 has lower surface area (14.48%), total pore volume, and micropore volume than the ratio 10:30:0. MnO introduced into activated carbon, was blocked the pores of activated carbon, thus decreasing the surface area. The micropore fraction of activated carbon was around 55.15-59.19%. Adding $KMnO_4$ has not significant effect in micropore fractions. The average pore diameters at ratios 10:10:5, 10:30:0, and 10:30:5 were 1.75 nm, 1.63 nm, and 1.65 nm, which were in a range of micropores size. The adsorption isotherm can be characterized as type I isotherm that can be implied that the activated carbon was microporous solid. FESEM images of the product exposed the development of pore and porous structure from rubberwood sawdust char. Thus, the micropore of rubberwood sawdust char can be generated by using $KOH/KMnO_4$ as an activation agent. Furthermore, the properties of activated carbon derived from rubberwood sawdust are suitable to be used as electrode material in supercapacitor application, and the next study will prove the effectiveness of $KMnO_4$ on the specific capacitance.

References

[1] Gao Y, Yue Q, Gao B and Li A 2020 Insight into activated carbon from different kinds of chemical activating agents: A review *Science of The Total Environment* 746 141094
[2] Danish M and Ahmad T 2018 A review on utilization of wood biomass as a sustainable precursor for activated carbon production and application *Renewable and Sustainable Energy Reviews* 87 1–21
[3] Lin Z, Goikolea E, Balducci A, Naoi K, Taberna P L, Salanne M, Yushin G and Simon P 2018 Materials for supercapacitors: When Li-ion battery power is not enough *Materials Today* 21 419–36
[4] Fu Y, Shen Y, Zhang Z, Ge X and Chen M 2019 Activated bio-chars derived from rice husk via one- and two-step KOH-catalyzed pyrolysis for phenol adsorption *Science of The Total Environment* 646 1567–77
[5] Malarvizhi R and Ho Y-S 2010 The influence of pH and the structure of the dye molecules on adsorption isotherm modeling using activated carbon *Desalination* 264 97–101
[6] del Mar Gómez-Tamayo M, Macías-García A, Díaz Díez M A and Cuerda-Correa E M 2008 Adsorption of Zn(II) in aqueous solution by activated carbons prepared from evergreen oak (Quercus rotundifolia L.) *Journal of Hazardous Materials* 153 28–36

[7] Peng C, Yan X, Wang R, Lang J, Ou Y and Xue Q 2013 Promising activated carbons derived from waste tea-leaves and their application in high performance supercapacitors electrodes *Electrochimica Acta* 87 401–8

[8] He X, Ling P, Qiu J, Yu M, Zhang X, Yu C and Zheng M 2013 Efficient preparation of biomass based mesoporous carbons for supercapacitors with both high energy density and high power density *Journal of Power Sources* 240 109–13

[9] Rufford T E, Hulicova-Jurcakova D, Zhu Z and Lu G Q 2008 Nanoporous carbon electrode from waste coffee beans for high performance supercapacitors *Electrochemistry Communications* 10 1594–7

[10] Villota S M, Lei H, Villota E, Qian M, Lavarias J, Taylan V, Agulto I, Mateo W, Valentin M and Denson M 2019 Microwave-Assisted Activation of Waste Cocoa Pod Husk by H_3PO_4 and KOH—Comparative Insight into Textural Properties and Pore Development *ACS Omega* 4 7088–95

[11] Zhang L, Xu C (Charles) and Champagne P 2010 Overview of recent advances in thermo-chemical conversion of biomass *Energy Conversion and Management* 51 969–82

[12] Açıkyıldız M, Gürses A and Karaca S 2014 Preparation and characterization of activated carbon from plant wastes with chemical activation *Microporous and Mesoporous Materials* 198 45-9

[13] Srinivasan R, Elaiyappillai E, Pandian H P, Vengudusamy R, Johnson P M, Chen S-M and Karvembu R 2019 Sustainable porous activated carbon from Polyalthia longifolia seeds as electrode material for supercapacitor application *Journal of Electroanalytical Chemistry* 849 113382

[14] Taer E, Deraman M, Talib I A, Umar A A, Oyama M and Yunus R M 2010 Physical, electrochemical and supercapacitive properties of activated carbon pellets from pre-carbonized rubber wood sawdust by CO2 activation *Current Applied Physics* 10 1071–5

[15] Liu Y, Zhang J and Hu R 2017 One-pot hydrothermal synthesis and supercapacitive performance of nitrogen and MnO co-doped hierarchical porous carbon monoliths *Ceramics International* 43 4427–33

[16] Liao Q, Li N, Cui H and Wang C 2013 Vertically-aligned graphene@MnO nanosheets as binder-free high-performance electrochemical pseudocapacitor electrodes *J. Mater. Chem. A* 1 13715

[17] Guan T, Zhao J, Zhang G, Zhang D, Han B, Tang N, Wang J and Li K 2018 Insight into controllability and predictability of pore structures in pitch-based activated carbons *Microporous and Mesoporous Materials* 271 118–27

[18] Kim S J, Bai B C, Kim M I and Lee Y-S 2020 Improved specific capacitance of pitch-based activated carbon by KOH/KMnO4 agent for supercapacitors *Carbon Letters* 30 585–91

[19] Mistar E M, Alfatah T and Supardan M D 2020 Synthesis and characterization of activated carbon
from Bambusa vulgaris striata using two-step KOH activation *Journal of Materials Research and Technology* 9 6278–86

[20] Yagmur E, Gokce Y, Tekin S, Semerci N I and Aktas Z 2020 Characteristics and comparison of activated carbons prepared from oleaster (Elaeagnus angustifolia L.) fruit using KOH and ZnCl2 *Fuel* 267 117232

[21] Karnan M, Subramani K, Srividhya P K and Sathish M 2017 Electrochemical Studies on Corncob Derived Activated Porous Carbon for Supercapacitors Application in Aqueous and Non-aqueous Electrolytes *Electrochimica Acta* 228 586–96

[22] Shaaban A, Se S-M, Ibrahim I M and Ahsan Q 2015 Preparation of rubber wood sawdust-based activated carbon and its use as a filler of polyurethane matrix composites for microwave absorption *New Carbon Materials* 30 167–75

[23] Nuilerd T, Pongyeela P and Chungsiriporn J 2018 Pellet activated carbon production using parawood charcoal from gasifier by KOH activation for adsorption of iron in water 7

[24] Yurdakal S, Garlisi C, Özcan L, Bellardita M and Palmisano G 2019 (Photo)catalyst

Characterization Techniques *Heterogeneous Photocatalysis* (Elsevier) pp 87–152

Acknowledgments
The authors are appreciated to PSU Energy System Research Institute for research funds, Prince of Songkla University for financial support (Grant no. ENG6201078S year 2019), Southern Biomass Fuel company for the raw material and Department of Chemical Engineering, Prince of Songkla University for instruction and providing equipment.

IOP Conf. Series: Materials Science and Engineering 1163 (2021) 012020 doi:10.1088/1757-899X/1163/1/012020

Inactivation of *Escherichia coli* by several types of gas bubbles

P. Naewkanya, A. Petiraksakul
Department of Chemical Engineering, Faculty of Engineering, King Mongkut's University of Technology North Bangkok, Bangkok, Thailand

E-mail: pattama.n@eat.kmutnb.ac.th

Abstract. The ability of gas bubbles to inactivate *Escherichia coli* suspended in distilled water at varied temperature and several types of gas, namely carbon dioxide, nitrogen and argon, was investigated. The results of the experiment showed that carbon dioxide gas bubbles gave the better inactivation of *Escherichia coli* than other gasses and the number of surviving *Escherichia coli* was reduced by approximately 0.76-log from the treatment of carbon dioxide gas bubbles at 4-10°C and gas flowrate of 2.5 ml/min for 40 min. The pH of solution after treatment with carbon dioxide gas bubbles for 40 min was 3.8. Furthermore, the number of surviving *Escherichia coli* was reduced by approximately 1.62-log when the solution temperature was increased to 50°C with the same condition of the treatment of carbon dioxide gas bubbles at 4-10°C. It was found that the addition of carbon dioxide gas bubbles together with increasing the solution temperature can significantly reduce the number of *Escherichia coli*. The most suitable of inactivation kinetics reaction was first order and the rate constant, k, at 4-10°C, 30°C and 50°C was 0.0220, 0.0013 and 0.0475 min^{-1}, respectively. The decimal reduction time D at 4-10°C, 30°C and 50°C was 45.45, 769.23 and 21.05 min, respectively.

1. Introduction

Bacteria are the most abundant forms of life on earth and are a natural component of lakes, rivers, and streams. Most of these bacteria are harmless to humans but a relatively small number of species causes disease. However, certain bacteria inhabited in body of organisms such as the intestinal tract of warm-blooded animals and a human stomach have the potential to cause sickness and disease in humans [1].

Escherichia coli is the most common gram-negative pathogen, causing urinary tract infections, septicaemia and new-borns meningitis [2]. It was first isolated from the stools of children by Theodor Escherich in 1885 [3]. *Escherichia coli* is classified in the family *Enterobacteriaceae* and is a rod-shaped bacterium, commonly found in terminal small intestine and large intestine of human and many warm-blooded animals [4-6]. The optimum temperature for *Escherichia coli* growth is 37°C but growth can occur between 4-45°C as well. The optimal pH for growth is in a range of pH 6-8. It can survive as low as pH 4 and as high as pH 9 [3,6,7].

Most strains of *Escherichia coli* are harmless. However, *Escherichia coli* O157:H7 can cause several diseases [8, 9]. The infective dose of *Escherichia coli* O157:H7 as low as 10 cells can be caused disease that lower than infective dose of most foodborne pathogens is greater than 10,000 cells [8,9]. Waterborne transmission of *Escherichia coli* O157:H7 is an emerging concern to human health.

Water is essential to life. Improving quality of safe drinking water has a profound effect on health [10]. *Escherichia coli* has usually been used as test microorganisms because it was a hygiene indicator bacterium [11]. There are several inactivation methods of *Escherichia coli* in water, such as increasing temperature, pressure, treating with chemicals and dissolved gases, and exposure time. [12-15]. Kobayashi F. *et al.*, 2016 [14] reported that the number of surviving *Escherichia coli* cells in physiological saline was decreased 1-log following micro-bubble carbon dioxide treatment at 35°C and 1.0 MPa for 10 min and the number of surviving *Escherichia coli* cells was decreased with increasing pressure or temperature. Adrian *et al.*, 2019 [16] found that dissolved carbon dioxide solution contained with 0.17 M NaCl could decreased 0.6-log of surviving *Escherichia coli* cells at pH of 4.1, temperature of 38°C and atmospheric pressure. In addition, argon and nitrogen gas were also reported an inhibitory effect on inactivation of bacteria in modified atmosphere packaging for maintaining the quality of meat and fruit [17-19]. Hence, this research aimed to study the effects of gases, temperature and contact time for *Escherichia coli* inactivation by using several types of gas bubbles and the inactivation kinetics of *Escherichia coli* were also investigated.

2. Materials and methods

2.1 Inoculum preparation and inoculation procedure

A pathogenic strain of *Escherichia coli* TISTR 780 (ATCC 8739) was used as the test strains. The cell culture was obtained from the Thailand Institute of Scientific and Technological Research (TISTR), Thailand. The activated cells were cultured individually in nutrient broth (Himedia laboratories pvt.ltd.) and then incubated at 37°C for 48 h before using as the working inoculum. The final bacterial concentrations in the inoculum samples were approximately controlled in the range of 8 log CFU/ml in the nutrient broth.

2.2 Washing conditions

2.2.1 Preparation of dissolved gas solutions. Each dissolved gas solution was prepared by filling 200 ml of distilled water in a 300 ml beaker placed in a water-ice mixture bath. N_2 (99.99 %, Bangkok Industrial Gas Co., Ltd., Thailand), Ar (99.995 %, Bangkok Industrial Gas Co., Ltd., Thailand), CO_2 (99.5%, Bangkok Industrial Gas Co., Ltd., Thailand), $50CO_2$:$50N_2$ v/v or $50CO_2$:50Ar v/v were dissolved into distilled water via a gas diffuser at flowrate of 2.5 ml/min and 20 min to final saturation. The flowrate of gases was controlled by a flowmeter with a needle valve and gas mixtures between CO_2 and N_2 or Ar were mixed in a tube before feeding it into the distilled water. Gas and gas mixtures were flushed into the distilled water until the dissolved oxygen was nearly zero measured by the YSI 52 DO/BOD meter. The temperature of the deoxygenated water was measured by a thermometer and was controlled in the range of 4-10°C, 30°C or 50°C.

2.2.2 Inactivation. After preparing the dissolved gas solution, 2 ml of *Escherichia coli* inoculum was dropped into the solution and the solution was then mixed by the flow of gas via a gas diffuser until the end inactivation period. The initial concentration of *Escherichia coli* in solution was approximately controlled in the range of 6 log CFU/ml. After the inactivation time at 0, 10, 20, 30 and 40 min, one milliliter of each sample was taken from each solution for measuring the number of *Escherichia coli.*

2.3. Microbial analysis and enumeration

One milliliter of diluted solution was made serial dilution (1:10), adding 1 ml of the solution to 9 ml of water, before measurement. Then, one milliliter of each diluted sample was dropped in a commercial count plate, a 3M *Escherichia coli* count plate and all plates were incubated at 35 (+1°C) for 24 ± 2 h (AOAC, 2002) [20]. The amount of *Escherichia coli* was counted after incubation and was calculated the number reverse before dilutions. The colony counts are expressed as log CFU/ml. All experiments were performed in duplicate, and the data presented are averaged from the results of duplicate experiments.

2.4 Inactivation kinetics
The kinetic of inactivation model was developed based on the basic kinetic equation for chemical reaction rate. The first-order kinetic model assumes an exponential reduction of the number of survivors over treatment time [20]. The model for the inactivation of microorganisms can be written as follows:

$$\frac{d(N/No)}{dt} = -k\,(N/No) \qquad (1)$$

where N and No are the surviving number and the initial number of the microorganisms, respectively, t is time (min) and k is first-order of rate constant (min^{-1}). The integration from $t = 0$ to t = t in equation (1) can be written as follows:

$$N/No = e^{-kt} \qquad (2)$$

The values of k in the model were used to estimate the time needed to inactivation of microorganisms and the time at a given temperature to reduce the population by 1 log cycle (D value) will be equal to

$$D = \frac{2.303}{k} \qquad (3)$$

and substitution of equation (3) into equation (2) yields the first-order survival model:

$$\log\frac{N}{No} = -\frac{t}{D} \qquad (4)$$

where D is the decimal reduction time (min). When one log reduction is achieved, it means 90% of relevant microorganisms have been killed [21].

3. Result

The results of colony counting of *Escherichia coli* showed only dark blue spots and some gas bubbles (Fig. 1) after incubating for 24hours same as the results reported by Gangar, *et.al.* 1999[22]. The effect of gas bubbles on the inactivation of *Escherichia coli* in distilled water is shown in Fig. 2. The number of surviving *Escherichia coli* was reduced by approximately 0.76-log from the treatment of carbon dioxide gas bubbles at 4-10°C and gas flowrate of 2.5 ml/min for 40 min, 0.46-log reductions, 0.11-log reductions by 50%CO_2:50%N_2 v/v and N_2, respectively, while Ar and 50%CO_2:50%Ar v/v increased the surviving *Escherichia coli*. The pH of solution after carbon dioxide gas bubble treatment at 40 min was 3.8 while the optimum pH for *Escherichia coli* growth is in the range of 6-8, however, growth can occur as low as pH 4 and as high as pH 9 as well [3,6,7]. It seems that the reduction of pH after treating with carbon dioxide gas bubbles could reduce the number of *Escherichia coli*. Adrian *et al.*, 2018 [23] reported that the pH of dissolved carbon dioxide solution with 0.17 M NaCl at room temperature was dropped from 5.9 to 4.1 and the reduction of *Escherichia coli* added to the solution was found only 0.05-log. Satoshi I. *et al.*, 2010 [24] explained that the critical pH of some bacteria including *Escherichia coli* was approximately 4.7. The synergistic effects of the pH below the critical point and dissolved carbon dioxide could have the *Escherichia coli* inactivation. Dissolved carbon dioxide in solutions easily diffused into the phospholipid bilayer of the membranes [16]. These results were supported by others [25, 26] who reported that dissolved carbon dioxide in solutions easily diffused into bacterial cell, resulting in a decrease in intracellular pH and cell death.

Fig. 3 shows the effect of temperature on inactivation of *Escherichia coli* in distilled water by carbon dioxide gas bubble. The number of surviving *Escherichia coli* was reduced by approximately 1.62-log from the treatment of carbon dioxide gas bubbles at 50°C and gas flowrate of 2.5 ml/min for 40 min, in comparison with the number reduction of *Escherichia coli* by 0.76-log and 0.05-log at 4-10°C and 30°C, respectively. This result was confirmed by the previous works [3,6,7] that *Escherichia coli* can survive between 4-45°C. The inactivation at 50°C could be explained by thermal inactivation mechanism resulting the reduction of surviving *Escherichia coli* more than at 30°C. In comparison with results of Stephane Guyot, *et.al.* 2014 [12] who reported that the addition of carbon dioxide gas and the increase of temperature can significantly reduce the number of *Escherichia coli*. On the other hand, the reduction of temperature in the inactivation process to 4-10°C gave better results than the inactivation at 30°C. This may explain that the *Escherichia coli* cells are halted with 'cold shock response', indicating that this strain has no capacity to efficiently adapt to low temperatures [27, 28].

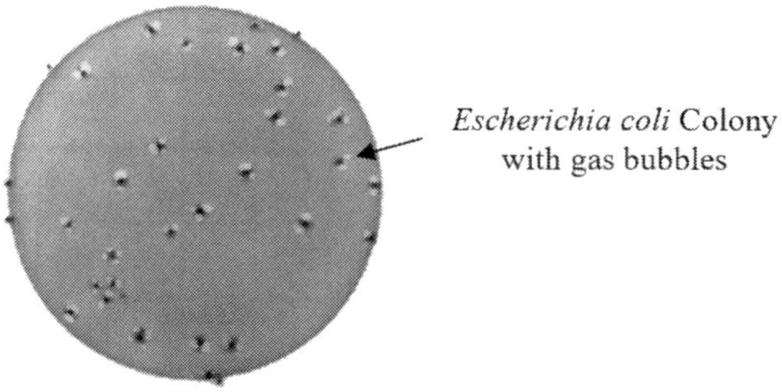

Figure 1. Colony of *Escherichia coli* after incubating for 24 hours at inactivation by carbon dioxide gas bubbles at 30°C and 30 min.

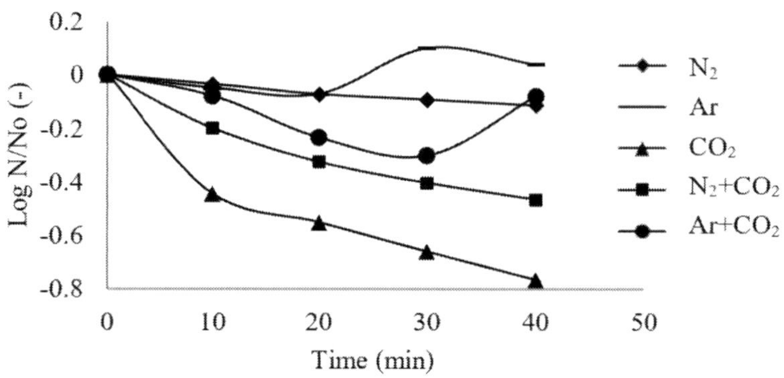

Figure 2. Effect of gas bubbles on inactivation of *Escherichia coli* in distilled water by gas bubble of N_2(♦), Ar(-), CO_2(▲), 50%N_2:50%CO_2v/v (■) and 50%Ar:50%CO_2v/v (●) at 4-10°C

Kinetic models are an important tool for inactivation of microorganisms. The inactivation rate can be usually expressed using the values of k and, a decimal reduction time, (D). Each microorganism strain is a specific D value for each inactivation process. Fig. 3 showed the inactivation of *Escherichia coli* for this study. A linear regression analysis from Equation (4) showed in straight lines and the results of kinetic parameters were shown in Table 1. The rate constant k at 50°C was more than at 4-10°C and at 30°C. The decimal reduction time D implied the retention time required to achieve the inactivation of 90% of the microorganisms, which showed that *Escherichia coli* was inactivated at 50°C more than at 4-10°C and at 30°C, which was consistent with the results above.

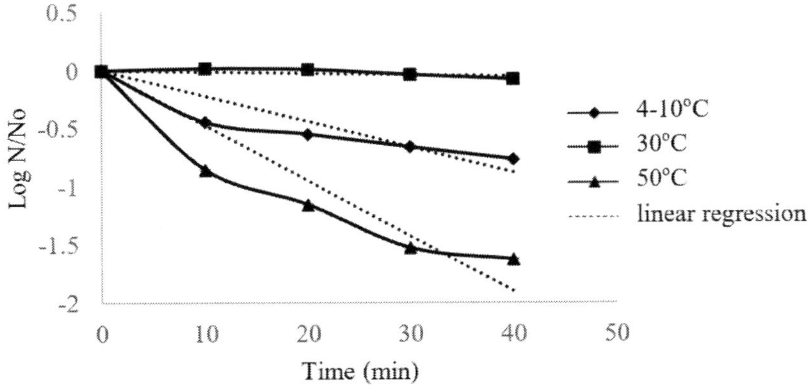

Figure 3. Effect of temperature on inactivation of *Escherichia coli* in distilled water by carbon dioxide gas bubbles at 4-10°C (◆), 30°C (■), 50°C (▲)

Table 1. kinetic parameters after solving (eq. 2, eq. 3 and eq.4) to describe inactivation of *Escherichia coli* in distilled water by carbon dioxide gas bubbles.

Temperature	k (min^{-1})	D (min)	R^2
4-7°C	0.0220	45.45	0.9503
30°C	0.0013	769.23	0.6406
50°C	0.0475	21.05	0.9617

4. Conclusion

Gas bubbles were effective in the reduction of *Escherichia coli* in distilled water and its effectiveness varied concomitantly with temperature, time and types of gas in the solution. In addition, the number of surviving *Escherichia coli* was reduced by approximately 1.62-log from the treatment of carbon dioxide gas bubbles at 50°C and gas flowrate of 2.5 ml/min for 40 min. The rate constants k for the inactivation model at 4-10°C, 30°C and 50°C were 0.022, 0.0013 and 0.0475 min^{-1}, and 45.45, 769.23 and 21.05 min for the decimal reduction time D, respectively. These results suggest that carbon dioxide gas bubble treatment may become a future inactivation method and it can be used in washing process for cleaning of fruit, vegetables, fish and poultry in the future.

References

[1] Russell J B, Diez-Gonzalez F and Jarvis G N 2000 Symposium: farm health and safety Invited Review: Effects of Diet Shifts on *Escherichia coli* in Cattle *J. Dairy Sci.* **83(4)** 863-73
[2] Sarff L D, McCracken G H, Schiffer M S, Glode M P, Robbins J B, Orskov I and Orskov F 1975 Epidemiology of *Escherichia coli* K1 in healthy and diseased newborns *The Lancet* 1099-101
[3] Escherich T 1885 Die darmbakterien des neugeboren und sauglings *Fortschritte der Medizin* **3(2)** 515-22
[4] Mackie R I, Sghir A and Gaskins H R 1999 Developmental microbial ecology of the neonatal gastrointestinal tract *Am J Clin Nutr.* **69(5)** 1035-45
[5] Gilligan P H 2013 *Identification of pathogens by classical clinical tests The Prokaryotes* (Berlin Heidelberg: Springer) chapter 4 pp 57-89
[6] Serres M H and Riley M 2006 *Volume 1: Symbiotic associations, biotechnology, applied*

microbiology, genomics and metabolism in Escherichia coli The Prokaryotes (Berlin Heidelberg: Springer) chapter 1 pp 261-74

[7] Doyle M P and Schoeni J L 1984 Survival and growth characteristics of *Escherichia coli* associated with hemorrhagic colitis *Appl Environ Microbiol.* **8(4)** 855-56

[8] Armstrong G L, Hollingsworth J and Morris Jr J G 1996 Emerging foodborne pathogens: *Escherichia coli* O157:H7 as a model of entry of a new pathogen into the food supply of the developed world *Epidemiologic Reviews* **18** 29-51

[9] Shallow S, Daily P and Rothrock G 1997 Foodborne diseases active surveillance network 1996 *Morbidity and Mortality Weekly Report* **46(12)** 258-261

[10] WHO (World Health Organization) 2008 *Guidelines for Drinking-water Quality Incorporating 1st and 2nd Addenda* Vol 1 Recommendations 3rd ed WHO Geneva Switzerland

[11] Damar S and Balaban M O 2006 Review of dense phase CO_2 technology: microbial and enzyme inactivation, and effects on food quality *J. Food Sci.* **71** R1-R11

[12] Stéphane G, Laurence P, Alain H, Mélanie R, Julia H T, Paul M, Eric F and Patrick G 2014 Extremely rapid acclimation of *Escherichia coli* to high temperature over a few generations of a fed-batch culture during slow warming *Microbiologyopen.* 52-63

[13] McWilliam L E C and Stewart C S 2002 Susceptibility of *Escherichia coli* O157:H7 and non-O157 isolates to lactate *Lett Appl Microbiol.* **35** 176-80

[14] Kobayashi F, Odake S, Miura t and Akuzawa R 2016 Pasteurization and changes of casein and free amino acid contents of bovine milk by low-pressure CO_2 microbubbles *LWT-Food Sci and Tech.* **71** 221-26

[15] Hongmei L, Zhong K, Liao X, and Hu X 2014 Inactivation of microorganisms naturally present in raw bovine milk by high-pressure carbon dioxide *Inter J. Food Sci Tech.* **49** 696-702

[16] Adrian G S , Richard P and Barry N 2019 Virus and bacteria inactivation by CO_2 bubbles in solution *JPN Clean Water* **5** 1-9

[17] Spencer K C 1995 The use of argon and other noble gases for the MAP of foods *International Conference on Modified Atmosphere Packaging and Related Technologies* Campden & Chorleywood Research Association, Chipping Campden, UK. 6-7

[18] Chung H S and Moon K D 2009 Browning characteristics of fresh-cut 'Tsugaru' apples as affected by pre-slicing storage atmospheres *Food Chem.* **114** 1433-37

[19] Ulrike H, Sonja R, Meik A K and Judith K 2013 Comparison of argon-based and nitrogen-based modified atmosphere packaging on bacterial growth and product quality of chicken breast fillets *Poultry Science* **92** 1348-56

[20] AOAC 2002 Association of Official Analytical Chemist, Official Method 998.08 *Escherichia coli* Count in Poultry, Meat and Seafood *J. AOAC Inter.* **82** 73

[21] Hayriye B, Jairus R D D, Ryan J T, D Scott L and P Michael D 2016 Thermal Inactivation Kinetics of Sporolactobacillus nakayamae Spores, a Spoilage Bacterium Isolated from a Model Mashed Potato–Scallion Mixture *J. Food Prot.* **79(9)** 1482-89

[22] Gangar V, Curiale M S, Lindberg K and Lenarz S G 1999 Dry rehydratable film method for enumerating confirmed *Escherichia coli* in Poultry, Meat, and Seafood: Collaborative Study *J. AOAC Inter.* **82(1)** 73-78

[23] Garrido A, Pashley R M and Ninham B W 2018 Water sterilisation using different hot gases in a bubble column reactor *J. Envir Chem Engi* **6** 2651-59

[24] Satoshi I, Katsuhisa K and Satoshi H 2010 Effects of pH on Bacterial Inactivation in Aqueous Solutions due to Low-Temperature Atmospheric Pressure *Plasma Process and Polymer* **7** 33-42

[25] Garcia G L, Geeraerd A H, Spilimbergod S, Elsta K, Van L G, Debevere J, Van I J F and Devlieghere F 2007 High pressure carbon dioxide inactivation of microorganisms in foods: the past, the present and the future *Inter J Food Microbiology* **117** 1-28

[26] Kobayashi F, Daisuke S, Tetsuya T and Hiromi I 2012 Inactivation of *Lactobacillus fructivorans* in physiological saline and unpasteurised sake using CO_2 microbubbles at ambient temperature

and low pressure *Inter J Food Sci and Tech.* **47** 1151-57

[27] Asif S, Shafiqul I, Md Fahad H and Talha B E 2019 Cold Shock and Thawing Effect on the Growth of *Escherichia coli EC Microbiology* **15(1)** 36-43

[28] Beran R K and Simons R W 2001 Cold-temperature induction of Escherichia coli polynucleotide phosphorylase occurs by reversal of its autoregulation *Molecular Microbiology* **39(1)** 112-125

IOP Conf. Series: Materials Science and Engineering 1163 (2021) 012021 doi:10.1088/1757-899X/1163/1/012021

PREDICTION OF VEGETARIAN FOOD PREFERENCES FOR THE AGING SOCIETY

Athakorn Kengpol

Advanced Industrial Engineering Management Systems Research Center,

Faculty of Engineering,

Department of Industrial Engineering,

King Mongkut's University of Technology North Bangkok, Bangkok, Thailand

athakorn@kmutnb.ac.th

Wilaitip Punyota

Faculty of Engineering,

Department of Industrial Engineering,

King Mongkut's University of Technology North Bangkok, Bangkok, Thailand

vilaitip.p@gmail.com

Abstract. The objective of this research is to predict vegetarian food preferences from chronic disease among the elderly by using a hybrid method that includes both an artificial neural network (ANN) and particle swarm optimization (PSO), called ANN-PSO. ANN is a mathematical model that mimics the human brain that is intelligent in learning, prediction, recognition, classification by practice, and solving complex problems. In this study, data collection of vegetarian food preferences, including gender (male and female), a chronic disease selected from the diseases that are common among the elderly, and a vegetarian menu suitable for the chronic disease. Data were collected by interviewing 100 elderly people. Then, the data were analysed using artificial neural networks and applied the particle swarm optimization method to determine the appropriate parameters (weights) for the neural network. The results indicate that the application of PSO along with ANN can accurately predict vegetarian preferences for the aging society. The accurate vegetarian prediction model resulted in increasing consumption of vegetarian food and allowed manufacturers to produce meals or present menus tailored to the individual preferences of the elderly.

1. Introduction

The number of elderly, destined to be 60 or older, is growing in almost all countries and this global trend is likely to increase [1]. An essential component of health in the aging population and influencing the aging process is nutrition. Nutrition for the elderly however has not been clearly defined. Also, no food manufacturer has produced foods that are especially suitable for the elderly. Since the metabolic rate of the elderly decreases with age and the energy requirements of the elderly decreases as well [2]. Ninety-five percent of the elderly have chronic disease problems consisting of high blood pressure, osteoarthritis, cataract, high cholesterol, diabetes, dementia, cardiovascular, depression, stroke and emphysema. The main factor causing these diseases is dietary behavior. If older adults consume healthy foods, this can help to reduce the risk of chronic disease [3].

As the World Health Organization (WHO) mentioned, many diseases that occur in the elderly are a result of dietary factors. These factors are part of the natural transition in the aging process. There are, for example, dietary fat is likely to be related to colon pancreatic, and prostate cancer. Additionally, vascular disease is caused by blood pressure, increased blood lipids, and glucose intolerance, all of which are affected by dietary factors, leading to cardiovascular disease [4]. Changing diet affects the level of risk factors and may have a greater impact on the elderly. Reducing saturated fats and reducing salt intake, which can lower blood pressure, can have a significant effect on lowering the risk of cardiovascular disease (CVD). Additionally, vegetarian dietary patterns are known to reduce CVD mortality and the risk of coronary heart disease (CHD) by 40% [5]. The consumption of vegetarian diets, such as fruits and vegetables, legumes and whole grains, and beans have been consistently associated with a lower risk for many chronic degenerative diseases and increased longevity [6]. Therefore, plant-based vegetarian diets are better protected against chronic disease than meat-based diets [7].

In recent years, the elderly have become more attentive to their health issues and try to consume healthy and suitable vegetarian food for them, but no models have yet recommended a suitable vegetarian food for the elderly. If the elderly are advised to choose a suitable vegetarian food, they are more likely to consume vegetarian food and may keep them in good health. Therefore, the objective of this research is to predict vegetarian food among the elderly preferences using an artificial neural network (ANN) hybrid with particle swarm optimization (PSO) or called ANN-PSO. An artificial neural network (ANN) is effective in classification, prediction. PSO is one of the powerful algorithms for finding the optimal parameter. PSO can increase efficiency and speed in the neural network model when the PSO is hybridized with ANN. PSO was used to find the structure and to calculate the optimal initial weight of the neural network model to obtain a more accurate vegetarian prediction model. The experiments were performed to compare the performance between the ANN model and the ANN-PSO model. The accurate vegetarian prediction model resulted in increasing consumption of vegetarian food and allowed manufacturers to produce meals or present menus tailored to the individual preferences of the elderly.

The paper is as following, section 2 discusses related studies; section 3 is comprised of the details of methodology research; section 4 is an analysis of the experimental results; and finally, the conclusion can be found in section 5.

2. Related Studies

As mentioned in part 1, the objective of this research is to predict vegetarian food among the elderly preferences using an artificial neural networks (ANN) hybrid with particle swarm optimization (PSO) to find the optimal structural and initial weight calculation of the artificial neural network model. Therefore, the research related to this research is as following.

2.1. Vegetarian food

Vegetarian food is defined as a meal that does not contain meat. Most vegetarian diets are made up of plants, vegetables and fruits, as well as using vegetable seasonings. There are studies involving fruit and vegetable consumption as follows: Yuan-Ting (2016) studied of older people who consumed more fruits and vegetables and found that they used less medical services [8]. Kahleova et al. (2018) studied a vegetarian diet for a 40 percent reduction in cardiovascular risk and coronary heart disease risk. In addition, it reduces the risk of the cardiac muscle by more than 80 percent according to nutrition [5].

2.2. *An artificial Neural Network*

An artificial neural network is a machine learning model that is primarily used to classify or predict information. ANN is a model of the human brain. The human brain is able to adapt to changing situations and can learn very quickly in the right context. The model architecture is built with data and learning rules and is trained through algorithms to predict the later dataset [9]. ANN is organized into layers consisting of an input layer, hidden layer and output layer as shown in Figure 1. The hidden layer is the root of the actual computed ANN of the network. The network operates upon receiving the input dataset and the output layer produces the desired results. The weight of the neurons entering the network can be different. In the same way, the transfer function of different neurons may be different (usually the same). Therefore, ANN training is required to achieve proper networking [10].

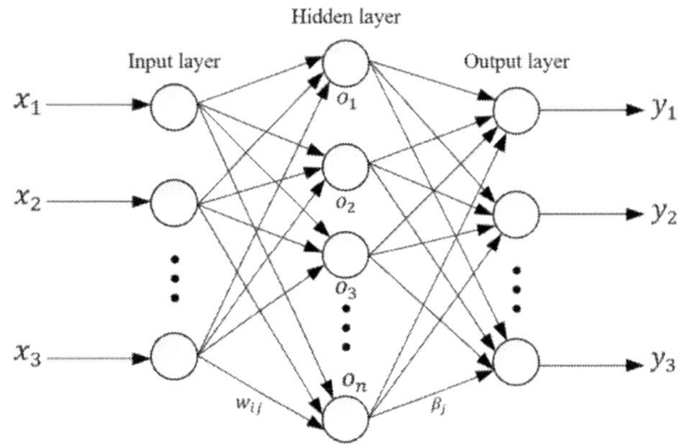

Figure 1. Architecture of a single hidden layer feedforward neural networks.

The input data is transmitted from the input layer to the output layer and is carried out using a feedforward technique, that is, the data is fed from the first layer and then flows one-way through and through, all the way to the last layer without backward previous class. If the input data is transmitted from the input data layer to the output layer, and back-propagation (BP) techniques are used, a method that optimizes the connected weight of the nodes based upon the difference between the network results and the actual results [11]. This distributed processing system is mathematically defined by equation 1:

$$y_1 = \sum\nolimits_{i=1}^{n} w_{ij} x_j + \beta_j \qquad (1)$$

where, x_j is the input training node, w_{ij} are the connection weights associated with the input, hidden and layer nodes, β_j is the bias of the hidden and the output layer nodes and n is the number of input signals, i is the source node ($i = 1,...,n$) and j is the destination node ($j = 1,...,n$) [12]. Learning the ANN process is typically using a back-propagation algorithm (BP). BP is a complex gradient algorithm used to optimize the ANN by changing the weight of each node and bias conditions until the output values at the output layer most closely predict the actual results, resulting in less error training [13]. In this work, PSO is applied as a training algorithm.

Flow chart of the ANN training process can be described as follow. First of all, the input and output parameters of the ANN model are specified. Second, define appropriate learning algorithms and input-output datasets. The most appropriate neural network architecture and training parameters are then selected. Finally, the trained networks can be tested and used for simulations and predictions after the neural network training as shown the figure 2 [14].

2.3. *Particle Swarm Optimization*

Particle swarm optimization (PSO), an algorithm using swarm intelligence proposed by Kennedy and Eberhart [15]. The approach was inspired by birds, fish, and other similar social behaviours found in nature [16], by regrouping, redirecting, and speed suddenly to simulate the movement of the group. The principle idea of this algorithm is that it tries to find an optimised area, where each space has a degree of possibility for a candidate solution [17]. PSO starts with a random particle population and the algorithm finds the optimal population by updating generations [18].

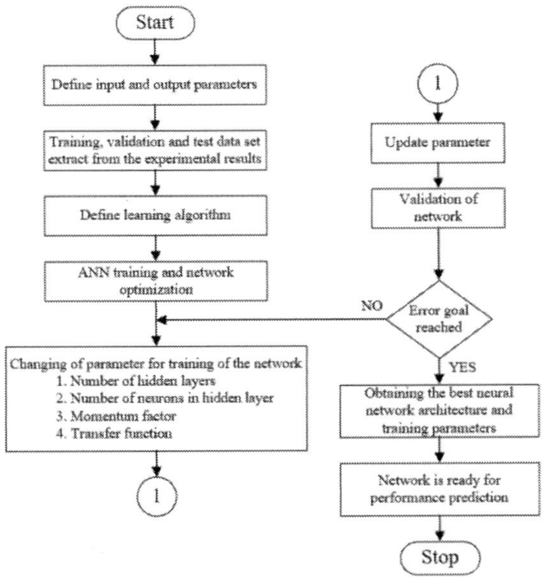

Figure 2. Flow chart of the ANN training process [18].

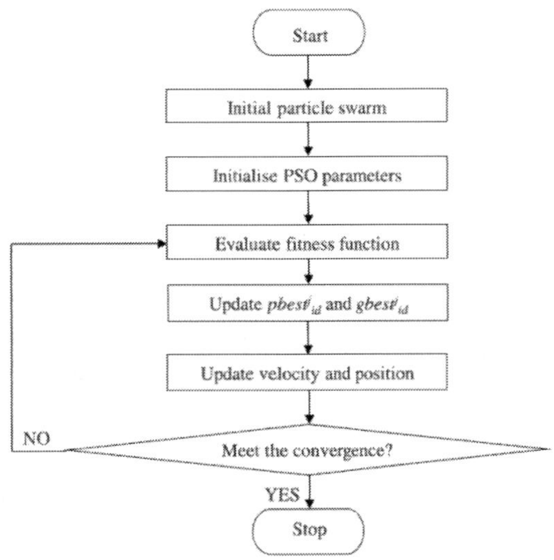

Figure 3. Flow chart of a Particle Swarm Optimization (PSO) [14].

The flowchart of PSO technique is shown in Figure 3 [14]. The operation of the PSO is described as follows 1. initialisation: determination of the position and velocity of random particles 2. evaluate fitness function: the fitness value is calculated for each particle with its position and speed updated 3. update $pbest^j_{id}$ and $gbest^j_{id}$: the fitness value of each particle compared to the best individual $pbest^j_{id}$ if the new fitness value is better than $pbest^j_{id}$, this value is given to $pbest^j_{id}$ and the current position of particle X^j_{id}. Among all the particles, the best fitness value is defined as $pbest^j_{id}$ or the global best value 4. update velocity and position: as defined by equation 2 and 3. 5. Meet the convergence: steps 2-4 are repeated until the stopping threshold is successful in order to achieve the best fitness value or the maximum number of iterations. The convergence means that the results from the model are close to the actual values. The more convergence, the closer the results from the model to the actual values. In this experiment, it can be observed from the loss function, which convergence near zero [19].

Updated velocity of particle

$$V_{id}^{j+1} = wV_{id}^{j} + c_1 r_1 (pbest^j_{id} - X^j_{id}) + c_2 r_2 (gbest^j_{id} - X^j_{id}) \quad (2)$$

Updated position of particle

$$X_{id}^{j+1} = X_{id}^{j} + V_{id}^{j+1} \quad (3)$$

Where V_{id}^{j+1} = updated velocity of particle i in dimension d search region
V_{id}^{j} = velocity of particle i at iteration j
X_{id}^{j+1} = updated position of particle i in dimension d search region
X_{id}^{j} = position of particle i at iteration j
c_1, c_2 = acceleration factors
r_1, r_2 = random constant between 0 and 1
w = inertia weight

The PSO algorithm is used to develop the weight of a multilayer back forward of artificial neural network. The position of the particles in the loop is described as the particle with the coordinates as the connecting weight. Throughout the training process, the above equations (Equations (2) and (3)) refine the network weights until meeting the criteria. PSO uses a randomization process in the search area of the problem to bring the particles in the population to the optimal position. The optimum values obtained from the PSO were then used to testing the ANN model. The steps of the PSO training algorithm for ANN are shown in figure 4. The PSO algorithm is applied to provide the ANN model to have better predictive results.

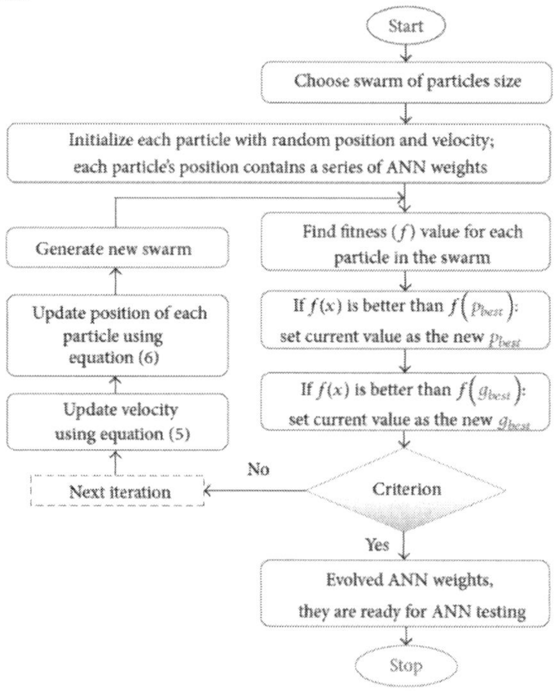

Figure 4. PSO-based algorithm flowchart in optimization of the weights of ANN [13].

The application of ANN and PSO are, for example, Kengpol (2006) assessing customer satisfaction in choosing the right fragrances for each customer group by using ANN [20]. Somchai et al. (2014) used to identify populations at high risk for hypertension and to correctly identify patients with cervical cancer. In classification, the data was 90.95 percent on average by using ANN [21]. Ozerdem et al. (2017) modelled the problem of predicting short-term burden using the particle swarm optimization loads for the feedforward neural network. The result shown that the neural network is optimized for the particle group and results in faster convergence in less time [22]. Bensingh (2019) applied a hybrid neural network (ANN) and particle swarm optimization (PSO) techniques to predict the optimal process parameters of the biosphere lens injection molding process. In this experiment, the efficiency of the ANN-PSO and ANN-GA (Genetic algorithm) hybrid was compared. It was found that PSO converged faster than GA within 40 to 50 iterations for all the five test runs [23]. Al-Majidi (2020) predicted the peak power point of solar cell arrays using feedforward artificial neural network (ANN) and particle swarm optimization (PSO) techniques to improve the accuracy of the ANN model. The result shown that the PSO technique, used in conjunction with feedforward ANN, can accurately predict the maximum power point. In addition, the converging speed of the proposed method has been enhanced under a transient state [12]. Zhang et al. (2020) predicted the stability of roads in tunnels and underground areas using the PSO optimized ANN model, abbreviated to the ANN-PSO model. Additionally, five other methods were developed and compared, including multiple linear regression (MLR), hybrid neural fuzzy inference system (HYFIS), support vector machine (SVM), classification and regression tree (CART), and conditional inference tree (CIT) in conjunction with the proposed ANN-PSO model to provide a comprehensive assessment. All five method comparisons show that ANN-PSO is most effective when measured by Mean Absolute Error (MAE), Root Mean Squared Error (RMSE) and R^2 [24].

Prediction is the analysis of historical data to find patterns of relationships in a data set that can be used as models for predictions, predicting outcomes or things that may happen in the future. Predictive results help a person or organization to make more effective decisions. From a related study on the prediction of food, Giuliani et al. (2015) predict daily food intake as a model to measure and assess differences in reaction and food [25]. Hunt et al. (2018) predict food insecurity in communal areas. Food safety links are examined among multiple demographics, socioeconomic, social risk factors, and food access variables. The findings of this research can help provide information on appropriate interventions by recommending food assistance programs to those in need [26]. Giacalone et al. (2019) study the relationship between the suitability of the situation and the product selection for food and beverages. A simple way to hedonic data is that of item-by-use (IBU) "appropriateness" or "situational fitness". Consumers are given a list of possible consumption situations and are asked to specify how well the product fits each of them. This study shown that the perceived suitability of consumers in food and beverage choices significantly affects their response to consumption [27]. Ruby et al. (2019) predict an adult consumer's willingness to handle safe food in the home, where there was no previous research on safe food handling and preparation among adult consumers. Ruby, therefore, predicts the effectiveness of using the theory of planned behavior (TPB) model to describe the objectives of safe food handling and preparation and to analyze the effect of food safety knowledge on consumer attitudes. The findings suggest that food safety knowledge affects consumer attitudes and can also help develop effective intervention strategies for improving safe food handling at home [28]. In the food industry, food safety, suitability, and quality are key issues directly related to health and society. Predicting or recommending food consumption to consumers is one of the factors that make consumers decide to consume food. Most consumers look for food products that are more reliable and expect manufacturers to provide quality products that are suitable for themselves [29]. Increasing consumer awareness of food safety and quality issues has led to the development of new, more complex techniques for choosing the right food for the consumer. Prediction of food that is appropriate for the consumer is a promotion of the health of the consumer. From the perspective of the manufacturer, being able to produce food that meets the needs of consumers is also promotion.

From the research mentioned above, the artificial neural network is capable of learning, recognition, classification, prediction. PSO can increase efficiency and speed in artificial neural network models. When PSO is hybridized with ANN, the neural network model is optimized as the PSO determines the optimum weight to feed the ANN model, which can be used to predict it more efficiently and reduce processing time. Therefore, the objective of this research is to apply the artificial neural network to predict vegetarian food preferences among the elderly and apply ANN hybrid with PSO to calculate the optimal weight of the neural network by comparing the efficacy between the ANN model and hybrid ANN with PSO (ANN-PSO). PSO was used to optimize the weights between neurons and each layer bias in the ANN model. The performance of the model can be considered by analyzing loss function (cross entropy loss) and accuracy.

3. Methodology

The method of operation is separated into three steps, data collection, data analysis, and validation, as shown in Figure 5. Each of the steps is detailed as following.

3.1. Data collection

In this study, all 10 of the most consumed vegetarian dishes, which have been chosen by vegetarian restaurants. These dishes have nutritional properties as determined by the Institute of Nutrition in Thailand [30]. Ten food menus are included: 1. chili sauce with tamarind 2. red curry chicken with chakram 3. spice vegan soup 4. spicy fruit salad 5. Thai rice and herb salad 6. deep fried spring roll 7. Thai spicy mixed mushroom salad 8. stir-fired tempeh with cashew nuts 9. spicy winged bean salad 10. fried ivy gourd leaf with spicy cream dressing [31]. Each menu is made from ingredients that offer preventive, mitigating and curing benefits. For example, ivy gourd is a gentle, nutritious green fruit and a good source of protein, calcium, fiber and beta-carotene, vitamin A. Ivy Gourd can preserve eyesight, and improve vision, and lowers blood sugar levels by 20 percent [32]. Consuming certain mushrooms or consuming specific ingredients from them may reduce the risk of certain diseases. Mushrooms have been proposed as a potential breast cancer risk reduction agent [33]. These vegetarian foods are suitable for all 5 medical including high blood pressure, osteoarthritis, cataracts, high cholesterol, and diabetes. These are the most common diseases among the elderly in Thailand [1].

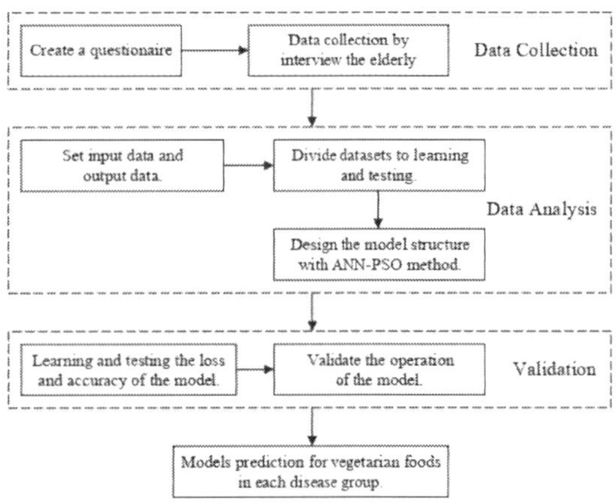

Figure 5. A structure of prediction preference vegetarian food.

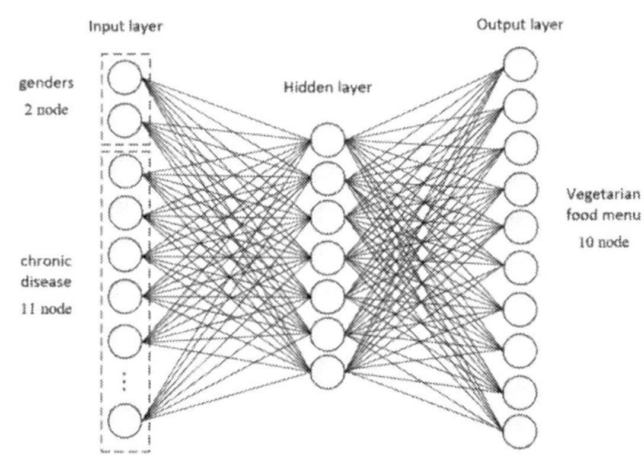

Figure 6. Neural network model for predict vegetarian food preferences.

This research creates a questionnaire and gather data from a sample of the elderly, which is a pilot study. Data is gathered from 100 customers who bought vegetarian food in vegetarian restaurants. The data are collected in 2 parts as follows: part 1 is individual questionnaires for vegetarian food consumers consisted of gender and 11 chronic diseases, setting the nominal scale and part 2 is a vegetarian menu preference questionnaire, where all 10 vegetarian menus are defined.

3.2. Data Analysis

Design a data analysis using Python program version 3.7 to create model an artificial neural network. The experiment is performed by assigning the input layer to 13 nodes (gender 2 node, chronic disease 11 node included high blood pressure, osteoarthritis, cataract, high cholesterol, diabetes, dementia, cardiovascular, depression, stroke, emphysema and other) and the output layer 10 nodes (vegetarian food menu 10 node). In this trial, PSO is used to determine the optimal weight of a backpropagation neural network. Define the activation function of the hidden layer and output layer as sigmoid function. The sigmoid function, which produces a result between 0 - 1, is suitable for using in this task where the probability result is 0 = no and 1 = yes, as shown in Equation 4. The training cycles are set equal to 1,000 epochs. The learning rate and momentum are in the range [0, 1] [34].

$$Sigmoid\ Function = \frac{1}{1+e^{-x}} \tag{4}$$

3.3. *Validation*

This step determines how close the simulation is to the real value. By considering the loss function value using cross entropy loss, which is used for multi-class problem classification. Cross entropy loss is a calculation of the error that the result of the model is different from the actual result. The lower the cross entropy loss value, the more accurate the classification of the data can be. The assessment of accuracy is the ratio of the total number of correct predictions calculated from Equation (5) by True Positive (TP) rate, True Negative (TN) rate, False Positive (FP) rate, False Negative (FN) rate for each type of dataset [35].

$$Accuracy = \frac{TP+TN}{TP+FP+TN+FN} \tag{5}$$

4. Experiments and results

In this experiment, a survey of vegetarian food preferences among 100 elderly people in Bangkok, Thailand was initiated by collecting data on gender, congenital disease, and vegetarian food that the elderly enjoyed. This research is a pilot study to survey the preferences of vegetarian food in a restaurant with a small number of elderly people. Therefore, the researcher collects data from only 100 elderly people, regardless of the proportion of the elderly by age, because if considering the proportion of the elderly according to the age, the data can be collected less and the data is insufficient for training. The majority of the sample was 78% female, 22% male, and had a chronic disease. The most common disease in the sample was high blood pressure 31.71%, osteoarthritis 20.72%, cataract 18.94%, diabetes 15.39%, and high cholesterol 6.18%, respectively. The results of a survey of vegetarian food menu for the elderly are shown in Table 1.

The experiment is conducted on an Intel (R) Core (TM) i7-10510U CPU@1.80Ghz 2.30 GHz RAM 8GB. This experiment is designed and conducted using a Python programming language to create an artificial neural network model based upon a vegetarian food preference survey database. In this section introducing experiments and results, these trials predict the elderly preferences for vegetarian food by analysing data comparison between an artificial neural network (ANN) and an artificial neural network hybrid with particle swarm optimization (ANN-PSO). In terms of ANN, it divides the 80% training dataset and 20% testing data of all data. The neural network has input layer 13 nodes and output layer 10 nodes output. The neural network is designed and tested for the number of hidden layers, which found that one hidden layer of 7 nodes yielded the best results as shown in Figure 6. In addition, adjusting the learning rate 0.5 and the momentum value 0.9, and set the training cycle equal to 1000 epochs, resulting in optimal results.

In terms of ANN-PSO, the experiments adjust r_1, r_2, w which are random within the range [0, 1] and the ratio of c_1, c_2, which determines the importance given to local best and global best to find the best way to predict vegetarian preferences [36]. Finally, the optimum parameter $w = 0.5, c_1 = 0.4, c_2 = 0.5, r_1 = 0.1, r_2 = 0.1$ provide the best prediction model results. The prediction results for vegetarian diets were obtained from the output layer of the neural network model. By comparison in terms of the Cross Entropy Loss. What confirms that the neural network model works correctly is that as training repetitions increase, the Cross Entropy Loss value decreases.

From the simulation, it is found that the ANN neural network alone has a loss function (cross entropy loss) of training data 0.15 and 0.19 of testing data and that the neural network analysis method hybrid with PSO had a loss function of training data 0.03 and test data 0.023 as shown in Table 2.

Table 1. The results of a survey of vegetarian food menu for the elderly.

Chronic disease	Favorite food menu
High blood pressure	(M) Chili sauce with tamarind (F) Red curry chicken with chakram
Osteoarthritis	(M) Spice vegan soup (F) Spice vegan soup
Cataract	(M) Spicy fruit salad (F) Spice vegan soup
High cholesterol	(M) Thai rice and herb salad, chili sauce with tamarind, spice vegan soup, spicy fruit salad (F) spicy fruit salad
Diabetes	(M) Spicy fruit salad (F) Thai spicy mixed mushroom salad
Dementia	(M) Chili sauce with tamarind (F) Red curry chicken with chakram
Depression	(M) Spice vegan soup
Stroke	(F) Stir-fired tempeh with cashew nuts

Table 2. The results of the vegetarian diet prediction experiment

Parameter	ANN		ANN-PSO	
	train ANN	test ANN	train ANN-PSO	test ANN-PSO
Cross Entropy Loss	0.15	0.19	0.03	0.02

The simulation test show that the Cross Entropy Loss decreased with the increase in the number of training and testing cycles, and the accuracy of ANN training data 0.92 and testing data 0.90. As for the ANN-PSO method, training accuracy was 0.95 and for testing data 0.96. From the experiment, the Cross Entropy Loss and the accuracy are consistent, that is, when the number of training cycles increased, the lower the Cross Entropy Loss (less error), the greater the accuracy, meaning the ANN-PSO can actually be used to predict preferences as shown in Figure 7.

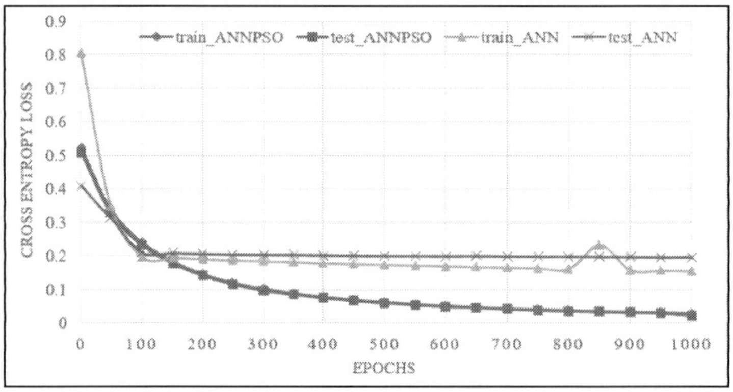

Figure 7. Relationship Cross Entropy Loss and Epochs

5. Conclusion

In this research, a prediction of vegetarian food preferences among the elderly is presented using a hybrid method that included an artificial neural network (ANN) and particle swarm optimization (PSO) or called ANN-PSO. Based upon a survey of elderly vegetarian food preferences from a vegetarian restaurant in Thailand, where the majority of Thai people consume streamed rice along with dished or condiments. The survey found that most of the elderly liked the spice vegan soup the most followed by spicy fruit salad, chili sauce with tamarind. Most of the elderly male prefer chili sauce with tamarind and spicy fruit salad the most. The elderly female prefer red curry chicken with chakram, spice vegan soup, and spicy fruit salad respectively. In the survey, most of the elderly choose to consume food that they like, regardless of the benefits or properties of the ingredients on the menu and their health. In terms of the vegetarian food prediction experiment, the results show that the artificial neural network method, hybrid with particle swarm optimization, which was used to determine the optimum weight of the neural network, performed better than that of an artificial neural network alone. Besides, this model has high accuracy up to 0.95, which can be applied to the elderly. This research is a tool that can predict vegetarian food preferences for the elderly, but whether the elderly choose to consume it or not, depending upon themself. It also allows producers to recognize the elderly preferences for vegetarian food and can offer food menus or produce foods based upon the preferences of the elderly. This research has explored and predicted preferences for only vegetarian food and the elderly in Thailand. For future work, this model can be applied to survey and predict the preferences of foreigners who like vegetarian food in Thailand or to explore and predict vegetarian preferences in other countries where different preferences can be tested. Moreover, it can be surveyed in an application on a mobile phone or a website so that the elderly or vegetarian food producers can comfortably predict vegetarian food preferences from each elderly person.

References

[1] United Nations 2019 Global and regional trend in population aging *World Population Ageing 2019* https://www.un.org/en/sections/issues-depth/ageing/

[2] Ahmed T and Haboubi H 2010 Assessment and management of nutrition in older people and its importance to health *Clin Interv Aging.* **5** 207-16

[3] Ministry of Public Health 2017 *Health at a Glance Thailand 2017* Bangkok http://bps.moph.go.th

[4] World Health Organization 2010 Burden: mortality, morbidity and risk factors *Global status report on noncommunicable diseases 2010*

[5] Kahleova H, Levin S and Barnard N D 2018 Vegetarian Dietary Patterns and Cardiovascular Disease *Prog. Cardiovasc. Dis.* **61** 54-61

[6] Sabate J 2020 The contribution of vegetarian diets to health and disease: a paradigm shift? *Am. J. Clin. Nutr.* **78** 502S–7S

[7] Willett W C 1999 Convergence of philosophy and science: the Third International Congress on Vegetarian Nutrition *Am. J. Clin. Nutr.* **70** 434S–8S

[8] Yuan-Ting C Lo, Wahlqvist M L, Huang Y and Lee M 2016 Elderly Taiwanese who spend more on fruits and vegetables and less on animal-derived foods use less medical services and incur lower medical costs *Br J. Nutr.* **115** 823-33

[9] Scapino L, Zondag H A, Diriken J, Rindt C C M, Bael J B and Sciacovelli A 2019 Modeling the performance of a sorption thermal energy storage reactor using artificial neural networks *Appl. Energ.* **253** 15

[10] Xue Y, Bai C, Qiu D, Kong F and Li Z 2020 Predicting rockburst with database using particle swarm optimization and extreme learning machine *Tunnelling and Underground Space Technology.* **98**

[11] Das G, Pattnaik K P and Padhy K S 2014 Artificial Neural Network trained by Particle Swarm Optimization for non-linear channel equalization *Expert Syst Appl.* **41** 3491-96

[12] Al-Majidi S D, Abbod M F and Al-Raweshidy H S 2020 A particle swarm optimisation-trained

feedforward neural network for predicting the maximum power point of a photovoltaic array *Eng. Appl. Artif Intell.* **92**

[13] Sivapathasekaran C and Sen R 2013 Performance evaluation of an ANN–GA aided experimental modeling and optimization procedure for enhanced synthesis of marine biosurfactant in a stirred tank reactor *J. Chem. Technol. Biotechnol.* **88** 794-99

[14] Xiaoxia W, Tao W and Bingshu W 2008 Hybrid PSO-BP Based Probabilistic Neural Network for Power Transformer Fault Diagnosis *Int. Symp. on Intelligent Information Technology Application* (China: Shanghai) vol 1 p 545–9

[15] Kennedy J and Eberhart R 1995 Particle swarm optimization *Proc. Int. Conf. on Neural Networks* vol 4 p 1942-8

[16] Reyes-Sierra M and Coello C A 2006 Multi-Objective Particle Swarm Optimizers: A Survey of the State-of-the-Art *Int. J. Comput Intell. Res.* **2** 287–308

[17] Vasumathi B and Moorthi S 2012 Implementation of hybrid ANN – PSO algorithm on FPGA for harmonic estimation *Eng. Appl. Artif Intell.* **25** 476–83

[18] Lazzús J A 2010 Optimization of activity coefficient models to describe vapor–liquid equilibrium of (alcohol + water) mixtures using a particle swarm algorithm *Comput Math Appl.* **60** 2260–9

[19] Lei F and Szeri A Z 2007 Inverse analysis of constitutive models: biological soft tissues *J. Biomech.* **40** 936–40.

[20] Kengpol A and Wangananon W 2006 The expert system for assessing customer satisfaction on fragrance notes: Using artificial neural networks *Comput Ind Eng.* **51** 567-584

[21] Somchai J, Cheawchanwattana S, Sunat K and Meungkod N 2014 Extreme Learning Machine for Pre-Hypertension Classification *Kasetsart Engineering Journal.* **89** 89-100

[22] Ozerdem O C, Olaniyi E O and Oyedotun O K 2017 Short term load forecasting using particle swarm optimization neural network *Procedia Computer Science.* **120** 382-93

[23] Bensingh R J, Machavaram R, Boopathy S R and Jebaraj C 2019 Injection molding process optimization of a bi-aspheric lens using hybrid artificial neural networks (ANNs) and particle swarm optimization (PSO) *Measurement.* **134** 359-74

[24] Zhang X, Nguyen H, Bui X, Le H A, Nguyen-Thoi T, Moayedi H and Mahesh V 2020 Evaluating and Predicting the Stability of Roadways in Tunnelling and Underground Space Using Artificial Neural Network-Based Particle Swarm Optimization *Tunnelling and Underground Space Technology.* **103**

[25] Giuliani N R, Tomiyama A J, Mann T and Berkman E T 2015 Prediction of daily food intake as a function of measurement modality and restriction status *Psychosom Med.* **77** 583–90

[26] Hunt B R, Benjamins M R, Khan S and Hirschtick J L 2018 Predictors of Food Insecurity in Selected Chicago Community Areas *J. Nutr. Educ. Behav.* **51** 287-99

[27] Giacalone D and Jaeger S R 2019 Consumer ratings of situational ('item-by-use') appropriateness predict food choice responses obtained in central location tests *Journal of Food Quality and Preference.* **78**

[28] Ruby G E, Abidin U F U Z, Lihan S, Jambari N N and Radu S 2019 Predicting intention on safe food handling among adult consumers: A cross sectional study in Sibu district, Malaysia *Food Contr.* **106**

[29] Gori A, Cevoli C, Fabbri A, Caboni M F and Losi G 2012 A rapid method to discriminate season of production and feeding regimen of butters based on infrared spectroscopy and artificial neural networks *J. Food. Eng.* **109** 525-30

[30] Chanchanakit K 2017 Food and Nutrition for the Older adults *Journal of Mahidol R2R.* **3**

[31] Kengpol A and Punyota W 2020 A Survey of Vegetarian Food Menu for the Elderly *National Conference of Industrial Operations Development.* (Thailand: Bangkok) 7-12

[32] Sargunam H 2017 Ivy Gourd – Medicinal and Nutritional Values *Int. J. Curr. Res.* **9** 47604-7

[33] Feeney M J 2014 Mushrooms and Health Summit Proceedings *J. Nutr.* **144**

[34] Sutskever I, Martens L, Dahl G and Hinton G 2013 On the importance of initialization and

momentum in deep learning *Proc. Int. Conf. on Machine Learning* (USA: Georgia) vol 28 p 1139-47

[35] Suwarno S 2019 Performance Evaluation of Artificial Neural Network Classifiers For Predicting Cesarean Sections *International Journal of Scientific & Technology Research.* **8** 1843-46

[36] Jayaprakasam S, Rahim S K A and Leow C Y 2015 PSOGSA-Explore: A new hybrid metaheuristic approach forbeampattern optimization in collaborative beamforming *Applied Soft Computing.* **30**

The Degradation of Austenitic Stainless Steel at High Temperature in Simulated Carbon Monoxide Containing Atmosphere of Biomass-to-Liquid Plants

Paweena Treewiriyakitja[1], Penpisuth Thongyoug[2], Suwijak Pokwitidkul[1], Jennarong Tungtrongpairoj[1*]

[1]High Temperature Corrosion Research Centre, Department of Materials and Production Technology Engineering, Faculty of Engineering, King Mongkut's University of Technology North Bangkok, Bangkok 10800, Thailand
[2]Department of Industrial Engineering and Management, Faculty of Engineering and Industrial Technology, Silpakorn University, Nakhon Pathom 73000, Thailand

Jennarong.t@eng.kmutnb.ac.th

Abstract. Biomass fuel is effective renewable energy and being used for replacing fossil fuel energy. It can be produced from synthesis gas containing a high percentage of carbon monoxide (CO) and hydrogen (H_2) in biomass-to-liquid plants. Austenitic stainless steel AISI 316L (Cr17% Ni 10% Mo 2%) is used for equipment parts in chemical and petrochemical industries due to good corrosion resistance at various operating conditions. The corrosion resistance of stainless steel may be degraded by the reduction reaction of the passive film and carbide formation from carbon diffusion, which leads to the intergranular corrosion on the steel surface. This research aims to study the degradation of stainless steel AISI 316L in a simulated carbon monoxide containing atmosphere at 15- 45%CO and a sensitizing temperature of 800 °C. Before the test, the samples were preoxidized in the air at 800 °C for 6 hours. An electrochemical reactivation (EPR) technique was used to analyze for detecting sensitization. The mass change of AISI 316L slightly increased after the reduction test. Besides, the high carbon diffusion was shown on the steel surfaces as chromium carbides at the high percentage of carbon monoxide.

1. Introduction

Biomass fuel is effective renewable energy and being used for replacing fossil fuel energy and utilizing agricultural and rubber wastes. Hydrocarbon compounds from the wastes are transformed to biomass pyrolysis products such as water, acetone, ethanol, formic acid, and char at a temperature of 500 - 600°C. Synthesis gases containing carbon monoxide(CO), carbon dioxide(CO_2), hydrogen(H_2), and hydrocarbon-containing gases(C_xH_y) such as methane(CH_4) are produced when heated at the temperature between 500 - 1400°C in gasification processes. The syngas with a high percentage of carbon monoxide (CO) and hydrogen (H_2) (Table 1) is further cleaned up and adjusted the proportion of carbon monoxide and hydrogen for producing liquid oil in the next process [1–4]. AISI 316L austenitic stainless steels (Cr17% Ni 10% Mo 2%) uses for various applications in corrosion environment due to the protective-oxide film formation. However, the carbide formation of $Cr_{23}C_6$ and

Cr_7C_3 were found at grain boundaries after gas carburization at high temperature [5–7]. The carbide formation and the deterioration of passive film when using in a reduction atmosphere can decrease the corrosion resistance, which leads to intergranular corrosion on steel surfaces (Equation 1-4) [8–10].

$$Cr_xO_y(s) + 2yC(s) = Cr_xC_y(s) + yCO(g) \quad (1)$$
$$\tfrac{3}{2}Cr + C = \tfrac{1}{2}Cr_3C_2 \quad (2)$$
$$\tfrac{7}{3}Cr + C = \tfrac{1}{3}Cr_7C_3 \quad (3)$$
$$\tfrac{23}{6}Cr + C = \tfrac{1}{6}Cr_{23}C_6 \quad (4)$$

Table 1. Composition of synthesis gas

Substance	Composition (%)
H_2	20-40
CO	35-40
CO_2	25-35
CH_4	0-15
N_2	2-5

This research aims to study the degradation of austenitic stainless steel in simulated carbon monoxide containing atmosphere of biomass-to-liquid plants between 15-45%CO at 800°C. The effect of different percentages of carbon monoxide on the degradation and carbide formation can be investigated. The degree of sensitization (%DOS) with EPR double loop technique can be analyzed chromium depleted zone caused by chromium carbide formation [11, 12]. The result can be used to predict the appropriate application of AISI 316L in the biomass-to-liquid plants.

2. Experimental method

The AISI 316L specimens were cut with dimension 20 mm x 20 mm x1 mm and ground to 1000 grit SiC paper, then cleaned by acetone. The samples were pre-oxidized at 800°C for 6 hours and measured weight change. After the preoxidization, the samples were heated to 800°C for 1 hour as shown in Fig.1. $CO\text{-}N_2$ gas mixtures were fed into the furnace atmosphere with different percentages of 15%CO, 30%CO, and 45%CO, relating to the flow rate of 3, 4.5, and 6 L/min, respectively. The cross-sectional microstructure of the reduced AISI 316L samples were investigated by optical microscope after etching with glyceregia. Electrochemical Potentiodynamic Reactivation (EPR) technique was used to analyzed carbide on steel surfaces. The electrochemical technique is related to the ASTM 108 standard using reagent water (1 L of 0.5 M H_2SO_4 +0.01 M KSCN and distilled water or deionized water) as electrolytes. The testing samples were anodically polarized from open circuit potential to a passivation potential of +300mV versus SCE (saturated calomel electrode) and reference electrode (Platinum), using sweep rate of 1.67 mV/s. The applied potential was then reversed and calculate the potentiokinetic reactivation ratio values of activation current peak (Ia) and reactivation current peak (Ir) or % degree of sensitization to evaluate chromium depleted zones on steel surfaces.

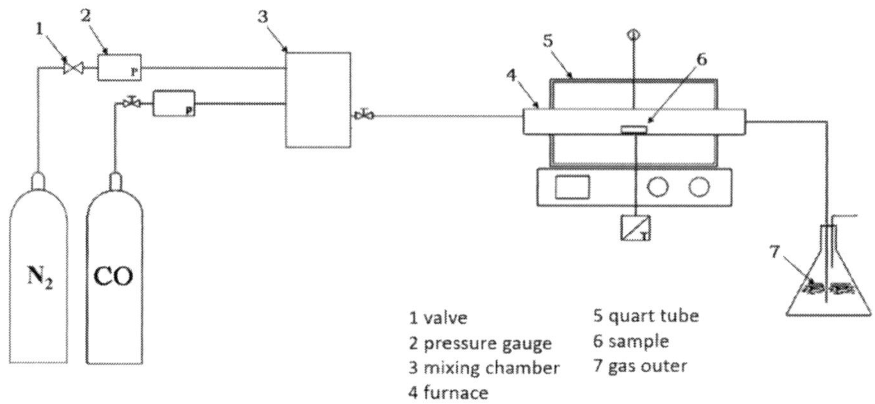

Figure 1. A schematic diagram of the reduction test.

3. Results and Discussion

The average weight of AISI 316L samples slightly increased around 10.36 μg/mm^2 after the preoxidation test in the air atmosphere at 800 °C for 6 hours. After the reduction test at 800 °C, the mass gain of preoxidized samples were increased insignificantly about 0.01144, 0.00802, and 0.02283 μg/mm^2 in the simulated carbon monoxide containing atmospheres at 15, 30, and 45%CO, respectively, as shown in Figure 2. Comparing the mass gain at different carbon monoxide percentages, it showed the highest amount at the reduction of the 45%CO-55%N_2 atmosphere, which was more than the previous two conditions around two times. Furthermore, the microstructures of the preoxidized and reduced specimens in the cross-sectional areas were examined by optical microscope, as illustrated in Figure 3 and Figure 4. The sensitized microstructure with chromium carbides at grain boundaries was clearly presented in every specimen after the preoxidation and reduction test but did not exist in the as-received specimen before the tests. The carbide network formation increased when increasing the percentage of carbon monoxide in mixture gas, especially on the surface area of steel specimens contacting the simulated atmosphere. The carbon from carbon monoxide gas can diffuse into the steel substrate following the reverse Boudouard reaction as the decomposition of carbon monoxide at a temperature lower than around 900 °C (Equation 1) [13]. Besides, the diffused carbon can react with the chromium and iron to be chromium and iron carbides following the equation 2 and 3.

$$2\text{CO} = \text{CO}_2 + \text{C} \tag{1}$$

$$x\text{Cr} + \text{yC} = \text{Cr}_x\text{C}_y \tag{2}$$

$$\text{Fe} + 3\text{C} = \text{Fe}_3\text{C} \tag{3}$$

From a thermodynamics point of view, the stability of chromium carbide is higher than that of iron carbide, which means the chromium carbide can easily occur. The chromium carbide can possibly be formed into $Cr_{23}C_6$, Cr_7C_3, and Cr_3C_2 following the Gibbs free energies ($\Delta G°$) of around -750, -625, and -500 kJ/mol of C at 800 °C, respectively [14, 15]. For this reason, it can be presumed the $Cr_{23}C_6$ was initially presented and transformed to be Cr_7C_3, and Cr_3C_2. In addition, the carbide formations are in good agreement of the previous works of T. Turpin et al. and H.M. Tawancy et al., which observed the $Cr_{23}C_6$ and Cr_7C_3 after gas carburizing of martensitic stainless steels and AISI 310 at the high temperature around 900 °C [5, 6].

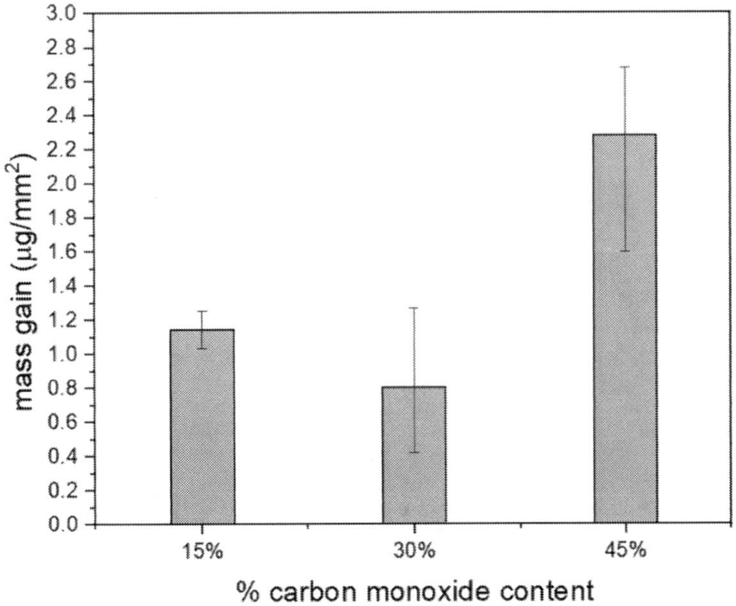

Figure 2. Mass change of preoxidized AISI 316L after reduction test with different percentage of CO in gas mixtures.

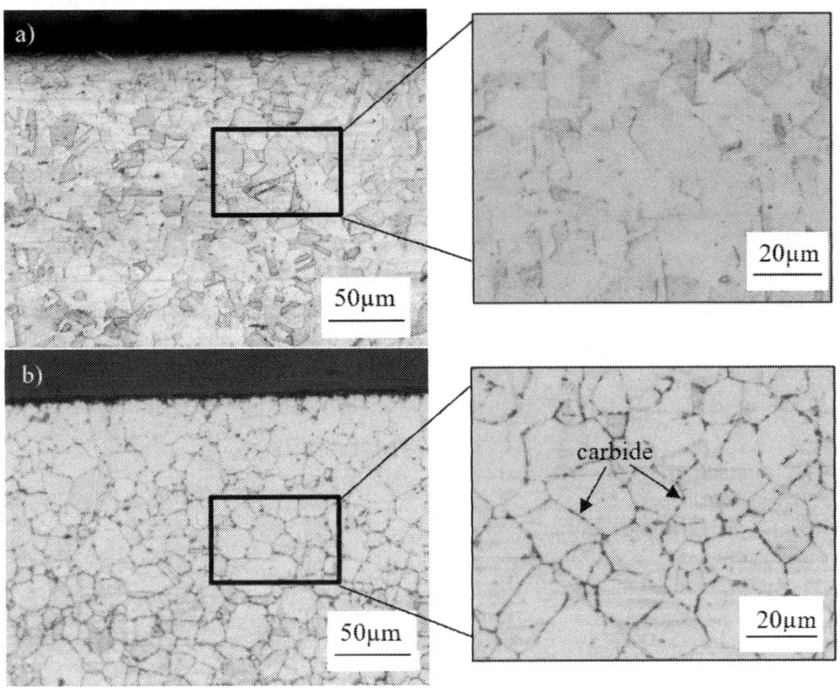

Figure 3. The cross-section of AISI 316L steel (a) before (b) after exposure at 800 °C for 6 hours in air atmosphere.

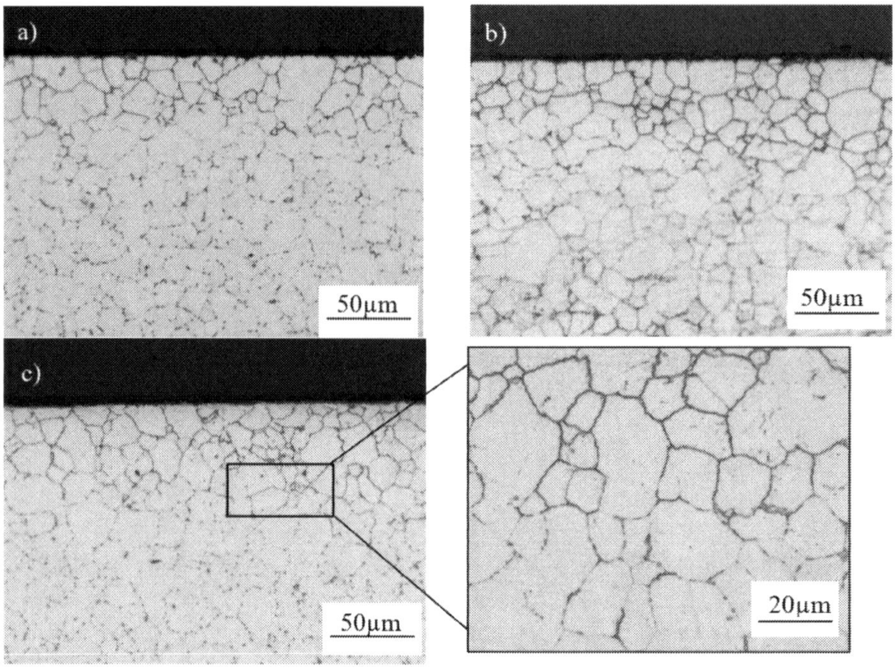

Figure 4. The cross-section of preoxidized AISI 316L steel after the reduction test in the simulated carbon monoxide containing atmospheres at (a)15%, (b) 30%, and (c) 45% CO.

The AISI 316L specimens were further analyzed by Double-loop electrochemical potentiokinetic reactivation (DL-EPR) analysis after sensitizing in the simulated carbon monoxide containing atmosphere to observe the intergranular corrosion (IGC) susceptibility by etching with glyceregia. The activation and reactivation current density profiles of those specimens at different test conditions were shown in Figure 5. The high-skewed reactivation current (Ir) profile which is indicated the reformation of the passive film layer, was detected in all sensitized specimens and raised after feeding more fraction of CO gas. Besides, the ratio of reactivation and anodic current densities (Ir/Ia) was more than one percent, which can indicate the sensitivity of IGC. The degree of sensitization of the sensitized specimens was also plotted in Figure 6, which is reasonably related to the ratio (Ir/Ia). It can be concluded the degree of sensitization will increase around 2 times when adding 15 percent carbon monoxide which may promote the high sensitivity of intergranular corrosion in AISI 316L and cause to failure at high temperature. After the DL-EPR test, the specimens were characterized to present the microstructure when passive film breakdown, as shown in Figure 7.

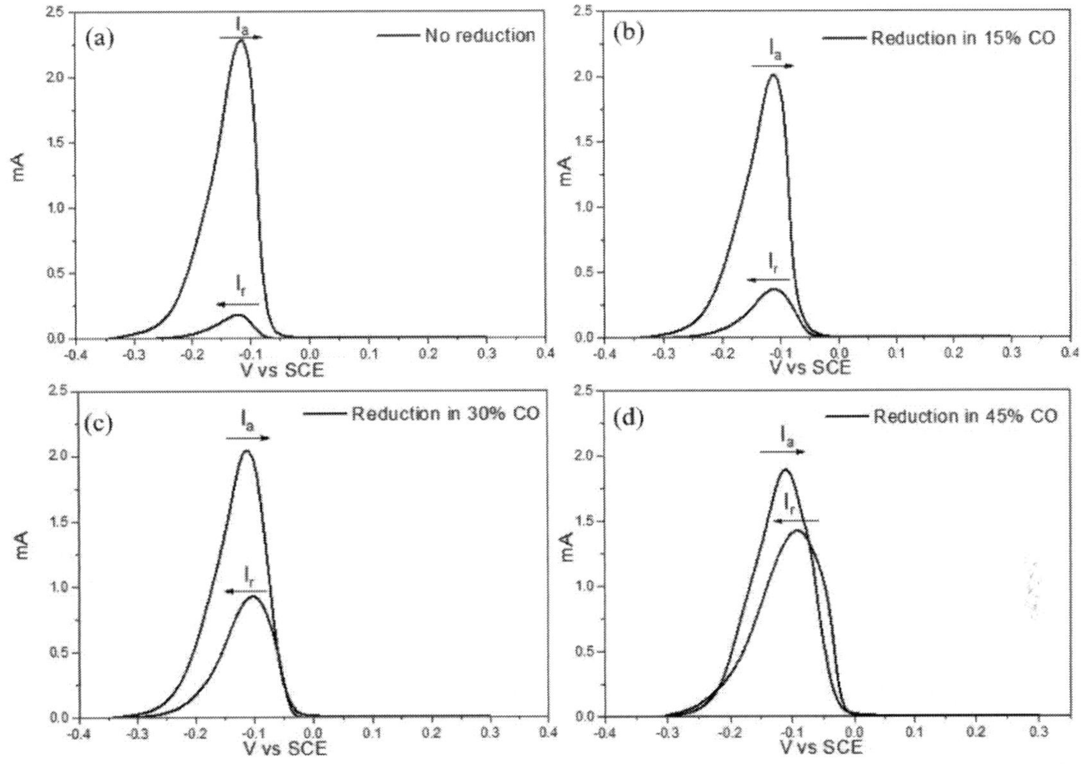

Figure 5. The activation and reactivation current density profile of the preoxidized AISI 316L specimens (a) after exposure in 15%CO (b), 30%CO (c), and 45%CO (d) atmospheres.

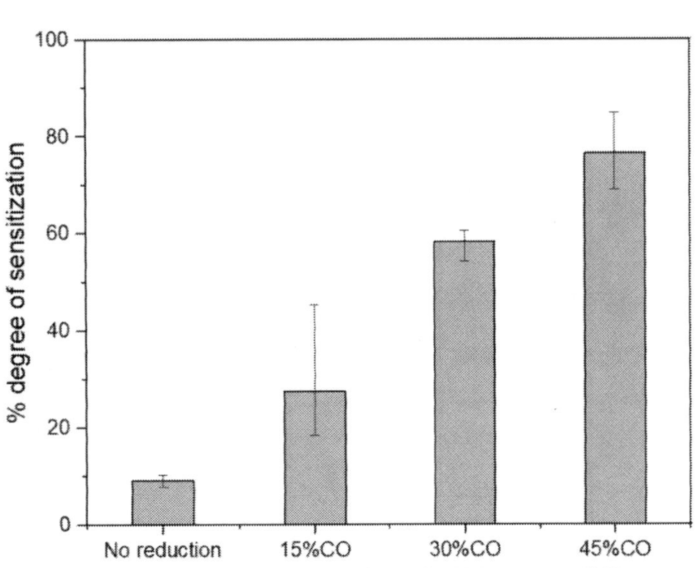

Figure 6. The degree of sensitization of AISI 316L at different testing conditions.

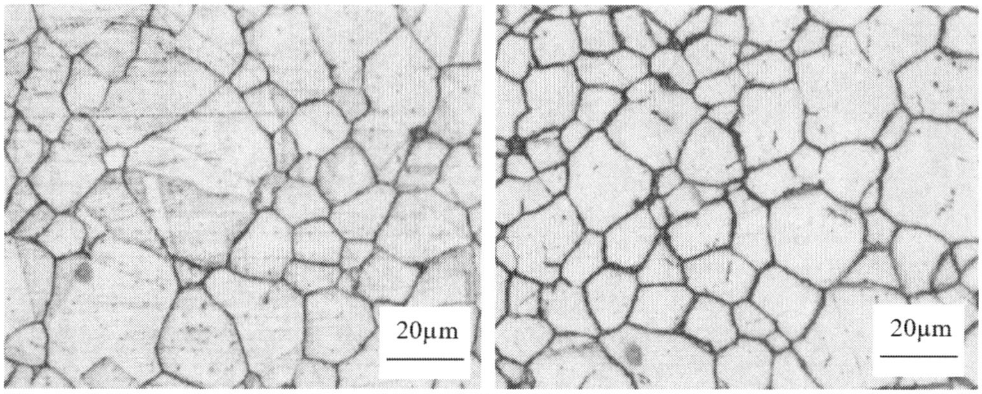

Figure 7. The cross-section of preoxidized AISI 316L steel (a) exposed in the 45%CO-55%N_2 atmosphere (b) after the DL-EPR test.

4. Conclusion

- The mass gain of preoxidized AISI 316L seemed to be not changed and unaffected after the reduction at 800 °C in various simulated carbon monoxide atmospheres.
- The carbide network can be formed after heating to the sensitization temperature and increased after feeding more percentages of carbon monoxide gas in the simulated atmosphere.
- The $Cr_{23}C_6$, Cr_7C_3, and Cr_2C_3 can possibly be formed at the grain boundaries due to the high stability formation with the more negative change of Gibbs free energy.
- The carburizing reaction plays an essential role in the carbon monoxide atmospheres following the high carbon diffusion and carbide formation on steel surfaces.
- The degree of sensitivity relating to the severity of intergranular corrosion in AISI 316L is double when the carbon monoxide increases in the atmosphere by around 15%.

Acknowledgment
This research is funded by King Mongkut's University of Technology North Bangkok with contract no. KMUTNB-64-NEW-12.

References

[1] Vernon, P.D., Green, M.L., Cheetham, A.K. and Ashcroft, A.T. 1990 Partial oxidation of methane to synthesis gas *Partial oxidation of methane to synthesis gas. Catalysis Letters* **6(2)** 181–6

[2] Speight J G (ed) 2011 *Handbook of Industrial Hydrocarbon Processes* (Elsevier)

[3] Wender I 1996 Reactions of synthesis gas *Fuel Processing Technology* **48** 189–297

[4] Niu L, Liu X, Liu J, Liu X, Wen X, Yang Y, Xu J and Li Y 2019 Tuning carburization behaviors of metallic iron catalysts with potassium promoter and CO/syngas/C2H4/C2H2 gases *Journal of Catalysis* **371** 333–45

[5] Tawancy H M 2020 On the comparative degradation of selected Cr2O3-forming alloys by high-temperature carburization with a case study involving 310 stainless steel *Engineering Failure Analysis* **110** 104402

[6] T. TURPIN, J. DULCY, and M. GANTOIS 2005 Carbon Diffusion and Phase Transformations during Gas Carburizing of High-Alloyed Stainless experimental study and theoretical modeling. Steels: Experimental Study and Theoretical Modeling *Metallurgical and Materials Transactions* **36** 2751–60

[7] Karabelchtchikova O Gas Carburizing Fundamentals: Kinetics and Thermodynamics of Mass Transfer in the Carburizing Atmosphere, Gas-Steel Interface and Carbon Diffusion in Steel: Thesis submitted to Worcester Polytechnique Institute, 2007

[8] Mori, Tomoyuki, Jian Yang, and Mamoru Kuwabara 2007 Mechanism of carbothermic reduction of chromium oxide *ISIJ international* **47** 1387–93

[9] Liu Z, Hou Q, Li C, Li X and Shao J 2018 Correlation between grain boundaries, carbides and stress corrosion cracking of Alloy 690TT in a high temperature caustic solution with lead *Corrosion Science* **144** 97–106

[10] Guo X, Wu S, Xu J and Du K 2011 Reducing gas composition optimization for COREX® pre-reduction shaft furnace based on CO-H2 mixture *Procedia Engineering* **15** 4702–6

[11] Laranjo R D 2019 Thermodtnamic Analysis in The Production of Chromium Carbide from The Reduction of Chromium Oxide with Methane-Containing Gas **4** 227–32

[12] G01 Committee Test Method for Electrochemical Reactivation (EPR) for Detecting Sensitization of AISI Type 304 and 304L Stainless Steels (West Conshohocken, PA: ASTM International)

[13] 1986 Institute of Ferrous Metallurgy of the Techniqueal University, RWTH, Aachen - IEHK *Steel Research* **57** 299–305

[14] Kühn G 1980 O. Kubaschewski, C. B. Alcock. Metallurgical Thermochemistry. 5th Edition revised and enlarged. International Series on Materials Science and Technology, Vol. 24, G. v. Raynor (ed.). Pergamon Press Oxford, New York, Toronto, Sidney, Paris, Frankfurt 1979 500 Seiten mit 180 Tabellen, 118 Abbildungen und 984 Literaturzitate. Preis: Broschiert US $ 20,00 *Krist. Techn.* **15** 176

[15] Kubaschewski O and Alcock C B 1979 *Metallurgical thermochemistry (Pergamon international library. international series on materials science and technology* vol.24*)* 5th edn (Oxford: Pergamon)

The Degradation of SA213-T22 steels Coated with Ni-based Alloy Containing WC-Co in Simulated Molten Salt Atmospheres

Penpisuth Thongyoug[1], Paweena Treewiriyakitja[2], Methawat Keereerakwattana[2], Jennarong Tungtrongpairoj[2*]

[1]Department of Industrial Engineering and Management, Faculty of Engineering and Industrial Technology, Silpakorn University, Nakhon Pathom 73000, Thailand

[2]High Temperature Corrosion Research Centre, Department of Materials and Production Technology Engineering, Faculty of Engineering, King Mongkut's University of Technology North Bangkok, Bangkok 10800, Thailand

Jennarong.t@eng.kmutnb.ac.th

Abstract. This work reveals the high-temperature oxidation resistance of WC-Co coated on SA213-T22 by High-Velocity Oxy-Fuel (HVOF) technique in simulated molten salt atmospheres and compare that with the uncoated Cr-containing steel. The corrosion behaviors of SA213-T22 coated and uncoated with WC-Co alloys were studied by immersion test under simulated molten salt atmospheres at 600 °C for 3 hours and investigated their corrosion current densities by electrochemical test. Scanning electron microscope equipped with EDS technique and X-ray diffraction were used to evaluate the surface morphology and phase constitution. The oxidation rate of the coated and uncoated steels was observed under simulated atmospheres at 750 °C for 120 hours before the immersion test. The mass gain scale increase with increasing the oxidation time and the oxidation rate of scale are limited by the WC-Co alloys coating layer, which acted as a protective layer. The low corrosion current density of the WC-Co self-fluxing coated steel after oxidation was shown 5 – 10 μA/cm^2 after the corrosion test in molten salt for 3 hours. After the immersion, the scale layer on the uncoated specimen was disappeared, but one-half of the oxidized coating layer on the coated specimen has remained.

1. Introduction

Molten salt is the one excellent heat transfer fluids (HTFs) for solar thermal power plants due to thermal stability at high temperatures. There are high viscosity, low vapor pressure, and high viscosity at high temperature. Molten salt is based on nitrates/nitrites ($NaNO_3$ 60% and $KNO_3$40%) that would be used for HTFs in the solar thermal power plant. It melts at 223°C, stable liquid phase up to 600°C [1, 2]. The molten nitrate salt contains an oxidizing agent as NO_3^-, the corrosion reaction of molten nitrate salt attacked the parts at high temperature followed by Equation 1-6 [3]. The commercial-grade were also dissolved chloride ion coming from these impurities may be concerned for increased corrosion rate.

$$NO_3^- + 2e^- \leftrightarrow NO_2^- + O^{2-} \quad (1)$$
$$Fe + O^{2-} \leftrightarrow Fe + 2e^- \quad (2)$$
$$3FeO + O^{2-} \leftrightarrow Fe_3O_4 + 2e^- \quad (3)$$
$$2Fe_3O_4 + O^{2-} \leftrightarrow Fe_3O_4 + 2e^- \quad (4)$$
$$2Cr + 3O^{2-} \leftrightarrow Cr_2O_3 + 6e^- \quad (5)$$
$$Fe + Cr_2O_3 + O^{2-} \leftrightarrow FeCr_2O_4 + 2e^- \quad (6)$$

To combat the high-temperature corrosion, the basic grade of the ferrous alloys is carbon steel, which gives iron oxides on the surface at high temperature. Cr can be added up to about 2-9 wt %.giving alloys T22, T23, T24, T91, and T92 to increase the use temperature to 550-600 °C [8–4]. The Cr addition can also be higher than 12 wt %.giving stainless steels, which can be used up to 675 °C for the boiler application [9, 10] . The effect of the oxide layer on the corrosion resistance of low carbon boiler steels in oxidation atmospheres were investigated by electrochemical test and showed a dense-double layer of Fe_2O_3 and Fe_3O_4 acts as the protective film[11] . The present study focuses on the application of T22 steel, which is the intermediate grade between the cheaper carbon steel and the more expensive stainless steel .Since the coating can be applied to increase the high-temperature corrosion resistance of the alloy, the T22 intermediate grade will be coated for such a purpose in this study.

The Cr-Ni alloy was applied by electroplating technique for the protective coating layer on T22 steel, which showed a cracking coating layer through the substrate.[12] Thermal spray coating processes are considered for coating the boiler components to increase corrosion and erosion resistance[14 ,13] , and High-Velocity Oxy-Fuel (HVOF) is commonly used because of its flexibility in a variety of selected materials, good efficiency, and cost-effectiveness. Thermally sprayed coatings were mainly attacked through oxides and voids at splat boundaries[15] . The NiCr alloy is suitable for high-temperature corrosion resistance [17 ,16]. Most literature revealed that the addition of WC-Co alloys improve the micro-hardness value of materials, wear and erosion resistance when compared with the other powders [19 ,18]. In addition, WC-based coatings usually showed higher wear resistance than Cr_3C_2-based coatings .[22–20] They found that the corrosion potential of WC-Co coating increased after the process of heat treatment .[23] This paper investigated SA213-T22 steel uncoated and coated with WC-Co self-fluxing alloy powder by HVOF spray system at the high-temperature oxidation of the air and water vapor environment at 750 °C, and under the simulated corrosive atmosphere of molten salts at 600 °C.

2. Experimental method

2.1. Test specimens and coating preparation

This study used ferritic steel with 2.25wt%.Cr (SA213-T22) in a power plant, and the chemical composition is shown in Table 1. The coating material used was a WC-Co self-fluxing alloy powder with the chemical composition shown in Table 2. The powder particles contained 88WC-12Co particles (Figure 1(a)) and self-fluxing alloy particles (Figure 1(b)) in the ratio of 50:50 by weight .The specimens were machined with dimensions of approximately 10 mm x 15 mm x 5 mm and polished with 600 grit SiC papers, then cleaned with acetone .The specimens for coating were blasted before by alumina grit, and the coating parameters were reported in Table 3 [24], and the coating layer is shown in Figure 1. The SEM-EDS analysis showed the phase rich in W, Co, and C in the white area and the phase rich in Ni. Cr, Fe, and S in the gray area.

Table 1. Chemical composition of SA213-T22 steel (wt%.).

C	Si	Mn	S	P	Cr	Mo	Fe
0.12	0.5	0.6	0.025	0.025	2.25	0.54	Bal.

Table 2. Chemical composition of WC-Co self-fluxing alloy coating powder (wt%.).

WC	Co	Ni	Cr	Fe	Si	B	C
44	6	36	8	2	2	1.7	0.3

Table 3. Process parameters for the HVOF spray coating.

Parameter	
Kerosene (gph*)	5.2
O_2/fuel ratio (scfh**/gph)	385
Spray distance (mm)	390
Powder feed (g/min)	120
Carrier gas (L/min)	23

*gph =gallon per hour, **scfh =standard cubic feet per hour

2.2. High-temperature oxidation tests

High-temperature oxidation tests were conducted on the uncoated and coated specimens in air (21%O_2-N_2 bal.) and 10 %water vapor containing condition in a quartz tube furnace at 750 °C .The water vapor was carried out by 50%O_2 and 40%N_2 (Figure 2.) .The exposure times were 24, 48, 72, 96, and 120 hours and cooling to room temperature by argon .The specimens were weighed before and after the exposure by using a five-digit analytical balance, respectively.

2.3. Corrosion in molten salt and electrochemical test

The coated and uncoated specimens were placed in quartz crucibles and heated in a horizontal furnace at 750 °C for 120 hours in the air (21%O_2-N_2 bal.) .After the oxide formation, the specimens were tested by a commercial molten salt with a composition of 60 %$NaNO_3$ and 40 %KNO_3 at 600 °C for three hours .Every one hour, the specimens were further measured corrosion-resistant properties by the potentiodynamic polarization technique .The electrochemical cell was introduced with the applied potential of 0.167 mVs^{-1} and 0.5 M of NaCl solution at room temperature, and then the corresponding cables of the potentiostat connected to the electrochemical cell for current intensity measurement.

Figure 1. SEM images of the cross-sectional morphology of WC-Co self-fluxing coating layer.

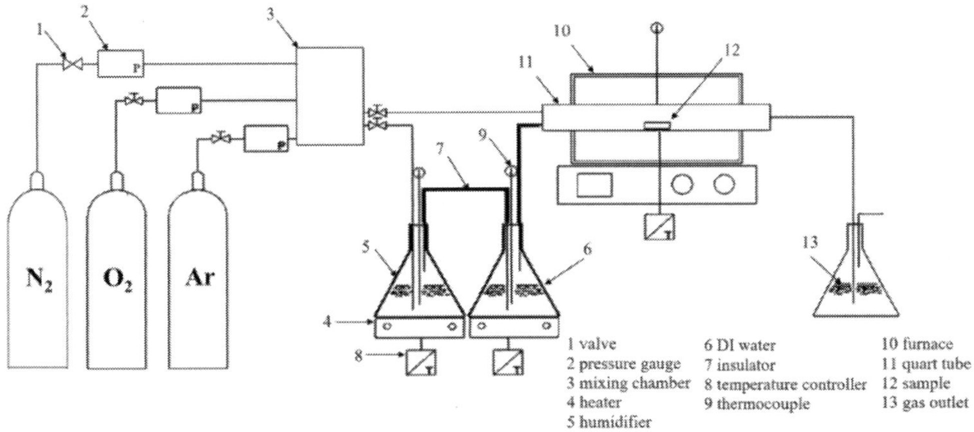

Figure 2. A schematic diagram of the high-temperature oxidation test.

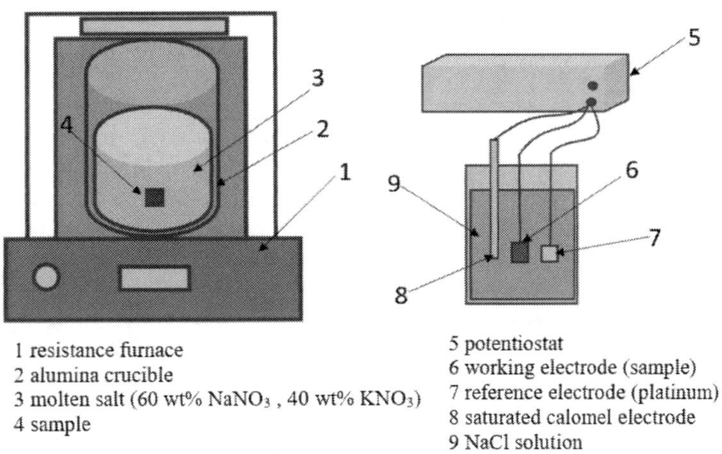

Figure 3. Corrosion in molten salt and electrochemical test setup.

3. Results and Discussion

The coated and uncoated specimens were evaluated in the simulated oxidation atmosphere at high temperatures as described in the previous section. The average mass gain of both samples is seemingly plotted as a parabolic curve, as shown in Figure 4. In Figure 5, the uncoated samples in the 10%H_2O-50%O_2-N_2 atmosphere showed intensive oxidation when compared with those exposed in 21%O_2-N_2 atmospheres .They were higher for every oxidation time and around 10 mg/cm^2 after 120 hours . Moreover, it was significantly higher than that from the previous works at 500 °C.[25] It can be noted that the mass gain of scale increased when increasing the oxidation time, and the percentage of oxygen and water vapor in the atmosphere at the same temperature, following the increase of oxidation kinetics .

After the WC-Co self-fluxing coating, the total mass gain -the accumulation of remain and spalling layers- of the coated specimens was less than that of the uncoated samples in both simulated atmospheres. The total mass gain in the water vapor-containing atmosphere was dramatically decreased at around twenty-two times and about three times for the air atmosphere. The oxide spallation started on the coated specimen after twenty-four hours, but it was slightly decreased after forty-eight hours. According to the past works, the water vapor may be reacted more rapidly with metals than oxygen provided that the formed oxide layer can grow by hydroxyl ion transportation [26–28]. In addition, the oxidation rate constant (k_p) decreased when compared with that of the uncoated specimens in both simulated atmospheres, as shown in Table 4 .The research works which observed the T22 steel in the air at 900 °C [29], and in steam (100%H_2O) at 725 °C showed exceptionally high k_p because of the effect of high temperature and high amount of water vapor .[30]

Besides, the specimens after oxidation were analyzed by XRD .The results of XRD analysis indicated peaks that were identified as scales in the uncoated samples .It was found that hematite (Fe_2O_3) and magnetite (Fe_3O_4) were the main phases on the surface of specimens in both conditions, and there was some wustite (FeO) in the water vapor containing condition (Figure 6(a)). For the coated specimens, NiO, Cr_3C_2, WO_2, and WC were shown in the coating layer before the exposure in both simulated atmospheres, and the Fe_2O_3, Cr_2O_3, and NiCr-spinel occurred after the exposure. However, the high intensity of W_2C peak was observed only in the water vapor containing condition, as shown in Figure 6 (b). It can be presumed that NiO, and Cr_3C_2 were oxidized to NiCr-spinel, Cr_2O_3, and some WO_2 during the oxidation test .From a thermodynamics point of view, the standard Gibbs free energy of Cr_2O_3 and WO_2 formations are lower than that of NiO formation. Therefore those former oxides can be stably formed after the oxidation .Furthermore, the Ni and Cr oxides showed excellent formation in the form of NiCr-spinel at high temperature in the coating layer on the steel.

The polarization curves of the uncoated and coated samples oxidized in air at 750 °C for 120 hours after the immersion test for 3 hours were shown in Figure 7. The corrosion potential of the uncoated sample is lower than that of a coated sample, and it can be discussed that the coated samples showed a higher corrosion resistance because of WC-Co self-fluxing coating layer. The corrosion potential decreases when immersion in molten salt for 1, 2, and 3 hours, respectively. In other words, the corrosion current density of the coated and uncoated specimens seemingly increases after immersion at different times, as shown in Figure 8 .Moreover, the corrosion current density of the coated samples is lower than that of the uncoated specimens at every time .The uncoated sample had a current density of around 30 $\mu A/cm^2$ before the immersion, but it was decreased after one hour and then rapidly increased to 85 $\mu A/cm^2$ after three hours .The scale layer on the uncoated specimen was disappeared after the test and revealed the strong iron peak on the surface (Figure 9) .However, the low corrosion current density was shown in the WC-Co self-fluxing coated steel which is around 5 $\mu A/cm^2$ after immersion for three hours. One-half of the oxidized coating layer on the coated specimen can remain after the test, and the layer was rich in Ni and W. It was deteriorated from 250 μm of thickness at the beginning to 153 μm after the tests .For this reason, it can be related to the low corrosion rate in the molten salt condition and concluded that the WC-Co self-fluxing layer explicitly promotes the high corrosion-resistant coating property. Consequently, the WC-Co self-fluxing layer showed an excellent protective layer .

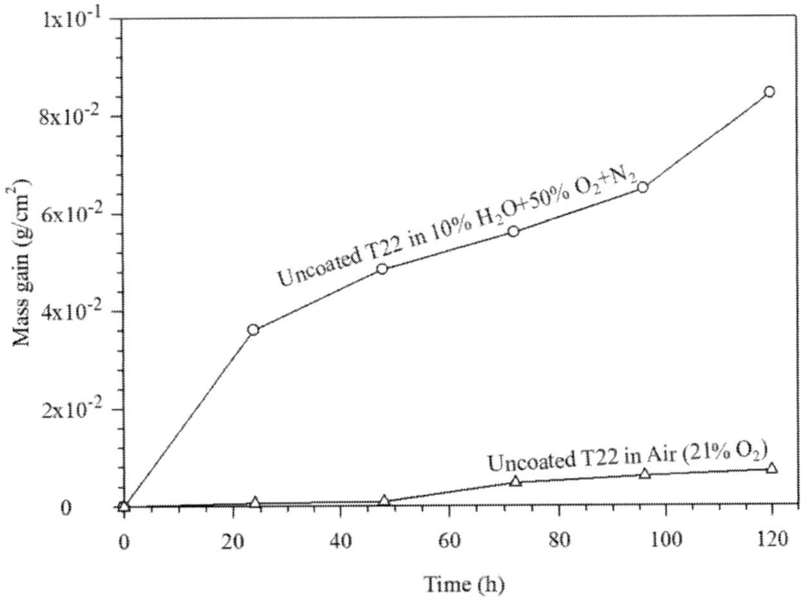

Figure 4. Mass gain of uncoated SA213-T22 after exposure in simulated atmospheres at 750°C for 120 hours.

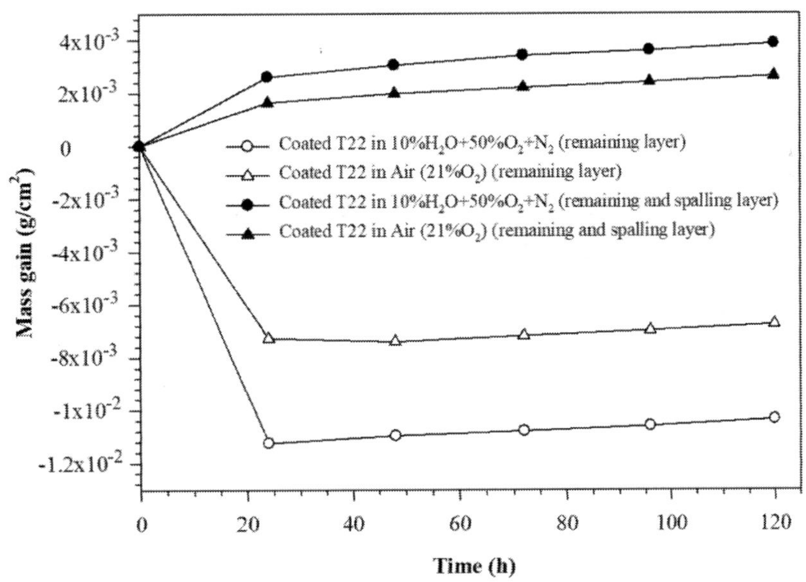

Figure 5. Mass gain of coated SA213-T22 with WC-Co self-fluxing after exposure in simulated atmospheres at 750°C for 120 hours.

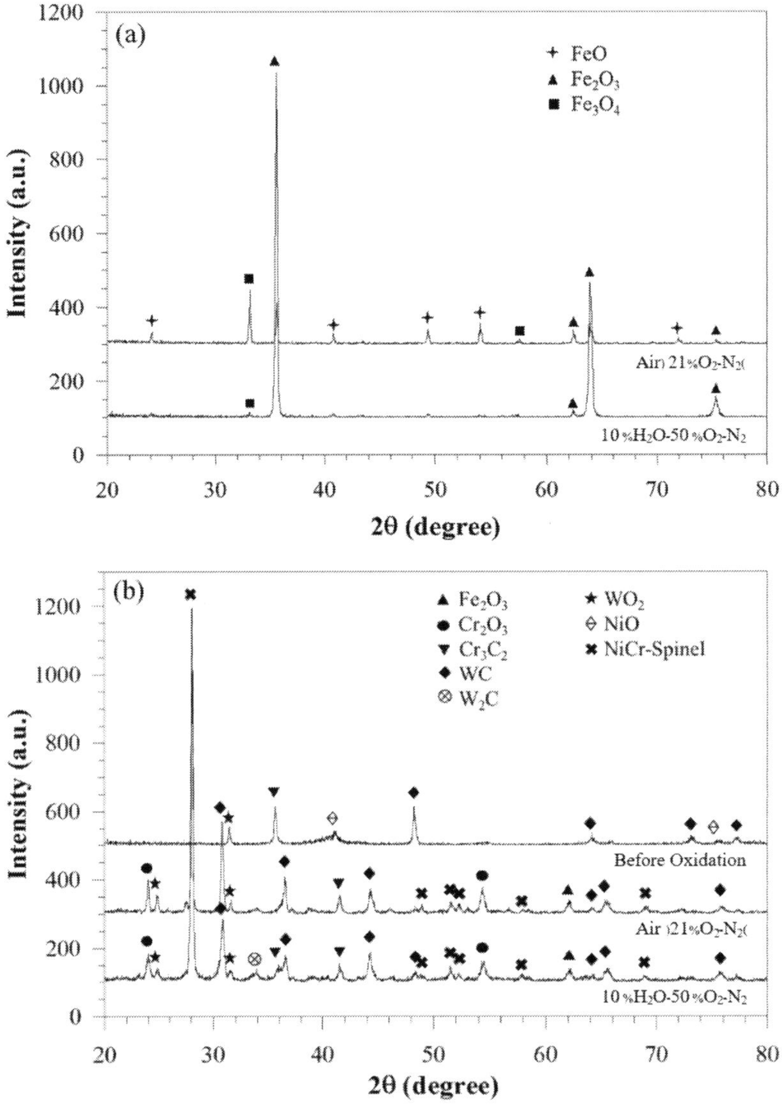

Figure 6 XRD spectrum of (a) uncoated and (b) coated SA213-T22 before and after exposure in simulated atmospheres at 750°C for 120 hours

Table 4. Calculated values of the parabolic rate constant (k_p) at 750 °C.

Descriptions	k_p (10^{-5} mg^2 cm^{-4} s^{-1})
Uncoated in Air	**12.60**
Uncoated in 10%H_2O containing	**1,487.14**
Coated in Air	**1.48**
Coated in 10%H_2O containing	**3.13**
Uncoated T22 at 900 °C [29]	**71,830***
Coated in 100%H_2O containing at 725 °C [30]	**6.67**
Uncoated T22 in molten salt at 900 °C [31]	**125,621.8**

- *k_p of uncoated T22 at 900 °C was 71,830x10^{-5} mg^2 cm^{-4} s^{-1} after exposure 24 hours.

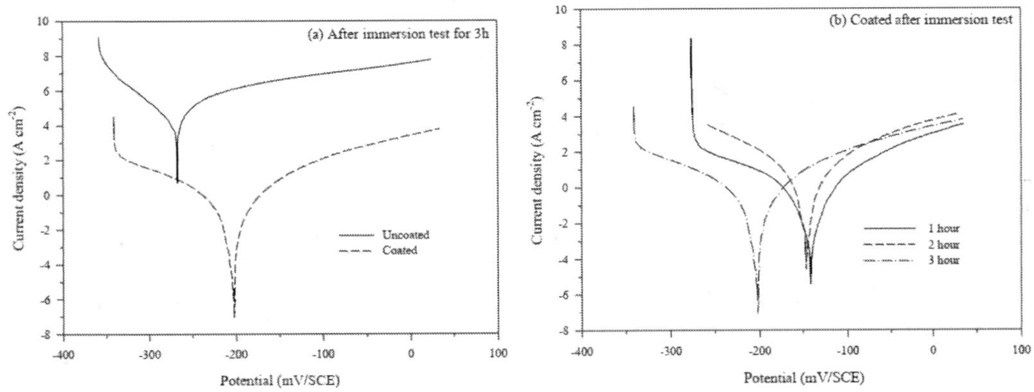

Figure 7. The polarization curves of uncoated and coated SA213-T22 exposed to 60%$NaNO_3$-40%KNO_3 after the immersion test for 3 hours (a) and at different immersion times (b).

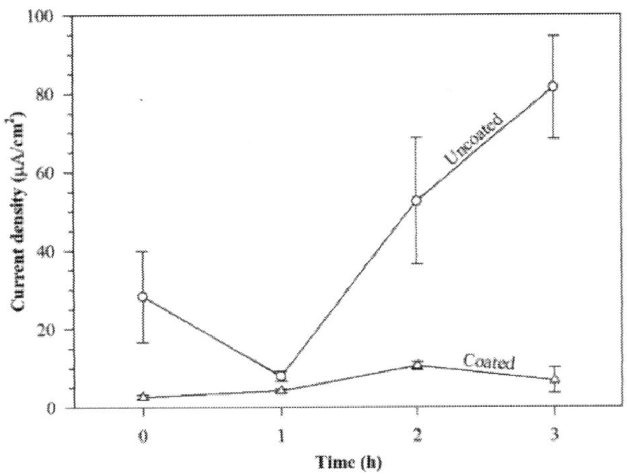

Figure 8. Current time series and cross-section of uncoated and coated SA213-T22 exposed to 60%$NaNO_3$-40%KNO_3 at 600 °C for 0 to 3 hours.

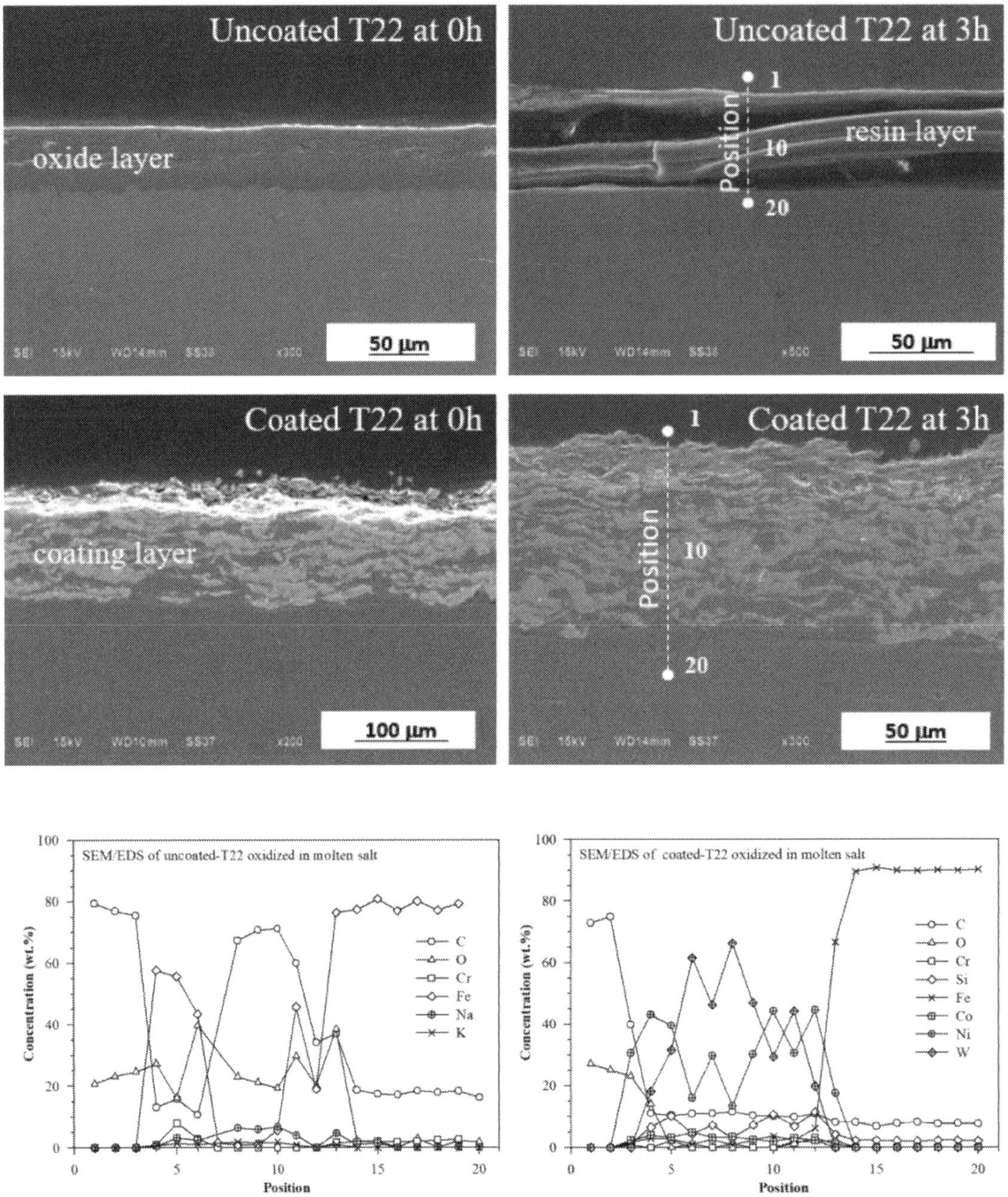

Figure 9. SEM/EDS of uncoated and coated SA213-T22 exposed to 60%$NaNO_3$-40%KNO_3 at 600 °C for 3 hours.

5. Conclusions

The WC-Co self-fluxing coating by using HVOF thermal spray method can be one optional method for protecting the oxidation of SA213-T22 steel at high temperature in thermal processing in molten salts for biomass power plants. The degradation behavior of SA213-T22 coated and uncoated with WC-Co alloys can be summarized as follows.

- The oxidation rate of SA213-T22 steel at 750 °C in water vapor containing condition (10% H_2O-50% O_2-N_2) was much higher than that found in air (21% O_2-N_2) because of the high percentage of oxygen and the inward transport of OH^- from water vapor, which promoting rapid scale cracking and accelerating oxidation.
- The total mass gain of WC-Co self-fluxing coated specimen was lower than that of the uncoated one by around 3 times in the air and 22 times in the water vapor containing condition.
- The spallation of the coating layer started at the beginning of the oxide formation mechanism before the NiCr-spinel formation.
- The formation of FeO and Fe_3O_4 prohibited on the WC-Co self-fluxing coated T22 steel at high temperature in simulated atmospheres, but a small content of Fe_2O_3 existed.
- The WC-Co self-fluxing coating was an excellent protective layer for exposure in air and water vapor conditions. It showed a good corrosion resistance in a simulated molten salt (60%$NaNO_3$-40%KNO_3) at high temperature due to the ability of NiCr-spinel, Cr_2O_3, WC, and W_2C on the high-temperature corrosion resistance and molten salt atmospheres.
- The NiCr-spinel was able to form Ni and Cr oxides after exposure at high temperature. Nonetheless, some of the coating layers started to deteriorate after oxidation for 24 hours and then decreased around 1.6 times after immersion in molten salts for 3 hours.

Acknowledgment

This research is funded by King Mongkut's University of Technology North Bangkok with contract no .ENG-63-73, and kindly technical supported by Department of Industrial Engineering and Management, Faculty of Engineering and Industrial Technology, Silpakorn University.

References

[1] Vignarooban K, Xu X, Arvay A, Hsu K and Kannan A M 2015 Heat transfer fluids for concentrating solar power systems – A review *Applied Energy* **146** 383–96

[2] Peng Q, Ding J, Wei X, Yang J and Yang X 2010 The preparation and properties of multi-component molten salts *Applied Energy* **87** 2812–7

[3] Soleimani Dorcheh A and Galetz M C 2016 Slurry aluminizing: A solution for molten nitrate salt corrosion in concentrated solar power plants *Solar Energy Materials and Solar Cells* **146** 8–15

[4] Wang Z-P, Zhang L, Liu J, Zhao Y-F, Geng B and Wang D-H 2004 *Failure analysis of high temperature steam oxidation of T22 and T91 tubes for power station* vol 25

[5] Quddus A, Ul-Hamid A and Saricimen H 2014 Carburization behavior of high velocity oxy fuel sprayed metallic coatings at 900°C *Anti-Corrosion Meth & Material* **61** 436–44

[6] Dudziak T, Hussain T, Orlicka D, Pokrywa A and Simms N J 2015 Fireside corrosion degradation of 15Mo3, T22, T23 & T91 in simulated coal-biomass co-fired environment *Materials and Corrosion* **66** 839–50

[7] Dudziak T, Łukaszewicz M, Simms N and Nicholls J 2016 Analysis of High Temperature Steam Oxidation of Superheater Steels Used in Coal Fired Boilers *Oxid Met* **85** 171–87

[8] Liu S W, Wang W Z and Liu C J 2017 Failure analysis of the boiler water-wall tube *Case Studies in Engineering Failure Analysis* **9** 35–9

[9] Lai G Y 2008, 2007 *High-temperature corrosion and materials applications* (Materials Park: ASM International)

[10] Holcomb G R 2011 Steam oxidation in steam boiler and turbine environments *Power Plant Life Management and Performance Improvement* (Elsevier) pp 453–89

[11] Kim Y-S and Kim J-G 2019 Improvement of corrosion resistance for low carbon steel pipeline in district heating environment using transient oxygen injection method *Journal of Industrial and Engineering Chemistry* **70** 169–77

[12] Kunyarong A and Fakpan K 2018 Cr–Ni alloy coating electrodeposited on T22 steel *Materials Today: Proceedings* **5** 9244–9

[13] Kaur M, Singh H and Prakash S 2008 A survey of the literature on the use of high velocity oxy-fuel spray technology for high temperature corrosion and erosion-corrosion resistant coatings *Anti-Corrosion Meth & Material* **55** 86–96

[14] Sidhu V P S, Goyal K and Goyal R 2017 An investigation of corrosion resistance of HVOF coated ASME SA213 T91 boiler steel in an actual boiler environment *Anti-Corrosion Meth & Material* **64** 499–507

[15] Rezakhani D 2007 Corrosion behaviours of several thermal spray coatings used on boiler tubes at elevated temperatures *Anti-Corrosion Meth & Material* **54** 237–43

[16] Somasundaram B, Kadoli R and Ramesh M R 2014 Evaluation of Thermocyclic Oxidation Behavior of HVOF Sprayed (Cr3C2-35% NiCr) + 5% Si Coatings on Boiler Tube Steels *Procedia Materials Science* **5** 398–407

[17] Singh H, Kaur M and Prakash S 2016 High-Temperature Exposure Studies of HVOF-Sprayed Cr3C2-25(NiCr)/(WC-Co) Coating *J Therm Spray Tech* **25** 1192–207

[18] Hawthorne HM, Arsenault B, Immarigeon JP, Legoux JG and Parameswaran VR 1999 Comparison of slurry and dry erosion behaviour of some HVOF thermal sprayed coatings *Wear* **225-229** 825–34

[19] Karaoglanli A C, Oge M, Doleker K M and Hotamis M 2017 Comparison of tribological properties of HVOF sprayed coatings with different composition *Surface and Coatings Technology* **318** 299–308

[20] Goyal D K, Singh H and Kumar H 2011 An overview of slurry erosion control by the application of high velocity oxy fuel sprayed coatings *Proceedings of the Institution of Mechanical Engineers, Part J: Journal of Engineering Tribology* **225** 1092–105

[21] Zhou W, Zhou K, Deng C, Zeng K and Li Y 2017 Hot corrosion behavior of HVOF-sprayed Cr3C2-WC-NiCoCrMo coating *Ceramics International* **43** 9390–400

[22] Vashishtha N and Sapate S 2019 Effect of Experimental Parameters on Wear Response of Thermally Sprayed Carbide Based Coatings *Mat. Res.* **22** 82

[23] Raghu Ram Mohan Reddy K, Ramanaiah N and Sarcar MMM 2017 Effect of heat treatment on corrosion behavior of duplex coatings *Journal of King Saud University - Engineering Sciences* **29** 84–90

[24] Villiers Lovelock H L de, Richter P W, Benson J M and Young P M 1998 Parameter Study of HP/HVOF Deposited WC-Co Coatings *J Therm Spray Tech* **7** 97–107

[25] Kahtan Kh.M. Al-Khazraji, Ahmed A.Moosa and Abbas Kh.Husssein 2009 Oxidation Properties in H2O Vapour of Steel-T22 Alloy Coated by Simultaneous Y-Doped Chromizing-Siliconizing Process **27**

[26] Galerie A, Wouters Y and Caillet M 2001 The Kinetic Behaviour of Metals in Water Vapour at High Temperatures: Can General Rules Be Proposed? *MSF* **369-372** 231–8

[27] Chandra-ambhorn S, Nilsonthi T, Wouters Y and Galerie A 2014 Oxidation of simulated recycled steels with 0.23 and 1.03 wt.% Si in Ar–20%H2O at 900 °C *Corrosion Science* **87** 101–10

[28] Wouters Y, Galerie A and Petit J-P 1997 Thermal oxidation of titanium by water vapour *Solid State Ionics* **104** 89–96

[29] Bala N, Singh H and Prakash S 2009 High-temperature oxidation studies of cold-sprayed Ni–20Cr and Ni–50Cr coatings on SAE 213-T22 boiler steel *Applied Surface Science* **255** 6862–9

[30] Dudziak T, Grobauer S, Simms N, Krupp U and Łukaszewicz M 2015 Metal Loss of Steam-Oxidized Alloys after Exposures at 675°C and 725°C for 500 Hours *High Temperature Materials and Processes* **34** 261

[31] Goyal K, Singh H and Bhatia R 2018 Cyclic high temperature corrosion studies of carbon nanotubes-Cr2O3 composite coatings on boiler steel at 900 ° C in molten salt environment *Anti-Corrosion Meth & Material* **65** 646–57

A study on activated carbon from Borassus flabellifer shell by chemical activation with potassium hydroxide for supercapacitor

C. Chaitammachok[1], *J. Taweekun[2], K. Maliwan[2], T. Theppaya[2]

[1]Energy Technology Program, Faculty of Engineering, Prince of Songkla University, 90112 Hat Yai, Songkhla, Thailand
[2]Department of Mechanical Engineering, Faculty of Engineering, Prince of Songkla University, 90112 Hat Yai, Songkhla, Thailand

*Juntakan Taweekun, Email address: jantakan.t@psu.ac.th

Abstract.
The supercapacitor is significant in daily life. It is an alternative option that can storage energy. This research focused on studying the effect of Potassium hydroxide concentration content increasing surface area of the *Borassus flabellifer* shell charcoal (BFC), power density and energy density of activated carbon as electrode and cycles. Borassus flabellifer shell as agricultural waste was carbonized with tube furnace at nitrogen atmosphere by heated at 400 °C for 1 hour. Activated carbon with KOH (the mass ratio of KOH: BFC; 1:1, 2:1, 3:1, 4:1 and 5:1) (BAC) was heated at 900 °C for 2 hours. The results of the research show that suitable concentration of KOH: BFC to activated with higher pore volume and higher surface area. BAC-5 (5:1) is the highest pore volume of 1679.61 m^2/g and the highest pore volume of 0.737 (cm^3/g) but BAC-4 (4:1) is the best because it has a similar surface area and is economical, the highest pore volume of 1607.092 m^2/g and the highest pore volume of 0.686 (cm^3/g). BAC-1, BAC-2, BAC-3, BAC-4 and BAC-5 have average pore diameters, respectively, around of 1.609 nm., 1.641 nm., 1.742 nm., 1.708 nm. and 1.754 nm. The average pore size of products is microporous.

Keywords: Activated carbon, Borassus flabellifer shell, Electrode, Microporous, Potassium hydroxide

1. Introduction

Activated carbon or activated charcoal is a charcoal that must first be treated with a chemical or physical method. To cause the physical structure of the charcoal to create large numbers of microscopic pores or cracks at the nanometer scale. The pore size of the activated carbon can increase the efficiency of adsorption. When there are many microscopic pores, the absorption of the smaller molecules increases. Currently, researchers are studies on the use of carbon as a component of electrode for supercapacitor. Research into the porous of activated carbon is widely used (Dubey, R. and Guruviah, V., 2019). The Surface area of porous is important in terms of capacitance inside. If it has surface area and pore size at the mesopore or micropore, it has more electrons can be stored (Oyedotun, K.O., 2018).

However, carbon is a fundamental element of living things. Which can be obtained from plants (shell, flower, leaf, coir pith, etc.) and animals (Hu, L., 2016, Niu, L., 2019) but in various researches it has seen the human food resource as important (Chaweewan, 2019). Therefore, initiated the introduction of waste crops from agricultural crops used to make activated carbon from the study area or waste from agriculture come to take advantage. There is a research on one species of plant and not yet widespread, which has similar characteristics with coconut. It is palmyra palm or *Borassus flabellifer*. Raj used *Borassus flabellifer* flower (male flower) into activated carbon it has surface area 474.99 $\mathbf{m^2/g}$, (Raj, F.R.M.S., Jaya, N.V., Boopathi, G., Kalpana, D. and Pandurangan, A., 2020). This plant is one kind of palm species. It is assumed to have originated in Africa. Later, it has been distributed in southern India and it spread to Southeast Asia including Thailand (Songkhla Provincial Agricultural Office, 1999). Most of the research uses all parts of the coconuts and some parts of palms that have similar species such as Lihong Yin used coconut fibers, it has surface area 2898 $\mathbf{m^2/g}$ (Yin, L., Chen, Y., Zhao, X., Hou, B. and Cao, B., 2016). T. Sesuk used coconut coir pith is a high surface area of 2056 $\mathbf{m^2/g}$ (Sesuk, T. et al., 2019). Anjali Jayakumar used coconut leaf sheath sheath -derived nitrogen doped carbon framework is developed and incorporated with nickel and cobalt metal (Jayakumar, A., Antony, R.P., Zhao, J. and Lee, J.M., 2018). Divyashree A. used coconut wastes, coconut fiber (Divyashree, A. et al., 2016). Activated carbon can be used as a heating fuel. Developed as a chemical adsorbent in water and supercapacitor. Many studies have been interested in using shell from this plant to make activated charcoal. Because of their high number of fibers and pores, they are well suited for the development for supercapacitor. Anukul K. Thakur used polypyrrole (PPY), Coconut shell activated carbon (AC) and Europium)III(oxide (Eu2O3). (Thakur, A.K., Choudhary, R.B., Majumder, M. and Gupta, G., 2017). Lin Peng mixed biomass wastes of coconut shell and sewage sludge, a high surface area is 3003 m²/g, a large pore volume 2.04 $\mathbf{cm^3}$/g (Peng, L et al., 2018). Kang Sun used coconut shell, a specific surface area is 1194.4 m²/g, total pore volume is 0.528 cm³/g, micropore percentage is 87.8 (Sun, K. et al., 2017) (etc. as shown in table 1).

Researchers then used the Borassus flabellifer shell (BF) in the research. In Thailand, it can be found a lot in many provinces, both the central and southern zones and it can also be found in many seasons. We using this resource because the price is very low. In addition to using the waste to be beneficial, it also helps reduce organic waste from the community. Another imperative thing is a substance that increases the porous surface area. There are 2 types of activation are physical reactivation (steam) (Misnon, I.I., 2015) and chemical activation (KOH, NaOH, $ZnCl_2$, K_2CO_3, H_2SO_4, etc.). Researchers use chemical activation because it can increase the porous surface area more than physical reactivation at same temperature. The optimum temperature is 600 - 900 °C (Sun, K. et al., 2017). Several studies have used strong bases to make it more micro porous. Potassium hydroxide is a very responsive substance. In this study, the effects of reaction conditions including Potassium hydroxide concentration content increasing surface area of the *Borassus flabellifer* shell charcoal (BFC).

Table 1 Comparison of activated carbons derived from various biowaste materials.

Bio-waste precursor	**Activation method**	**Mass ratio**	**Temperature (°C)**	**Surface area (m^2/g)**	**Reference**
Pitch fiber	NH3-gas	-	900	658	(Torchała, K.,2015)
Brussel sprout	KOH	2:1	800	2410	(Li, J., 2016)
Willow catkins	KOH	1:1	600	645	(Wang, K., 2015)
Willow	KOH	4:1	800	1775.7	(Xie, L., 2016)
Elm flower	KOH	1:1	700	2048.57	(Chen, H., 2018)
Soybean	KOH	-	700	1749	(Lin, G., 2018)
Palm shell	KOH	4:1	500	462.1	(Misnon, I.I., 2015)
Palm shell	steam	-	500	727	(Misnon, I.I., 2015)
Sunflower stalk	KOH	2:1	800	1505	(Wang, X., 2018)
Sakura	KOH	4:1	750	1433.76	(Ma, F., 2019)
Borassus flabellifer flower (male flower)	KOH	0.3:1	600	474.99	(Raj, F.R.M.S., 2019)
coconut coir pith	NaOH	3:1	700	2056	(Sesuk, T., 2019)
coconut shell and sewage sludge	KOH	3:1	700	3003	(Peng, L., 2018)
coconut shell	pyrolysis	-	900	1194.4	(Sun, K., 2017)
pet-cokes and palm kernel shell.	K_2CO_3	1.75:1	680	318	(Rashidi, N.A., 2020)
coconut fibers	KOH	4:1	850	2898	(Yin, L., 2016)

2. Experimental section

2.1. Preparation of biomass precursor

Borassus flabellifers (BF) were obtained from the local area of Sathing Phra District, Songkhla Province in Thailand. The as-received BF were cleaned with water to remove soil and other natural contaminants and then dried with an oven at *105* °C for *24* hours to prevent mold. After that, BF were packaged in a zip-lock bag to prevent moisture.

2.2. Carbonization procedure

Approximately, 3 Borassus flabellifers per burning 1 time were used and then burned to charcoal. BF were carbonized by heating with a rate of 6 °C/min to 400 °C and maintaining this temperature for 1 h with calcined in a tubular furnace under a nitrogen atmosphere. After the Borassus flabellifer charcoal (BFC) cooled, BFC were crushed to a smaller size with a stone mortar. BFC powder was packed in a zip-lock bag to prevent moisture. Then put it in a box with silica desiccant. The equation (1) used to calculate the yield of charcoal.

$$\text{The yield of char })\%\text{wt}(= \frac{\text{Weight of Char}}{\text{Weight of Raw material}} \times 100 \qquad)1($$

2.3. Activation procedure

Approximately 50.0 g of BFC was mixed with KOH)86.2%, AR grade(with a different mass ratio of KOH:BFC = 1:1, 2:1, 3:1, 4:1 or 5:1 in a beaker size 1,000 ml. product was stirred and immersed in beaker in closed box for 24 hrs. After slurry was filtered to leave only charcoal, it was activated by heating at a rate of 6 °C/min to 900 °C and maintaining this temperature for 2 h with calcined in a tubular furnace under a nitrogen atmosphere. followed by washing with dilute 0.1 MHCl solution and then deionized water until pH was adjusted to approximately 7. BAC was dried in oven at 105 °C for 24 hours. BAC was packed in a zip-lock bag to prevent moisture, then put in a box with desiccant silica. The final samples were denoted as BAC-x, where x represents the mass ratio of KOH to BFC. The equation (2) used to calculate the yield of activated carbon.

$$\text{The yield of activated carbon })\%\text{wt}(= \frac{\text{Weight of Activated carbon}}{\text{Weight of Raw material}} \times 100 \qquad)2($$

2.4. Physical and chemical characterization

Static volumetric nitrogen adsorption isotherms were measured on a surface characterization analyzer (Micromeritics, ASAP 2460 version 2.01, USA) at 77 K. The specific surface area and porosity calculated by Brunauere-Emmete-Teller (BET) (surface area and porosity analyzer) technique. The morphology observation was carried out with Field Emission Scanning Electron Microscope (FESEM) images were used to examine the morphological properties of activated carbon by Merlin compact, ZEISS Microscopy, Jena, Germany from the Center for Scientific and Technological Equipment, Walailak University, Thailand. The total pore volume was determined at P/Po = 0.99. The micropore volume and pore size distribution were determined by the Harkins and Jura equation and non-local density functional theory (NLDFT). EDX mapping exhibits the presence of sulfur and oxygen in carbon matrix with variable wt.% with increasing carbonization temperature.

3. Results and discussion

3.1 Textural properties of activated carbons

The activated carbons adsorption isotherm graph is exhibited in Fig. 1. BAC-1, BAC-2, BAC-3, BAC-4 and BAC-5 isotherm curves present similar trends and shapes until reach a relative pressure (P/Po) of 0.99. All isotherm curves may be related to type I physisorption isotherm (according to the IUPAC classification). Type I isotherms are encountered with microporous powders whose pore size does not exceed a few adsorbate molecular diameters. Gas molecule, when inside pores of these small dimensions, encounters the overlapping potential from the pore walls which enhances the quantity adsorbed at low relative pressures. At higher pressures the pores are filled by adsorbed or condensed adsorbate, indicating little or no additional adsorption after the micropores have been filled. Physical adsorption that produces the type I isotherm indicates that the pores are microporous and that the exposed surface resides almost exclusively within the micropores, which once filled with adsorbate, leave little or no external surface for additional adsorption. This indicates that all samples have broader range of pore size distribution which includes a wider range of micropores (pores diameter less than 2 nm) and possibly with narrow mesopores (pores diameter between 2-50 nm) (Thommes, M. et al., 2015) and it in accordance with the pore size distribution of BAC-1, BAC-2, BAC-3, BAC-4 and BAC-5 have average pore diameter was in a range of 1.609 - 1.754 nm. as shown in Fig. 4.

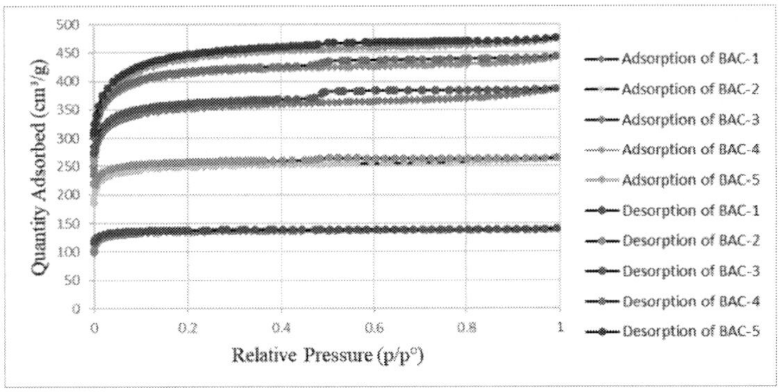

Fig. 1. Adsorption isotherm of activated carbons.

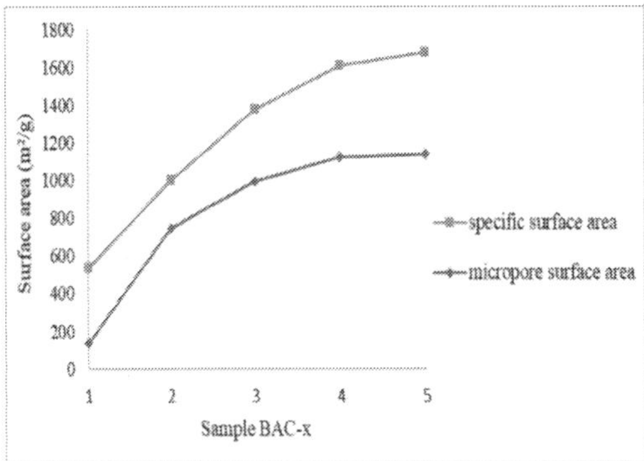

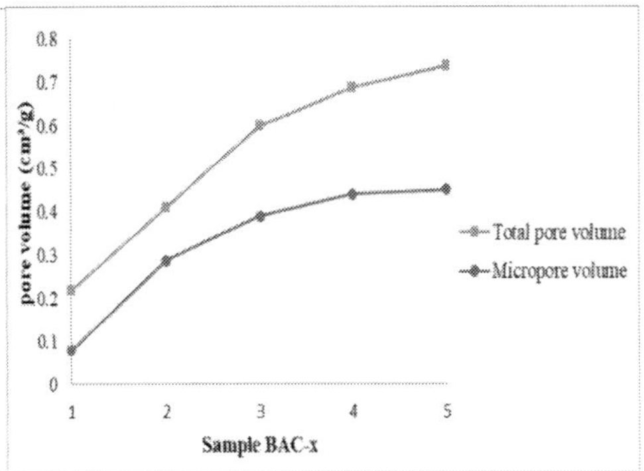

Fig. 2. Graph of the relationship of Concentration in the ratio of KOH: BFC of products to surface area and pore volume (Value obtained from Brunauere-Emmete-Teller (BET) technique as detailed in the table 2).

IOP Conf. Series: Materials Science and Engineering 1163 (2021) 012024 doi:10.1088/1757-899X/1163/1/012024

Table 2. List of parameters deduced from nitrogen adsorption/desorption analysis for different materials.

Sample	S_{BET} (m^2/g)	S_{micro} (m^2/g)	D_p (nm)	V_t (cm^3/g)	V_{micro} (cm^3/g)	% by Volume
BAC-1	537.436	135.33	1.609	0.216	0.079	36.85
BAC-2	997.861	744.18	1.641	0.409	0.286	69.93
BAC-3	1,375.770	990.96	1.742	0.599	0.389	64.94
BAC-4	1,607.092	1122.91	1.708	0.686	0.439	63.99
BAC-5	1,679.609	1140.18	1.754	0.737	0.450	61.03

$\mathbf{S_{BET}}$: specific surface area; $\mathbf{S_{micro}}$: micropore surface area; $\mathbf{D_p}$: Adsorption average pore diameter; $\boldsymbol{V_t}$: total pore volume; $\mathbf{V_{micro}}$: micropore volume.

3.2 Surface area and pore analysis

The overall external and internal surface area of porous and porosity were calculated by Brunauere-Emmete-Teller (BET) technique. Fig. 2 and table 1 were shown properties of activated carbons at different mass ratio of KOH:BFC. The experimental are carbonized at 400 °C and varied from 1 to 5 of impregnation ratio at 900 °C of activation temperature. BET surface area of derived activated carbon was discovered in a range of 537.436 -1,679.609 m²/g and total pore volume was in a range of 0.216 - 0.737 cm³/g. Result was increasing impregnation ratio resulted in improvement of surface area and total pore volume, which has a similar trend as shown in Fig. 2. The BET surface area of BAC-5 is the highest of 1,679.609 m²/g and the highest pore volume of 0.737 cm³/g. Adsorption average pore diameter was in a range of 1.609 - 1.754 nm. respectively, as shown in Table 1. It was shown increasing the KOH concentration used for activation results in the surface area and pore volume increase as well. KOH is a stimulant that causes BAC to have an increased pore content and causing a microporous (less than 2 nm.). Although BAC-5 has the highest surface area (as shown in table 2) but the best sample is BAC-4 (KOH:C; 4:1, heat at 900 °C) because it has a similar surface area and it is economical, the highest pore volume of 1607.092 m^2/g and the highest pore volume of 0.686 cm³/g.

3.3 Field Emission Scanning Electron Microscope analysis

The images obtained with FESEM can be used to provide a clear idea of the surface texture of porous materials. BAC-2, BAC-3, BAC-4 and BAC-5 depicted using FESEM at the magnification of 200 - 50000x. In Fig. 3 (a-d), it was magnified of 500x. BAC-2 was carbonized at 400 °C and mass ratio of KOH:BFC = 2:1, it was activated by heating at 900 °C. Fig. 3 (a) shows the nature of the pores of the BFC that are inherent in nature but there is corrosion caused by activation. Causing more holes in the outer surface. BAC-3, BAC-4 and BAC-5 were carbonized at 400 °C and mass ratio of KOH:BFC = 3:1, 4:1 and 5:1, there were activated by heating at 900 °C. Fig. 3. (b-d) shows a noticeably more porous surface, where the majority of porosities seen in the figure, micropore and radical pore size distribution as shown in Fig. 4. The results of FESEM were related to BET test, physical characteristics. Where more porosity tends to be in the same direction as the surface area and pore volume, as shown in Table 2.

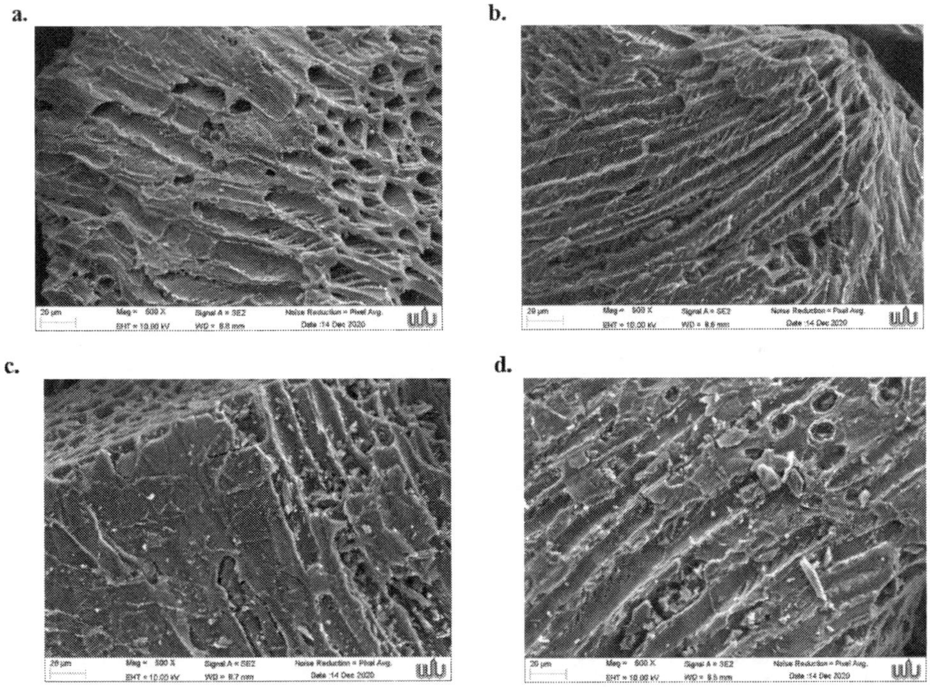

Fig. 3. FESEM images of the samples; a. BAC-2, b. BAC-3, c. BAC-4 and d. BAC-5.

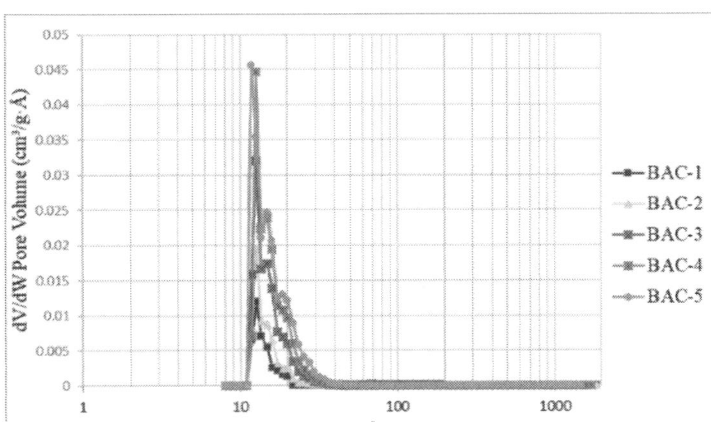

Fig. 4. The activated carbons pore size distributions

Table 3. Elemental composition of BAC-2, BAC-3, BAC-4 and BAC-5

Sample	C (%)	O (%)	K (%)	Si (%)	Mg (%)
BAC-2	95.31	3.90	0.56	0.14	0.09
BAC-3	97.15	1.95	0.62	0.29	<0.01
BAC-4	95.37	3.15	1.02	0.32	0.14
BAC-5	86.06	8.64	4.83	0.13	0.33

Measured by EDS analysis.

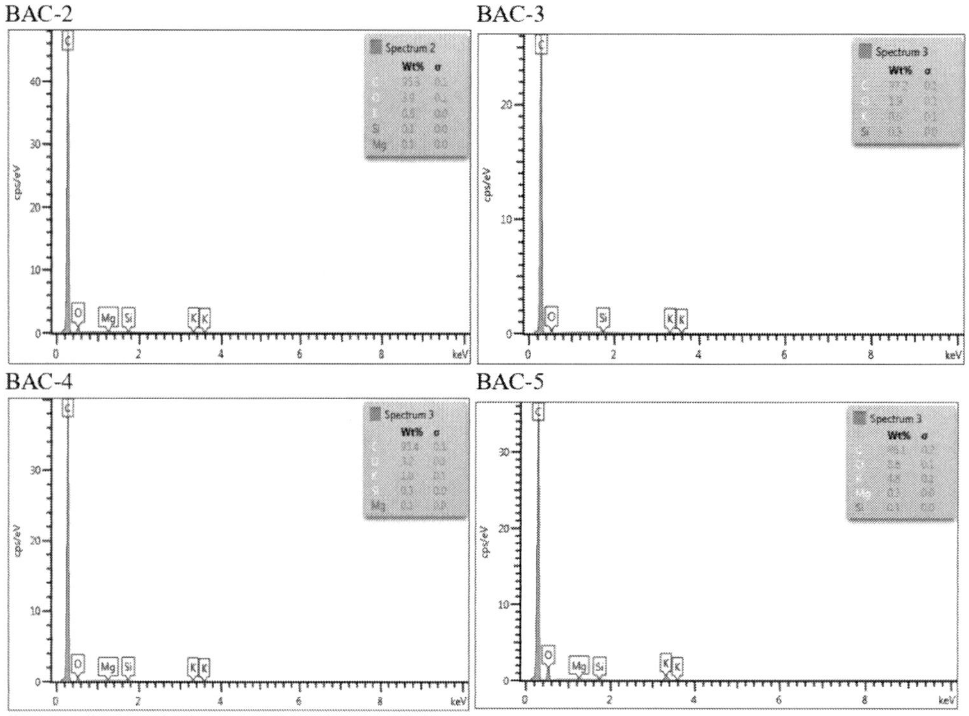

Fig.5. EDS spectrum

3.4 Energy-dispersive X-ray spectroscopy

Element analysis data for BAC-2, BAC-3, BAC-4 and BAC-5 were present in table 3. The carbon content on BAC-5 (86.06%) was lower than BAC-2 (95.31%), BAC-3 (97.15%) and BAC-4 (95.37%) because it has mass ratio of KOH more than others. BAC-3 was the highest carbon element. Carbon was important, when it was used the electrode. Increased purity of carbon affects the adsorption capacity of an activated carbon, the higher the carbon purity the greater the adsorption power.

3.5 The yield of Borassus flabellifer charcoal and activated carbon

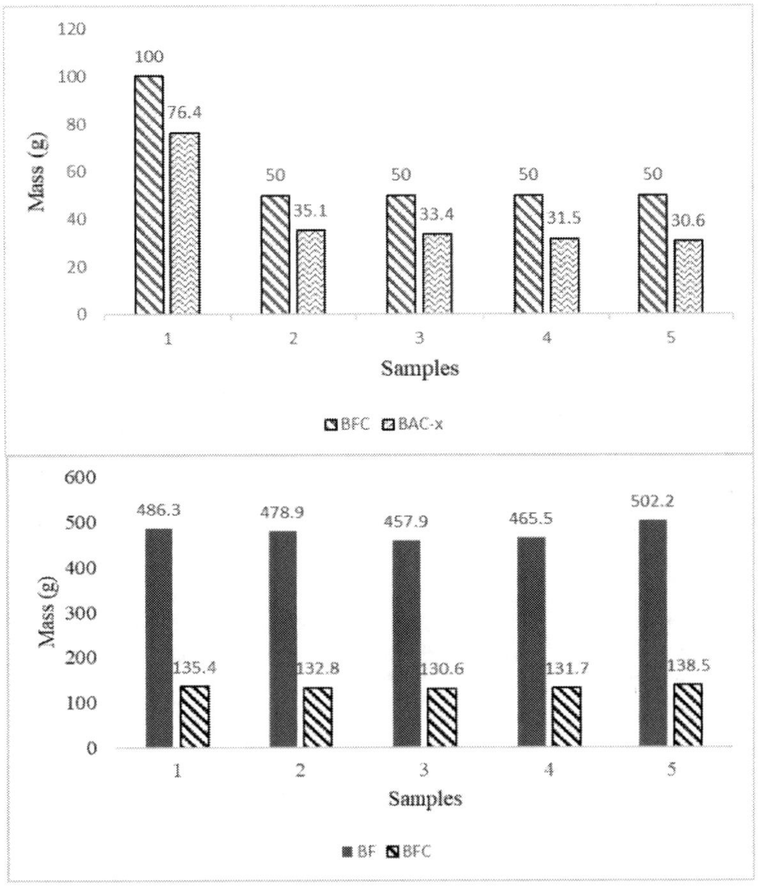

Fig. 6. yield of Borassus flabellifer charcoal and activated carbon diagram.

The Fig. 6 is the yield of Borassus flabellifer charcoal and activated carbon with 1:1,2:1, 3:1, 4:1 and 5:1 impregnation ratio at different activation temperatures. The equation (1) for the yield of char and equation (2) for the yield of activated carbon. Yield of Borassus flabellifer charcoal was in a range of 27.58% - 28.52% by weight. Yield of activated carbons was in a range of 16.88% - 21.27% by weight. BAC-5 is the lowest yield of activated carbons. Result is increased mass ratio of KOH:C, it reduced yield of activated carbons (Nuilerd, T., 2018).

4. Conclusions

The Borassus flabellifer activated carbons (BAC) have qualified to use for electrode. BAC-1, BAC-2, BAC-3, BAC-4 and BAC-5 isotherm curves may be related to type I physisorption isotherm (according to the IUPAC classification). The best sample is BAC-4 (KOH:C; 4:1, heat at 900 °C) because it has a similar surface area and it is economical, the highest pore volume of 1607.092 m^2/g and the highest pore volume of 0.686 cm^3/g. Adsorption average pore diameter was in a range of 1.609 - 1.754 nm. respectively, as shown in Table 1. It was shown increasing the KOH concentration used for activation results in the surface area and pore volume increase as well. KOH is a stimulant that causes BAC to have an increased pore content and causing a microporous (less than 2 nm.). Yield of Borassus flabellifer charcoal was in a range of 27.58% - 28.52% by weight. Yield of activated carbons was in a range of 16.88% - 21.27% by weight. The carbon content on BAC-5 (86.06%) was lower than BAC-2 (95.31%), BAC-3 (97.15%) and BAC-4 (95.37%) because it has mass ratio of KOH more than others. BAC-3 was the highest carbon element. In conclusion, BAC-4 should be used for electrode of supercapacitor and the microporous of Borassus flabellifer activated carbon can be developed by chemical activation with KOH.

Acknowledgments

The authors are appreciated to PSU Energy System Research Institute for research funds, Department of Chemical engineering, Prince of Songkla University for instruction and providing equipment.

References

[1] Burke, A. and Miller, M., 2010. Testing of electrochemical capacitors: Capacitance, resistance, energy density, and power capability. *Electrochimica Acta*, 55(25), pp.7538-7548.
[2] Chaweewan Pengpitak, 2019. Activated carbon. Division of Chemicals and Consumer Products. Department of Science Service.
[3] Chen, H., Yu, F., Wang, G., Chen, L., Dai, B. and Peng, S., 2018. Nitrogen and sulfur self-doped activated carbon directly derived from elm flower for high-performance supercapacitors. *ACS omega*, *3*(4), pp.4724-4732.
[4] Divyashree, A., Manaf, S.A.B.A., Yallappa, S., Chaitra, K., Kathyayini, N. and Hegde, G., 2016. Low cost, high performance supercapacitor electrode using coconut wastes: eco-friendly approach. *Journal of energy chemistry*, *25*(5), pp.880-887.
[5] Dubey, R. and Guruviah, V., 2019. Review of carbon-based electrode materials for supercapacitor energy storage. *Ionics*, 25(4), pp.1419-1445.
[6] Hu, L., Hou, J., Ma, Y., Li, H. and Zhai, T., 2016. Multi-heteroatom self-doped porous carbon derived from swim bladders for large capacitance supercapacitors. *Journal of Materials Chemistry A*, *4*(39), pp.15006-15014.
[7] Jayakumar, A., Antony, R.P., Zhao, J. and Lee, J.M., 2018. MOF-derived nickel and cobalt metal nanoparticles in a N-doped coral shaped carbon matrix of coconut leaf sheath origin for high performance supercapacitors and OER catalysis. *Electrochimica Acta*, *265*, pp.336-347.
[8] Kim, Y.I., Lee, Y.J., Yoo, J. and Kim, J.H., 2019. High-capacitance activated bio-carbons with controlled pore size distribution for sustainable energy storage. *Journal of Power Sources*, 438, p.226969.
[9] Li, J., Zan, G. and Wu, Q., 2016. Nitrogen and sulfur self-doped porous carbon from brussel sprouts as electrode materials for high stable supercapacitors. *RSC advances*, *6*(62), pp.57464-57472.
[10] Lin, G., Ma, R., Zhou, Y., Liu, Q., Dong, X. and Wang, J., 2018. KOH activation of biomass-derived nitrogen-doped carbons for supercapacitor and electrocatalytic oxygen reduction. *Electrochimica Acta*, *261*, pp.49-57.
[11] Lipu, M.H., Hannan, M.A., Hussain, A., Hoque, M.M., Ker, P.J., Saad, M.H.M. and Ayob, A., 2018. A review of state of health and remaining useful life estimation methods for lithium-ion battery in electric vehicles: Challenges and recommendations. *Journal of cleaner production*, 205, pp.115-133.
[12] Ma, F., Ding, S., Ren, H. and Liu, Y., 2019. Sakura-based activated carbon preparation and its performance in supercapacitor applications. *RSC advances*, *9*(5), pp.2474-2483.
[13] Misnon, I.I., Zain, N.K.M., Abd Aziz, R., Vidyadharan, B. and Jose, R., 2015. Electrochemical properties of carbon from oil palm kernel shell for high performance supercapacitors. *Electrochimica Acta*, *174*, pp.78-86.
[14] Niu, L., Shen, C., Yan, L., Zhang, J., Lin, Y., Gong, Y., Li, C., Sun, C.Q. and Xu, S., 2019. Waste bones derived nitrogen–doped carbon with high micropore ratio towards supercapacitor

applications. *Journal of colloid and interface science*, *547*, pp.92-101.

[15] Nuilerd, T., Pongyeela, P. and Chungsiriporn, J., 2018. Pellet activated carbon production using parawood charcoal from gasifier by KOH activation for adsorption of iron in water. *Songklanakarin Journal of Science and Technology*, *40*(2), pp.264-270.

[16] Oyedotun, K.O., 2018. Synthesis and characterization of carbon-based nanostructured material electrodes for designing novel hybrid supercapacitors (Doctoral dissertation, University of Pretoria).

[17] Peng, L., Liang, Y., Dong, H., Hu, H., Zhao, X., Cai, Y., Xiao, Y., Liu, Y. and Zheng, M., 2018. Super-hierarchical porous carbons derived from mixed biomass wastes by a stepwise removal strategy for high-performance supercapacitors. *Journal of power sources*, *377*, pp.151-160.

[18] Raj, F.R.M.S., Jaya, N.V., Boopathi, G., Kalpana, D. and Pandurangan, A., 2020. S-doped activated mesoporous carbon derived from the Borassus flabellifer flower as active electrodes for supercapacitors. *Materials Chemistry and Physics*, *240*, p.122151.

[19] Rashidi, N.A. and Yusup, S., 2020. Co-valorization of delayed petroleum coke–palm kernel shell for activated carbon production. *Journal of Hazardous Materials*, *403*, p.123876.

[20] Saikia, B.K., Benoy, S.M., Bora, M., Tamuly, J., Pandey, M. and Bhattacharya, D., 2020. A brief review on supercapacitor energy storage devices and utilization of natural carbon resources as their electrode materials. *Fuel*, 282, p.118796.

[21] Sesuk, T., Tammawat, P., Jivaganont, P., Somton, K., Limthongkul, P. and Kobsiriphat, W., 2019. Activated carbon derived from coconut coir pith as high performance supercapacitor electrode material. *Journal of Energy Storage*, *25*, p.100910.

[22] Sun, K., Leng, C.Y., Jiang, J.C., Bu, Q., Lin, G.F., Lu, X.C. and Zhu, G.Z., 2017. Microporous activated carbons from coconut shells produced by self-activation using the pyrolysis gases produced from them, that have an excellent electric double layer performance. *New Carbon Materials*, *32*(5), pp.451-459.

[23] Thakur, A.K., Choudhary, R.B., Majumder, M. and Gupta, G., 2017. In-situ integration of waste coconut shell derived activated carbon/polypyrrole/rare earth metal oxide (Eu2O3): a novel step towards ultrahigh volumetric capacitance. *Electrochimica Acta*, *251*, pp.532-545.

[24] Thommes, M., Kaneko, K., Neimark, A.V., Olivier, J.P., Rodriguez-Reinoso, F., Rouquerol, J. and Sing, K.S., 2015. Physisorption of gases, with special reference to the evaluation of surface area and pore size distribution (IUPAC Technical Report). Pure and Applied Chemistry, 87(9-10), pp.1051-1069.

[25] Torchała, K., Kierzek, K., Gryglewicz, G. and Machnikowski, J., 2015. Narrow-porous pitch-based carbon fibers of superior capacitance properties in aqueous electrolytes. *Electrochimica Acta*, *167*, pp.348-356.

[26] Verma, P., Maire, P. and Novák, P., 2010. A review of the features and analyses of the solid electrolyte interphase in Li-ion batteries. *Electrochimica Acta*, 55(22), pp.6332-6341.

[27] Wang, K., Zhao, N., Lei, S., Yan, R., Tian, X., Wang, J., Song, Y., Xu, D., Guo, Q. and Liu, L., 2015. Promising biomass-based activated carbons derived from willow catkins for high performance supercapacitors. *Electrochimica Acta*, *166*, pp.1-11.

[28] Wang, X., Yun, S., Fang, W., Zhang, C., Liang, X., Lei, Z. and Liu, Z., 2018. Layer-stacking activated carbon derived from sunflower stalk as electrode materials for high-performance supercapacitors. *ACS Sustainable Chemistry & Engineering*, *6*(9), pp.11397-11407.

[29] Wang, Y., Tian, J., Sun, Z., Wang, L., Xu, R., Li, M. and Chen, Z., 2020. A comprehensive review of battery modeling and state estimation approaches for advanced battery management systems. *Renewable and Sustainable Energy Reviews*, 131, p.110015.

[30] Xie, L., Sun, G., Su, F., Guo, X., Kong, Q., Li, X., Huang, X., Wan, L., Li, K., Lv, C. and Chen, C.M., 2016. Hierarchical porous carbon microtubes derived from willow catkins for supercapacitor applications. *Journal of Materials Chemistry A*, *4*(5), pp.1637-1646.

[31] Yin, L., Chen, Y., Zhao, X., Hou, B. and Cao, B., 2016. 3-Dimensional hierarchical porous activated carbon derived from coconut fibers with high-rate performance for symmetric supercapacitors. *Materials & Design*, *111*, pp.44-50.

IOP Conf. Series: Materials Science and Engineering 1163 (2021) 012025 doi:10.1088/1757-899X/1163/1/012025

Multi-criteria Flood Susceptibility Assessment Using Remote Sensing and GIS-based Approaches in Chi River Basin, Thailand

K Sakmongkoljit[1], N Homtong[1*], N Surakotra[1] and N Silaratana[1]

[1]Department of Geotechnology, Faculty of Technology, Khon Kaen University, Khon Kaen, 40002, Thailand

E-mail: nudth@kku.ac.th; Tel.: +66-94280-4848 (N.H.)

Abstract. The flood susceptibility map is a comprehensive resource for forecasting and preventing floods worldwide, particularly where floods occur on a regular basis, which is missing in numerous developing-country basins. The recurring flood hazard in Thailand is an endless problem as there is still a lack of understanding of the underlying causative factors. This research provides an answer of certain flood-causative parameters and the effects on flood susceptibility in Chi River Basin (CRB), Thailand, using an Analytical Hierarchical Process (AHP) method. As the result, total of seven hydrogeomorphological parameters can be ranked by its influence on flood hazard in the CRB, as follows, the distance from the river, elevation, geology, soil type, land use, drainage density, and geomorphology, respectively. Combination of all flood-causative parameters based on their weights and ranks provided the flood susceptibility map. The map was evaluated as accurate relative to previous satellite-based flood maps using a binary classification test, providing a robust flood susceptibility map. The findings of this work have contributed to a decent understanding of the present problem of floods in the CRB in the context of its actual causes, which is beneficial for a water management scheme.

1. Introduction

Flood perceived to be a frequent natural disaster inflicts damage worldwide [1-5]. Changing in flood frequency and magnitude is one of the impacts of global warming [6], which influences the flood control system downstream [7]. Climate change conditions cause these problems. Consequently, the water in the basin increases suddenly [6]. Flood disaster caused by both natural and anthropogenic influences [8], which raises economic agricultural and social losses, also affects housing and agricultural areas of riparian people [2,9-12]. However, there is no single solution provided in terms of a flood management plan, and the warning system is still a real-time foresee warning [13]. Thus, the determination of flood risk area is essential for water management of the basin [14], as the flood-prone area needed to be identified [15]. The flood susceptibility mapping in various techniques is a tool for spatial water management to enhance flood prevention and mitigation strategies [16,17], which is particularly essential for flood-prone areas [18].

The Analytical Hierarchy Process (AHP), as one among other multi-criteria analysis (MCAs) methods, is an efficient, cost-effective, less time-consuming, and convenient multi-criteria decision on which variables are ordered in a hierarchical structure [2,19,20], developed by Saaty [21]. Currently, remote sensing and Geographic Information Systems (GIS) techniques became the key elements for multi-criteria decision-making processes, particularly in water resource management [19]. MCA is a method that weights the physical parameters logically in the GIS study [18]. The combination of MCA (AHP method) and GIS is widely used in various research regarding water management issues [2,22,23], especially for flood susceptibility mapping. The AHP and GIS methods can solve problems of water management in various aspects such as identify, assess, and simulate flood-prone areas and its impacts [24], to reduce cost in lives and properties, especially for cities of developing countries [16].

The Chi River is one of the main rivers in Thailand, which flows through several provinces in the north-eastern region of Thailand [25], including Mahasarakham and Roi-Et Provinces. Floods often occurred in the basin [26], such as the flooding from the influence of the tropical storm Sonca in 2017. The floods caused damages to live and properties in the CRB [26]. However, the flood risk and susceptibility of the basin is still not identified. This study aims to assess hydrogeomorphological factors, which influence the occurrence of the flood. The spatial flood susceptibility map of the CRB in Mahasarakham and Roi-Et Provinces was developed and evaluated by utilizing the Analytical Hierarchical Process (AHP) and divided into different flood risk levels, spatially. This flood susceptibility map will be beneficial for flood warning and sustainable water management in the CRB, and further applied in other basins worldwide.

2. Methodology

2.1. The Case Study Area: Chi River Basin

The case study area is a part of the Chi River Basin (CRB) in the northeastern of Thailand, Mahasarakham, and Roi-Et Provinces; consists of 17 districts with the total area of 3,032.87 km^2 (Figure 1). Agriculture is a dominant land use activity in the research field. The study area is classified into three main categories according to geomorphological characteristics, which are the gentle plateau in the south and some areas in the north, floodplain in the central region along the Chi river, and the water bodies such as lakes and swamps throughout the area. According to the data from the Department of Mineral Resources of Thailand [27], in 2006, the area endured tropical monsoon conditions with 1,304.7 mm of a mean annual rainfall. Due to heavy rainfall in the study region, several small villages along the Chi River were dramatically affected by flood threats, both physically and economically.

2.2. Preparation of thematic layers

This study applied a GIS-based AHP technique for generating flood susceptibility map, as shown in Figure 2. Data for this study has been acquired from various sources such as government organizations and satellite images for delineating different flood hazard zones in the CRB. The 30 m-resolution of a Digital Elevation Model (DEM) acquired from the USGS online database and processed in ArcGIS Software version 10.5 for the CRB delineating, preparation of elevation, and geomorphology maps. The drainage density map has been calculated and generated through ArcGIS software version 10.5, using the hydrology tools. The DEM has been processed in the buffering function of the ArcGIS software version 10.5 for developing a distance from the river map with 500 m-interval. The Department of Mineral Resources (DMR) of Thailand was a geologic data source for preparing the CRB geologic map. The Landsat 8 images with 30 m-resolution from the USGS online database have been acquired for a supervised land use classification technique in ERDAS Imagine software to generate a land use map of the study area. Data for the soil type mapping has been compiled from the Land Development Department (LDD) database. After preparing all seven thematic layers, which converted to raster format with 30 m resolution, and reclassified using AHP ranks.

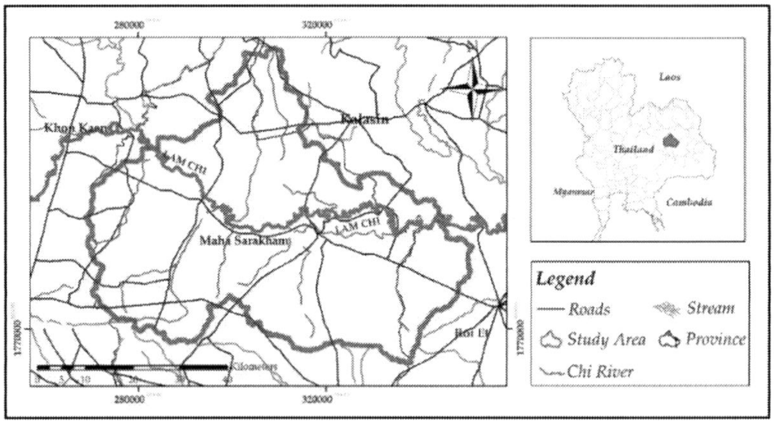

Figure 1. The study area of Chi River Basin, Mahasakham and Roi-Et Provinces, Thailand

2.3. Multi-Criteria Decision Analysis: Analytical hierarchy process (AHP)

The parameter weights can be assigned through the AHP technique which is multi-criteria decision analyses. The AHP method divides the problem into sections in the form of a hierarchical model by imitating the human brain's complex concepts to achieve the right decision based on the information available. The pairwise comparison process is the main part of AHP technique, that assigns numbers for representing data to determine each parameter weight. The values and criteria of the parameters are given for indicating its relative importance between two parameters. The next step is the weighted arithmetic mean for calculating the Eigenvalue. The goal of this process was to measure the weights in a pairwise comparison matrix. Each parameter weight was determined in a pairwise comparison matrix using the mean row approach.

3. Results

The results of this study consist of four sections. The first is a pairwise comparison, which provides the weight of each parameter as the results. The second section is thematic maps, which illustrate flood-causative parameter maps along with the result of rating scores of each parameter class. The third section is an assessment of a spatial flood risk via a flood susceptibility map of the CRB. The last section is an evaluation of the flood susceptibility map. The detail of each result section is explained as follows.

3.1. Pairwise comparison matrix

The pairwise comparison provides the weight of each flood-causative parameters for an AHP method of a flood susceptibility assessment. The results of weighted arithmetic mean are also shown in Table 1.

Table 1. The weight values and normalized rating of the flood-causative parameters from the pairwise comparison technique

Parameters	Distance from river	Elevation	Geology	Soil Type	Land-Use	Drainage Density	Geomor-phology	Weight
Distance from river	0.403	0.534	0.472	0.290	0.318	0.282	0.217	**0.360**
Elevation	0.134	0.178	0.283	0. 194	0.238	0.235	0.174	**0.205**
Geology	0.081	0.059	0.094	0.290	0.159	0.094	0.174	**0.136**
Soil Type	0.134	0.089	0.031	0.097	0.159	0.141	0.087	**0.106**
Land use	0.101	0.059	0.047	0.048	0.079	0.188	0.130	**0.093**
Drainage Density	0.067	0.036	0.047	0.032	0.020	0.047	0.174	**0.060**
Geomorphology	0.081	0.045	0.024	0.048	0.026	0.012	0.043	**0.040**

3.2. Thematic maps
The total of seven flood-causative thematic maps (Figure 2) The detail of each floods-causative parameter is discussed as follows; (1) the distance from the river: the thematic map was created and converted to the raster image (Figure 2a). The distance from the river considered the most influential factor for flood disasters in the study area with a normalized weight of 0.360, (2) elevation: the elevation is generally low in the middle of the area along floodplain next to the Chi river mainstream in the study area as illustrated in Figure 2b with the normalized weight of 0.205, (3) geology: Influence flood disaster in term of its permeability geologic distribution of the study area is illustrated in Figure 2c, which has a normalized weight of 0.136, (4) soil type: the clay content in soil causes a lower infiltration rate, which enhances flood risk of the area. The soil type map is shown in Figure 2d. The normalized weight of soil type had been assigned as 0.106, (5) land use: the CRB land use map is presented in Figure 2e, with the normalized weight of 0.093, (6) drainage density: the drainage density map is shown in Figure 2f. The drainage density parameter has a normalized weight of 0.060, and (7) geomorphology: Among these landforms, the floodplain along the mainstream has the highest flood risk than other landforms in the CRB. The geomorphological types map is illustrated in Figure 2g. The geomorphology has been assigned 0.040 for the normalized weight.

3.3. Assessment of flood susceptibility
All flood-causative parameter maps were combined using an overlay technique in ArcGIS software version 10.5 to generate the study area's flood susceptibility map. The parameters were overlaid by using a weight linear combination method according to each parameter weight and rank. The study area's flood susceptibility map was categorized into five different flood susceptibility levels using the natural break (Jenks) method. The spatial flood susceptibility levels are very high flooding (16.93%), high flooding (17.76%), moderate flooding (22.40%), low flooding (29.08%), and very low flooding (13.83%), that spread throughout the area as illustrated in Figure 3. The very high flooding mostly concentrates along the Chi river floodplain in the central part, indicated in a red area on the map. In contrast, dark green regions on the map represent very low flood susceptibility, spreading on the high elevation regions of the northern, southwestern, and south-eastern part of the study area, shown in Figure 3.

3.4. Evaluation of flood susceptibility map
The flood susceptibility map has been evaluated for its accuracy and reliability by comparing it with a compiled historical flood data of the study area, using a statistical evaluation method of the confusion matrix and binary classification test. The compiled historical floods of the study area in 2005-2016 has been acquired from the Geo-Informatics and Space Technology Development Agency of Thailand (GISTDA) [31]. The accuracy of the map shows a correct determination of flood and non-flood area as compared with an observation data (in this case is a compiled flood area map from GISTDA). The map has an accuracy of 73% according to a binary classification test.

4. Discussion
Although the AHP method was successfully applied for the flood susceptibility mapping in the Chi River Basin, Mahasarakham, and Roi-Et Provinces of Thailand as confirmed by a good accuracy, there are remarks, uncertainties, and limitations that need to be addressed for the study.

The first discussable point is a selection of flood causative parameters for the AHP analysis, as it is an adjustment of the experts. There are various numbers of parameters included in different studies of flood susceptibility determination using AHP. The number of parameters of four has been applied in the various studies, such as the studies of [8,22,23]; six parameters in [11,14,15,32]; seven parameters in [17,33]. A different list of parameters for flood susceptibility assessment should be determined according to specific physical characteristics of each basin [17], as well as social and economic aspects. The flood-causative parameters in our study can be grouped into categories as follows; (1) hydrography parameters: distance from the river, and drainage density, (2) hydrological parameters: geomorphology, and elevation, (3) permeability parameters: geology, land use, and soil type.

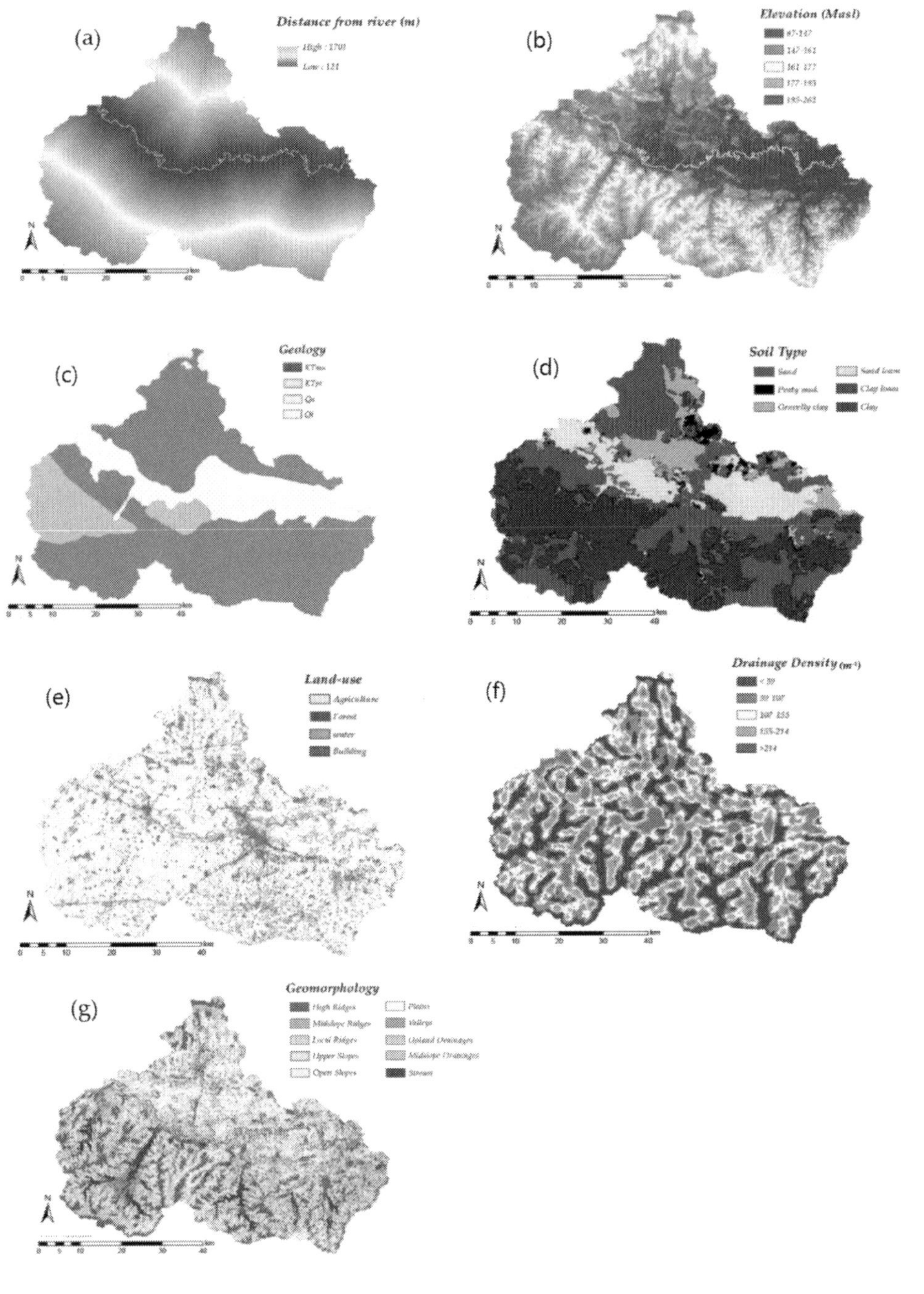

Figure 2. Flood-causative parameter thematic maps; (a) distance from river, (b) elevation, (c) geology, (d) soil type, (e) land-use, (f) drainage density, and (g) geomorphology.

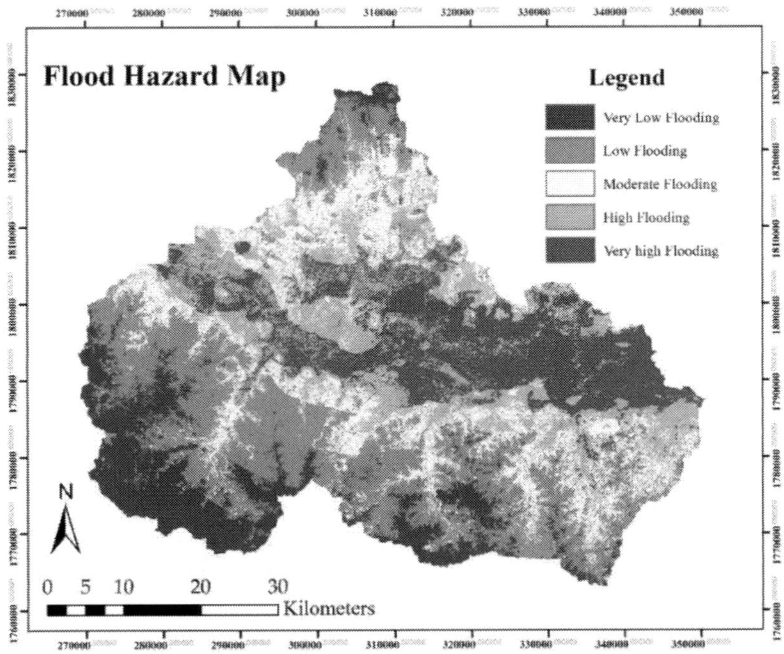

Figure 3. Flood susceptibility map of the Chi river basin, Mahasarakham and Roi-Et provinces; the red area indicates a very high flood susceptibility area, a dark green color indicates low flood susceptibility area.

In this research, seven flood-causative parameters are in categories (1) – (3), which involved only physical characteristics of the basin. All parameters were considered as static parameters which are not changeable or not like to change through time, reflecting physical flood susceptibility of the basin. Rainfall is a factor that has been excluded in the study since we considered it as a dynamic parameter which can change rapidly through time. This study determined a flood susceptibility spatially over the basin. However, the magnitude of flood disaster, i.e., flood depth; was not included in flood susceptibility levels. Thus, a combination of other various flood susceptibility assessments with the AHP is highly recommended for higher accuracy of flood risk studies. Final remark, the most effective flood susceptibility mapping can be generated by two factors, which are data and method. The AHP technique is a convenient and effective method for multi-criteria analysis. However, this flood susceptibility map can be improved if there are better data availability, as well as if other more effective methods can be evaluated as compared to the AHP.

5. Conclusions

The flood susceptibility evaluation using the Analytical Hierarchy Process (AHP) is successful in assessing the regions vulnerable to spatial floods in the Chi river basin, Mahasarakham and Roi-Et provinces, Thailand. The AHP is a cost and time-effective method. Seven hydrogeomorphologic parameters have been determined as the flood-causative parameters in the CRB; which are the distance from the river, drainage density, elevation, geology, soil type, land use, and geomorphology. The highest normalized weight in the study is a distance from the river, followed by elevation, geology, soil type, land use, drainage density, and geomorphology, respectively. The result reveals that 16.93 percentages of the entire study area have been identified as a very high flood zone in the central of the CsRB; along with 17.76 percentages of a high, 22.40 percentages of a moderate, 29.08 percentages of a low, and 13.83 percentages of a very low flood susceptibility regions. The fluvial flood is a dominant flood type in the study area because the weightiest parameter on flood susceptibility is a distance from the river. The flood susceptibility map has an accuracy of 73 percentages as compared to compiled previous flood maps from GISTDA by using the binary classification test. The AHP method is well-performed for the flood susceptibility mapping of the CRB, which can be utilized in other basins. However, an adjustment of parameters is recommended. The method can be further applied for other multi-criteria studies such as drought study, groundwater potential assessment, and so on.

References

[1] Al-Suhili R, Cullen C, Khanbilvardi R. An urban flash flood alert tool for megacities Application for Manhattan, New York City, USA. Hydrology. 2019;6(2).
[2] Das S. Geographic information system and AHP-based flood hazard zonation of Vaitarna basin, Maharashtra, India. Arabian Journal of Geosciences. 2018;11(19).
[3] Winsemius HC, Aerts Jeroen CJH, van Beek Ludovicus PH, Bierkens Marc FP, Bouwman A, Jongman B, et al. Global drivers of future river flood risk. Nature Climate Change. 2016;6(4):381-5.
[4] Wang Y, Li Z, Tang Z, Zeng G. A GIS-Based Spatial Multi-Criteria Approach for Flood Risk Assessment in the Dongting Lake Region, Hunan, Central China. Water Resources Management. 2011;25(13):3465-84.
[5] Yahaya S, Ahmad N, Abdalla R. Multicriteria analysis for flood vulnerable areas in Hadejia Jama'are River basin, Nigeria. European Journal of Scientific Research ISSN. 2010;42:1450-216.
[6] Hirabayashi Y, Mahendran R, Koirala S, Konoshima L, Yamazaki D, Watanabe S, et al. Global flood risk under climate change. Nature Climate Change. 2013;3(9):816-21.
[7] Wu C, Huang G, Yu H, Chen Z, Ma J. Impact of climate change on reservoir flood control in the upstream area of the Beijiang River basin, South China. Journal of Hydrometeorology. 2014;15(6):2203-18.
[8] Danumah JH, Odai SN, Saley BM, Szarzynski J, Thiel M, Kwaku A, et al. Flood risk assessment and mapping in Abidjan district using multi-criteria analysis (AHP) model and geoinformation techniques, (cote d'ivoire). Geoenvironmental Disasters. 2016;3(1):10.
[9] Hong H, Tsangaratos P, Ilia I, Liu J, Zhu AX, Chen W. Application of fuzzy weight of evidence and data mining techniques in construction of flood susceptibility map of Poyang County, China. Science of The Total Environment. 2018;625:575-88.
[10] A. Lagmay AMF, Racoma BA, Aracan KA, Alconis-Ayco J, Saddi IL. Disseminating near-real-time hazards information and flood maps in the Philippines through Web-GIS. Journal of Environmental Sciences. 2017;59:13-23.
[11] Gigovic L, Pamučar D, Bajić Z, Drobnjak S. Application of GIS-Interval Rough AHP Methodology for Flood Hazard Mapping in Urban Areas. Water. 2017;9:1-26.
[12] Forkuo EK. Flood Hazard Mapping using Aster Image data with GIS. Int J Geomat Geosci. 2011;1.
[13] Montz BE, Gruntfest E. Flash flood mitigation: Recommendations for research and applications. Environmental Hazards. 2002;4(1):15-22.
[14] Ouma YO, Tateishi R. Urban flood vulnerability and risk mapping using integrated multi-parametric AHP and GIS: Methodological overview and case study assessment. Water (Switzerland). 2014;6(6):1515-45.
[15] Kazakis N, Kougias I, Patsialis T. Assessment of flood hazard areas at a regional scale using an index-based approach and Analytical Hierarchy Process: Application in Rhodope–Evros region, Greece. Science of The Total Environment. 2015;538:555-63.
[16] Dano UL. Flash flood impact assessment in Jeddah city: An analytic hierarchy process approach. Hydrology. 2020;7(1).

[17] Vojtek M, Vojteková J. Flood Susceptibility Mapping on a National Scale in Slovakia Using the Analytical Hierarchy Process. Water. 2019;11(2):364.
[18] Seejata K, Yodying A, Wongthadam T, Mahavik N, Tantanee S, editors. Assessment of flood hazard areas using Analytical Hierarchy Process over the Lower Yom Basin, Sukhothai Province. Procedia Engineering; 2018.
[19] Kaliraj S, Chandrasekar N, Magesh NS. Identification of potential groundwater recharge zones in Vaigai upper basin, Tamil Nadu, using GIS-based analytical hierarchical process (AHP) technique. Arabian Journal of Geosciences. 2014;7(4):1385-401.
[20] Saaty TL. How to make a decision: The analytic hierarchy process. European Journal of Operational Research. 1990;48(1):9-26.
[21] Saaty TL. The Analytic Hierarchy Process. New York: McGraw-Hill; 1980.
[22] Rincón D, Khan UT, Armenakis C. Flood Risk Mapping Using GIS and Multi-Criteria Analysis: A Greater Toronto Area Case Study. Geosciences. 2018;8(8):275.
[23] Samanta S, Koloa C, Pal DK, Palsamanta B. Flood risk analysis in lower part of Markham river based on multi-criteria decision approach (MCDA). Hydrology. 2016;3(3).
[24] Komi K, Amisigo BA, Diekkrüger B. Integrated flood risk assessment of rural communities in the Oti River basin, West Africa. Hydrology. 2016;3(4).
[25] Chanlabut U, Gomontean B, Srifa A. Soil Organic Carbon Stocks across Hydrologic Schemes in Freshwater Wetlands of the Chi River Basin, Northeast Thailand. Wetlands. 2019.
[26] Arunyanart N, Limsiri C, Uchaipichat A. Flood hazards in the chi river basin, Thailand: Impact management of climate change. Applied Ecology and Environmental Research. 2017;15:841-61.
[27] DMR. Zoning for geology and geological resources management of Mahasarakham province. Bangkok: Department of Mineral Resources, Thailand; 2009.
[28] Olson DL. The Analytic Hierarchy Process. In: Olson DL, editor. Decision Aids for Selection Problems. New York, NY: Springer New York; 1996. p. 49-68.
[29] The Geologic Map of Mahasarakham Province [Internet]. Department of Mineral Resources, Thailand. 2007 [cited March 2019]. Available from: http://www.dmr.go.th/download/pdf/NorthEast/Mahasarakam.pdf.
[30] Soil Group Map of Mahasarakham province, Thailand [Internet]. Land Development Department of Thailand. 2019. Available from: http://gisinfo.ldd.go.th/cd_land_map.html?land_type=LAND_SOIL&province_id=044&amphur_id=&tambol_id=.
[31] Thailand Flood Monitoring System [Internet]. Geo-Informatics and Space Technology Development Agency of Thailand. 2019. Available from: http://flood.gistda.or.th/.
[32] Cabrera J, Lee HS. Impacts of Climate Change on Flood-Prone Areas in Davao Oriental, Philippines. Water. 2018;10.
[33] Elkhrachy I. Flash Flood Hazard Mapping Using Satellite Images and GIS Tools: A case study of Najran City, Kingdom of Saudi Arabia (KSA). The Egyptian Journal of Remote Sensing and Space Science. 2015;18(2):261-78.

Acknowledgments
This work was financially supported by the young researcher development project of Khon Kaen University.

REDUCTION OF CHLORIDE IN RAW WATER BY NANOFILTRATION AND REVERSE OSMOSIS

Suppakit Panwattanakul[1], Kasipong Kasikun[1], Charongpun Musikavong[1,2,3], Suthatip Sinyoung[1,3*]

[1]Department of Civil Engineering, Faculty of Engineering, Prince of Songkla University, Hat Yai, Songkhla 90110, Thailand
[2] Environmental Assessment and Technology for Hazardous Waste Management Research Center, Faculty of Environmental Management, Prince of Songkla University, Songkhla 90110, Thailand
[3]Center of Excellence on Hazardous Substance, Management (HSM), Bangkok, Thailand

E-mail: suthatip.sin@gmail.com

Abstract. This research is aimed at determining the reduction of chloride in the raw water of a water supply plant using nanofiltration and reverse osmosis. Raw water was prepared by adding NaCl at the concentrations of 300 and 600 mg/L. This work used a PACl at 15 mg/L as the pre-treatment for removing turbidity and DOC. Coagulated water was filtered with GF/C filter as the filtration process for minimizing membrane blockage. At chloride concentrations of 300 and 600 mg/L, the flux values were in the range of 66.7-70.0 and 62.8–65.0 L/m^2-h for NF-270 and 17.2–20.6 and 18.3-20.8 L/m^2-h for TW-30, respectively. The performance capabilities of NF-270 and TW-30 for reducing chloride were significantly different. NF-270 and TW-30 membranes were able to reduce chloride by 67 % and 92 % at the added dose of 300 mg/L, respectively. At the added amount of 600 mg/L, NF-270 and TW-30 membranes could reduce chloride by 63 % and 85 %, respectively. In the case of chloride in raw water at a concentration lower than 600 mg/L, the nanofiltration is suitable for removing chloride in regards to the low operation costs compared with the reverse osmosis. Nanofiltration could reduce chloride to lower concentration levels than the water supply standards of the Provincial Waterworks Authority.

Keywords: Chloride, Nanofiltration, Reverse osmosis, Surface water, Water supply

1. Introduction

Chloride naturally occurs as the major anion found in all-natural waters. It is the most abundant ion in seawater, with a concentration greater than 19,000 mg/L [1]. Salt or sodium chloride (NaCl) intake can lead to the condition known as salt poisoning, salt toxicity, hypernatremia, or water deprivation–sodium ion intoxication. Increased chloride concentrations in some environments have killed off native vegetation and allowed invasive salt-tolerant species to thrive [2].Chloride is corrosive to steel, thus it may corrode pipes in water treatment and industrial plants. Because it imparts a salty taste to water and is corrosive, elevated chloride levels in drinking water supplies can lead to increased treatment costs. [3,4]. The U.S. Environmental Protection Agency (USEPA) recommends a chronic criterion for aquatic life of a four-day average chloride concentration of 230 mg/L with an occurrence interval of once every three years [5]. The recommended acute criterion is 860 mg/L, which relates to a one-hour average concentration with a recurrence interval of less than once every three years [5]. In Thailand, the water supply standard for chloride concentration of the Provincial Waterworks Authority

(PWA) for consumption is 250 mg/L, but in areas of the world with water scarcity, drinking water can have considerably greater concentrations of chloride [6]. The main methods of desalination or chloride removal include thermal processes, such as distillation, electrodialysis, and membrane processes, such as reverse osmosis (RO). RO is the finest of all membrane filtration systems, with extremely small pores capable of removing particles as small as 0.1 nm. It is a process in which both dissolved organic matter, salts, and bacteria are removed from water using a different mechanism from distillation, ion exchange or activated carbon processes [7].

Chloride removal from seawater by the reverse osmosis process indicates the efficiency and applicability of RO technology [8,9]. However, due to the fine membrane construction, reverse osmosis not only removes contaminants present in the raw water, it also removes many of the desirable minerals from the raw water. Some researchers have been studying the long-term health effects of drinking demineralized water [10-12]. The NF membrane is a relatively recent technology that was developed mainly for potable water generation. NF delivers coarser filtration than RO, with the ability to remove particles as small as 0.002 to 0.005 μm in diameter. NF can remove harmful contaminants, such as pesticide compounds and organic macromolecules [13]. Nowadays, nanofiltration membranes have been used in the pretreatment of brackish water desalination worldwide due to their remarkable abilities to selectively reject different dissolved salts and provide high water flux at low operating costs and low energy consumption [14].

The U-Tapao Canal is the raw water source that produces the water supply for utility in the area of Songkhla province, Thailand. This canal receives wastewater and treated wastewater from community, agricultural, and industrial activities [15-17]. The Songkhla province area is close to the sea, so during the dry season, the sea level rises above the average level and the freshwater is contaminated by seawater causing chloride content ranges from 50 to 600 mg/l [18]. This research, therefore, focuses on investigating the potential of the NF membrane and RO process to remove chloride from the raw water of the water treatment plant in Songkhla province, Thailand. The challenge is in minimizing the operational costs and energy consumption as well as improving the quality of the water supply.

2. Material and methods

2.1 Materials and chemicals

Raw water samples were collected from the pumping station of the water treatment plant (WTP), Songkhla, Thailand, that is located adjacent to the U-Tapao Canal. The raw water samples were collected in November 2019. All samples were stored below 4°C until analysis and experiments were conducted. For the study of chloride removal, raw water was prepared by adding NaCl (99.99%) at the concentrations of 300 and 600 mg/L. In regards to membrane technology, this work focused on the mass transfer in the crossflow filtration process using two types of membrane which are NF-270 for nanofiltration membrane and TW-30 for reverse osmosis. All the properties of the membranes according to the manufacturers are shown in Table 1.

Table 1. Properties of nanofiltration (NF-270) and reverse osmosis (TW-30) membranes in this study

Properties	NF-270	TW-30
Manufacturer	DOW/Filmtec	DOW/Filmtec
Water permeability (L m^{-2} h^{-1} bar^{-1})	13.5	2
Molecular weight cutoff (Daltons)	170 to 300	100
Contact angle (°)	55.0	Not available
Zeta potential (mV)	-19	-20.6

2.2 Methods and analysis

2.2.1 Method

This study used polyaluminium chloride (PACl) at 15 mg/L as the pre-treatment for removing turbidity, suspended solids, and dissolved organic carbon (DOC). PACl was used in the coagulation experiments by the use of a jar-test apparatus (JLT leaching test jar-test - VELP scientific). Samples were placed in the jar-test apparatus and coagulants were injected into each of the samples .The jar-test apparatus was set on a rapid mixing speed of 100 rpm for 1 minute, and then on a slow mixing speed of 60 rpm for 8 minutes, 40 rpm for 8 minutes, and 25 rpm for 5 minutes, respectively.

The pH was controlled in the range of 6.8-7.2 by the addition of sodium hydroxide (NaOH) and sulfuric acid (H_2SO_4). After a 30-min rest period, the supernatant was measured for its turbidity and color. Coagulated water was filtered through a GF/C filter as the filtration process for minimizing membrane blockage. Then, the chloride was removed from the filtered water by the use of NF-270 and TW-30 membranes, operating at a 4 bar pressure and this was carried out for 8 hours. All experiments were conducted in triplicate.

2.2.2 Analysis

The physical and chemicals properties of the raw water were analysed following standard methods. The pH was directly measured using a HACH pH meter (accuracy of ±0.01pH unit). Alkalinity was determined following the standard method 2320 B (Titration method). Turbidity and color were measured by a turbidity meter (HACH, 2100 N) and a Genesys 10S UV/VIS spectrophotometer (Thermo Electron Corp. Madison, WI, USA). The chloride content was analyzed by the argentometric method following the standard methods for the examination of water and wastewater, APHA, AWWA & WEF, 23rd Edition, 2017[19]. DOC in water samples was determined by a combustion method (Standard Method 5310D) using a total organic carbon analyzer (TOC-V CSN, Shimadzu, Japan) [20].

3. Results and Discussion

3.1 Characteristics of raw water supply

The characteristics of the raw water from the pumping station at the U-Tapao canal in the rainy season are presented in Table 2. The pH value of raw water was close to neutral (6.8 ± 0.2). It can be stated that the pH of the raw water is suitable for coagulation process. Color and DOC values of 39 ± 3 Pt-Co and 3.9 ± 0.1 mg.C/L were determined for this sample. Turbidity is a significant parameter for representing the quality of a raw water supply. In this study, the turbidity was 34 ± 2 NTU. This was higher than 5 NTU and not adequately clear enough to allow for distribution to the public. The conventional process, which commonly consists of coagulation and filtration, is necessary in this case for removing turbidity as pre-treatment for minimizing membrane blockage before using membrane technology. Color, turbidity, and DOC in this work were higher than Provincial Waterworks Authority (PWA) standard [6] and results of other researchers [15,22]. This indicated that the characteristics of raw water in the canal varied according to the time of sampling. In regards to the characteristics of raw water, the water treatment plant must reduce the turbidity and color to be lower than the drinking water standards of 4 NTU and 15 Pt-Co, respectively [16]. Iron, manganese, and chloride content were detected in the raw water at 3.3 ± 0.2, 1.1 ± 0.6, and 19.6 ± 0.8 mg/L, respectively. This study is focused on chloride removal from raw water supply by membrane technology, therefore, raw water was prepared by adding NaCl at the concentration of 300 and 600 mg/L.

Table 2. Characterization of raw water

Parameters	Units	Results
pH	-	6.8 ± 0.2
Color	Pt-Co	39 ± 3
Turbidity	NTU	34 ± 2
DOC	mg C/L	3.9 ± 0.1
Alkalinity	mg/L	22 ± 0.2
Chloride	mg/L	19 ± 0.8
Iron	mg/L	3.3 ± 0.2
Manganese	mg/L	1.1 ± 0.6
Total Hardness	mg/L as $CaCO_3$	190 ± 5

3.2 Effect of chloride content on the flux of NF-270 and TW-30 membranes

This study looked at flux as a function of time using DI water as a feed solution of NF-270 and TW-30 membranes at a pressure of 4 bars. The DI water flux of NF-270 and TW-30 membranes were in the range of 69.4 to 73.9 L/m^2-h and 29.4 to 33.9 L/m^2-h., respectively. Then, the flux values of raw water containing various contents of chloride using NF-270 and TW-30 membranes were tested, and are shown in Figure 1 and 2. DI water (DI), raw water (RW), raw water and chloride at 300 and 600 mg/l (RW+Cl300 and RW+Cl600) were used as the feed solution.

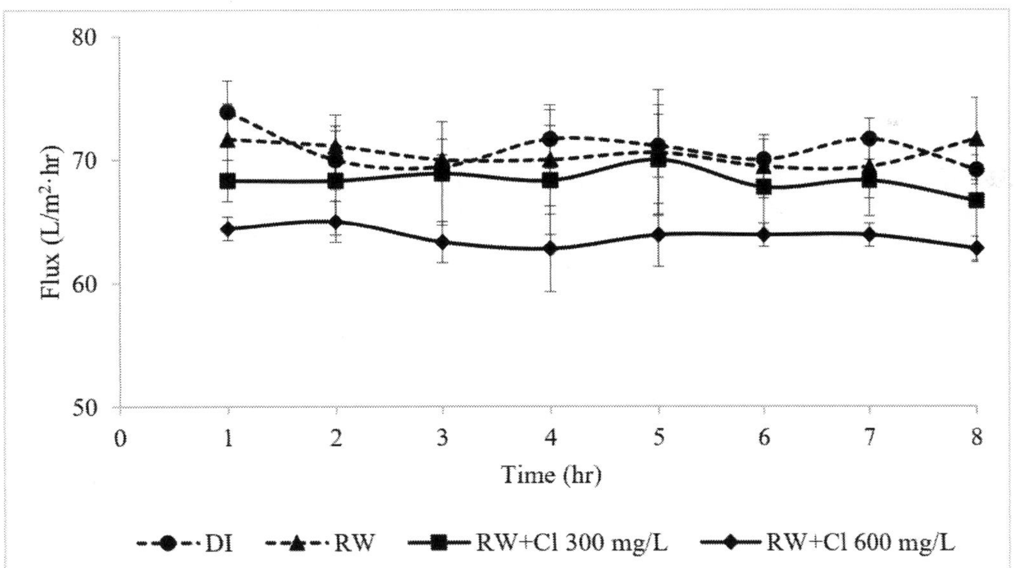

Figure 1. Flux as a function of time using DI water, raw water, and raw water with added NaCl as a feed solution of NF-270 membrane

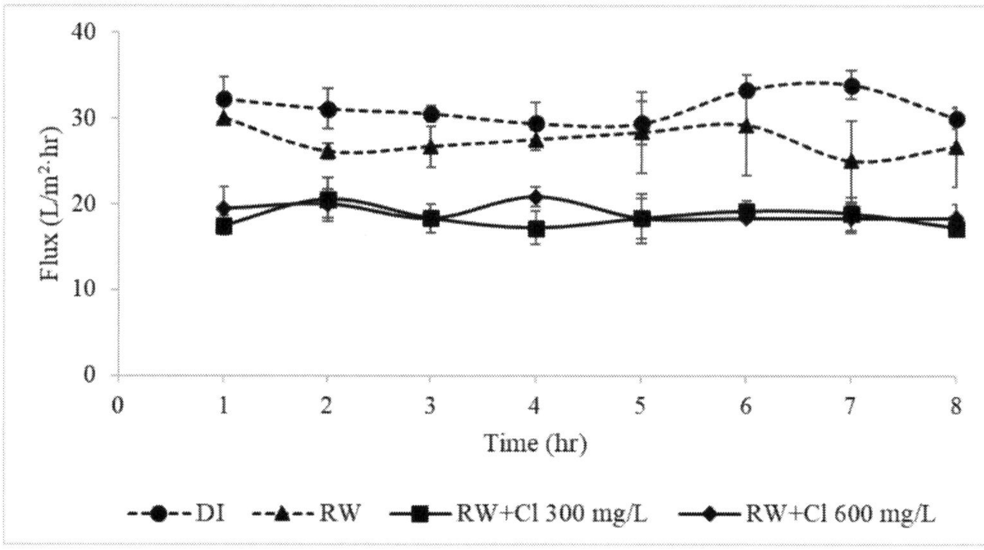

Figure 2. Flux as a function of time using DI water, raw water, and raw water with added NaCl as a feed solution of TW-30 membrane

In the case of NF-270, and raw water containing chloride content at 19, 300, and 600 mg/L, the flux values were in the range of 69.4 to 71.7, 68.3 to 70, and 62.8 to 65 L/m^2-h, respectively. When testing using the TW-30 membrane, the flux values were in the range of 25 to 30, 17.2 to 20.5 and 18.3 to 20.8 L/m^2-h, respectively. This indicated that the flux value of both systems is directly related to the initial content of chloride. This can be attributed to the raw water containing a high content of chloride which caused clogging. However, the results indicated that the flux value of the nanofiltration system was higher than the reverse osmosis system.

3.3 Reduction of chloride and efficiency of NF-270 and TW-30 membranes

Figure 3 shows the chloride content in raw water after using the NF-270 and TW-30 membranes at a pressure of 4 bars for 8 hours. For the NF-270 membrane, and initial concentration of chloride at 19, 300, and 600 mg/L, the chloride content in the raw water was 7.7, 97, and 214 mg/L, respectively. For the TW-30 membrane, and the initial concentration of chloride at 19, 300, and 600 mg/L, the chloride content in raw water was 6.3, 25, and 90 mg/L, respectively.

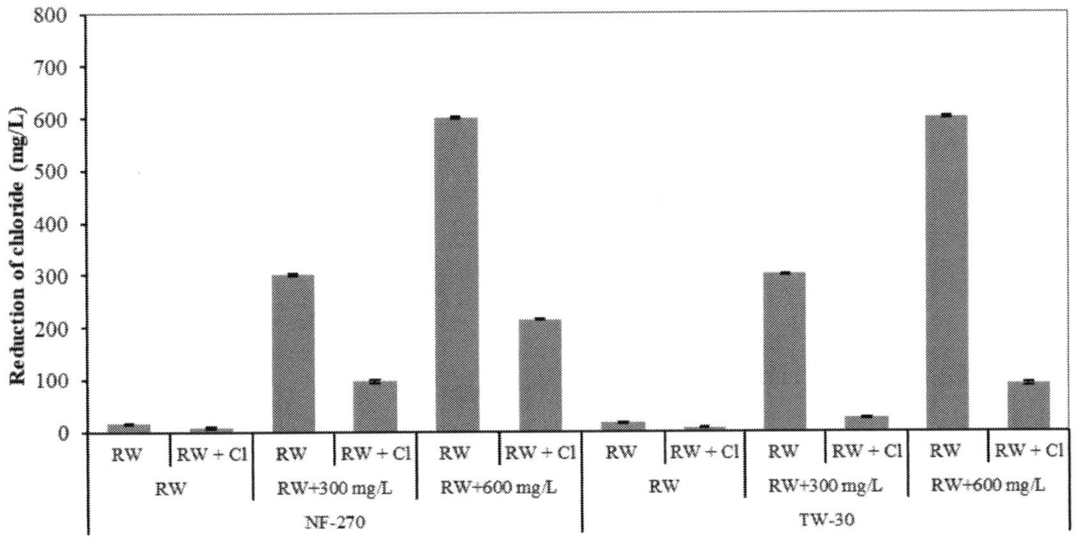

Figure 3. *Reduction of chloride at various initial concentrations using NF-270 and TW-30 membranes*

All the results showed that the chloride content in the raw water was lower 250 mg/L after using the NF-270 and TW-30 membranes which is lower than the water supply standard of the Provincial Waterworks Authority, Thailand for consumption [6].

In the comparison of chloride removal efficiency between the NF-270 and TW-30 membranes, the results indicated that the initial chloride concentrations at 300 and 600 mg/L, the chloride removal efficiency of the NF-270 was 67 and 63%, respectively. For TW-30, and the initial chloride concentrations at 300 and 600 mg/L, the chloride removal efficiency was 92 and 84%, respectively. The efficiency of chloride removal of the TW-30 is higher than the NF-270 membrane and the performance capability of TW-30 and NF-270 for reducing chloride was significantly different. Comparing the results obtained from this study and other research (Table 3), it was found that the nanofiltration membrane (NF-270) had the potential to eliminate chloride from water. The flux and efficiency of a nanofiltration membrane depended on the type and pressure in operation.

Table 3. Flux and Efficiency of NF270 membrane at various water sources and conditions

Source of water	Type of membrane	Initial Chloride Conc. (mg/L)	Flux (LMH)	% Removal	Pressure (bar)	References
Surface water	NF270	**300**	**71.7**	**66.9**	**4**	**This work**
		600	**70.0**	**62.5**	**4**	
Synthetic water	NF270	1000	113	59.1	10	Shin et al., 2020 [23]
	RO		170	99.8		
Pure water	NF270	1000	76	47.1	6	Chu et al., 2020 [24]
Seawater	NF270	5000	50.2	2.3-10.9	2	Hilal et al., 2005 [25]
		10000	66.5	11-29	9	
Seawater	NF270	60000	80	19-25	5	Istirokhatun et al., 2018 [26]

For the application purpose, the Tha Chin River is the main surface water source for water supply plant in Thailand. There are 9 monitoring station and reported that salinity did not change from 2012 to 2014, the while salinity increase significantly between 2015 to 2016.[27] This may affect the growth of some economic plants and the raw water of the water supply plants. Therefore, the results of chloride treatment by using nanomembranes obtained from this research and previous research confirms that the nanofiltration (NF-270) is suitable for removing chloride in the Tha Chin River and other water surface sources of Thailand.

4. Conclusion

This research shows that it is feasible to use nanofiltration membranes (NF-270) as reverse osmosis (TW-30) to reduce the chloride content from a raw water supply and the results can be concluded as follows:

- At chloride concentrations of 300 and 600 mg/L, the flux values were in the range of 66.7-70.0 and 62.8–65.0 L/m^2-h for NF-270 and 17.2–20.6 and 18.3-20.8 L/m^2-h for TW-30, respectively.
- NF-270 and TW-30 membranes were able to reduce chloride by 68 % and 92 % at the added dose of 300 mg/L, respectively.
- NF-270 and TW-30 membranes were able to reduce chloride by 63 % and 85 %, at the added dose of 300 mg/L, respectively.
- In the case of chloride concentrations lower than 600 mg/ L in raw water, nanofiltration is suitable for removing chloride.

5 .Acknowledgement

This research was funded by the Center of Excellence on Hazardous Substance Management)HSM(, Thailand under contract number HSM-PJ-CT-18-20. Also, the authors would like to thank the Engineering Faculty, Prince of Songkla University, Hatyai Campus for their assistance with the equipment.

References

[1] Stumm W and Morgan J J 1996 Aquatic chemistry chemical equilibria and rates in natural waters 3rd ed John Wiley & Sons Inc. New York p 1022

[2] Panno S V Nuzzo V Cartwright K Hensel B R and Krapac I G 1999 Changes to the chemical composition of groundwater in a fen-wetland complex caused by urban development. Wetlands 19(1): pp 236–245

[3] Environment Canada 2001 Priority Substances List Assessment Report: Road Salts. Health Canada http://www.ec.gc.ca/Substances/ese/eng/psap/final/roadsalts.cfm.

[4] Kaushal S S Groffman P M Likens G E Belt K T Stack W P Kelly V R Band L E and Fisher G T 2005 Increased salinization of fresh water in the northeastern United States. Proceedings of the National Academy of Sciences of the United States of America 102(38): pp 13517–13520

[5] USEPA 1988 Ambient water quality criteria for chloride-1988 440/5–88–001 U S Environmental Protection Agency Office of Water Regulations and Standards.

[6] Provincial Waterworks Authority (PWA) water supply standard 2011 PWA branch statistics. Available at: (accessed September 2020) https://en.pwa.co.th/download/pwastandard50-1.pdf

[7] Warsinger D M Nejati S Juybari H F 2020 Energy Efficiency Metrics in Membrane Distillation, book: Advances in Water Desalination Technologies World Scientific Publishing Company pp.272 – 282

[8] Garud R M Kore S V Kore V S and Kulkarni G S 2011 A Short Review on Process and Applications of Reverse Osmosis Universal Journal of Environmental Research and Technology Vol 1 Issue 3 pp 233-238

[9] Djebedjian B Gad H E Khaled I and About Rayan M 2020 Experimental and Analytical Study of A Reverse Osmosis Desalination Plant Mansoura Engineering Journal 34 (2) pp 71-92
[10] Schumann K Elsenhans B and Reichl FX 1993 Does intake of highly demineralized water damage the rat gastrointestinal tract Vet Hum Toxicol Vol 35 pp 28-31
[11] Rakhmanin Y A Mikhailova R I Filippova A V 1989 On some aspects of biological effects of distilled water. (In Russian.) Gig Sanit Vol 3 pp 92-93
[12] Kozisek F 2005 Health Risks from Drinking Demineralised Water National Institute of Public Health, Czech Republic. Chapter 12 in Water, Sanitation and Health Protection and the Human Environment; W.H.O
[13] Yin W Li X Suwarno S R Cornelissen E RChong T H 2019 Fouling behavior of isolated dissolved organic fractions from seawater in reverse osmosis (RO) desalination process Water Res. 159 p 385.
[14] Schaep J Van der Bruggen B Uytterhoeven S Croux R Vandecasteele C and Wilms D E 1998. Removal of hardness from groundwater by Nanofiltration Desalination 119(1-3) pp 295-301
[15] Musikavong C and Wattanachira S 2013 Identification of dissolved organic matter in raw water supply from reservoirs and canals as precursors to trihalomethanes formation, *Journal of Environmental Science and Health, Part A* 48 pp 760-771
[16] Phatthalung W N Musikavong C and Suttinun O 2016 The presence of aliphatic and aromatic amines in reservoir and canal water as precursors to disinfection by-products, Journal of Environmental Science and Health, Part A 51 pp 900-913
[17] Musikavong C Srimuang K Suksaroj T T and Suksaroj C 2017 Formation of Trihalomethanes of Dissolved Organic Matter Fractions in Reservoir and Canal Waters Journal of Environmental Science and Health Part A 51 pp 782–791
[18]Groundwater Resources Conservation and Recovery 2019 Department of Groundwater Resources, Ministry of Natural Resources and Environment. State Groundwater Situation Report of Thailand
[19] APHA 2017 Standard methods for the examination of water and wastewater APHA, AWWA & WEF 23rd Edition Argentometric Method
[20]APHA 1998 Standard Methods for the Examination of Water and Wastewater 20th Ed.; American Public Health Association AWWA and WEF: Washington DC
[21]Tongchang P Kumsuvan J Phatthalung W N Suksaroj W Wongrueng C and Musikavong C 2018 Reduction by Enhanced Coagulation of Dissolved Organic Nitrogen as a Precursor of N-Nitrosodimethylamine. Journal of Environmental Science and Health, PartA. 53 pp 583–593
[22] Sinyoung S Buadang R Songkaew W and Musikavong C 2020 Reduction of Chloride Contained in Groundwater by Membrane Technology Conference on Environmental Engineering, Science and Management, Environmental Engineering Association of Thailand (EEAT) , Chiang Rai, Thailand, pp.546-553
[23] Shin M G Kwon S J Park H Park Y I & Lee J H 2020 High-performance and acid-resistant nanofiltration membranes prepared by solvent activation on polyamide reverse osmosis membranes. Journal of Membrane Science 595 pp 117590
[24] Chu C H Wang C Xiao H F Wang Q Yang W J Liu N and Sun S P 2020 Separation of ions with equivalent and similar molecular weights by nanofiltration: sodium chloride and sodium acetate as an example. Separation and Purification Technology pp 117199
[25] Hilal N Al-Zoubi H Mohammad A W and Darwish N A 2005 Nanofiltration of highly concentrated salt solutions up to seawater salinity. Desalination, 184(1-3) pp 315–326
[26] Istirokhatun T Dewi M N Ilma H I and Susanto H 2018 Separation of antiscalants from reverse osmosis concentrates using nanofiltration Desalination 429 pp 105-110.
[27] Jirawat P and Ravipat L 2017 Study of Water Quality Changes of Lower Tha Chin River in Dry Season for Agriculture The 18th TSAE National Conference and The 10th TSAE International Conference: TSAE 2017 "Agricultural Engineering for the Digital Age", Bangkok, Thailand, pp.346-349

Comparison of machine learning algorithms for mortality prediction in intensive care patients on multi-center critical care databases

Thanakron Na Pattalung[1], **Sitthichok Chaichulee**[1,#]
[1] Institute of Biomedical Engineering, Department of Biomedical Sciences and Biomedical Engineering, Faculty of Medicine, Prince of Songkla University, Songkhla 90110, Thailand.

E-mail address: sitthichok.c@psu.ac.th

Abstract. Current scoring systems for mortality prediction in intensive care patients are usually applied once after 24 hours of admission, as all parameters needed for scoring are not yet available. In addition, several parameters are dynamic and may change according to patient conditions. It is hypothesized that mortality prediction should be made at the earliest when relevant information becomes available and continuously during patient stay. This study focuses on the development of algorithms for mortality prediction from vital signs and laboratory results based on the data from three recent critical care databases, i.e. the eICU collaborative research database, the Multiparameter Intelligent Monitoring in Intensive Care III (MIMIC-III) database, and the MIMIC-IV database. We employed logistic regression, *k*-nearest neighbours, neural networks and tree-based classifiers for such problem. Our models had the area under the receiver operating characteristic curve (AUROC) ranging from 0.67 – 0.95. Reliable mortality prediction can be made as early as the first 4 hours after ICU admission. We provided comprehensive analysis on different time frames used for prediction, models trained with top attributes, models trained with data combination, and missing values. Our results provide guidelines and benchmarks for the development of such algorithm in local narratives.

1. Introduction

Patients are admitted into the intensive care unit (ICU) generally following severe illness, accidents, or surgery. They require special medical attention and are at high risk of clinical deterioration and death. The patient's risk of death can be determined using demographic information, the severity of patient condition, vital sign measurements, and laboratory results [1]–[4]. Mortality prediction plays an important role not only for identifying high-risk patients and prioritizing patient care but also for administrative purposes, such as evaluating the performance of ICUs across health systems [5].

Conventional scoring systems for mortality prediction usually applied once after 24 hours of ICU admission as they require the aggregation of various clinical parameters in which many measurements are not yet available during the first hours of admission. The availability of clinical data stored in the modern hospital information system (HIS) allows the development of modern predictive tools for estimating patient mortality using machine learning. The tools could leverage the use of data for identifying patterns or relationship between clinical data and patient outcomes. They could make possible the early prediction of patient mortality that could benefit the effectiveness of patient care. Recently, the Massachusetts Institute of Technology (MIT) Laboratory for Computational Physiology released several public critical care databases [6]–[8], promoting the development of machine learning algorithms on high-volume datasets. The algorithms and techniques developed on large public datasets could be used as a blueprint for developing in-house decision support systems.

This study focuses on the comparison of different machine learning algorithms for mortality prediction on three different critical care databases, namely, the eICU collaborative research database [6], the Multiparameter Intelligent Monitoring in Intensive Care III (MIMIC-III) database [7], and the MIMIC-IV database [8]. To this end, our contributions are three-fold. Firstly, we present detailed benchmarking results using the data obtained from the first 4, 8, 12, 16, 20, and 24 hours following ICU admission. We show that the data obtained during the first 4 hours are enough to make the prediction of patient mortality with clinically acceptable performance. Secondly, we compare the availability of each clinical variables in different time frames across different databases. We found that the high frequency of recorded vital signs in the eICU database could contribute to the higher performance of such algorithms developed on the eICU database, compared to those developed on the MIMIC-III and MIMIC-IV databases. The frequency of recorded vital signs in the HIS, hence, may be one of the key contributors for the improvement of the algorithms. Finally, we employ the same preprocessing pipeline for all the databases in order to standardize the data to be in the same format prior to comparison. We, therefore, present benchmarking results, separately for each database, on the models trained with all the data combined. This is different from other studies.

Our paper is organized as follows. Section 2 presents related work regarding mortality prediction using both conventional scoring systems and machine learning algorithms. Section 3 describes the critical care databases used in our study. Section 4 explains methods for training and evaluating algorithms. Section 5 details the comparison of each algorithm of each database. Section 6 discusses the results and provides suggestions for the development of algorithms in local narratives. Section 7 concludes the contributions made by this study.

2. Related work

Predicting patient's clinical outcomes is an important yet challenging topic for research in critical care. Studies have shown that abnormalities in physiological observations are present in patients prior to their major clinical events such as infection, congestive heart failure, cardiac arrest and death. Eighty-four percent of patients had documented signs of physiological abnormalities within eight hours prior physiological derangement events [9]. Even though abnormalities in biochemistry results were shown to be inconsistent, abnormalities in vital signs were consistently observed. Several clinical scoring systems have been introduced to the ICU for predicting patient's risk of dying based on studying aggregated data from large cohorts of patients. In this section, we reviewed related work regarding mortality prediction: conventional scoring systems, machine learning algorithms, and challenges in processing critical care data.

2.1. Conventional scoring systems for mortality prediction

Various scoring systems has been introduced for mortality prediction in the ICU, for example: the Acute Physiology and Chronic Health Evaluation (APACHE) scores [1], [2]; the Simplified Acute Physiology Score (SAPS) [4]; and the Sequential Organ Failure Assessment (SOFA) score [3]. These clinical scoring systems are mostly based on patient demographics, physiological and biochemistry variables, obtained during the first 24 hours of ICU admission. They relied on a panel of clinical and statistical experts while developing such scoring systems. Each score has its specific purpose and been seen as complementary rather than mutually exclusive [5]. The APACHE [1] and SAPS [4] scores, for example, assess the severity of illness which has an impact on patient's mortality, while the SOFA score [3] provides an assessment of organ derangements. They obtained, in their original articles, the areas under the receiver operator characteristic curves (AUROC) between 0.68 and 0.88 [1]–[4].

2.2. *Machine learning techniques for mortality prediction*
Information stored in the HIS can be utilized through various techniques to observe the trends of disease progression in order to plan for prevention and intervention and to increase the effectiveness of treatment [10]. Machine learning is a tool to find a relationship of between input and output data. Many machine learning algorithms have been recently developed for predicting patient mortality following ICU admission based on large public critical care databases, such as MIMIC-II/III databases [7], [11]. The availability of data in the HIS together with algorithms developed could made possible the automated and real-time prediction of patient mortality in the ICU.

Recently, Awad *et al.* [12] proposed the Early Mortality Prediction for Intensive Care Unit (EMICU) patients, developed on the MIMIC II database [11], in which patient mortality can be predicted as early as 6 hours after ICU admission with an AUROC of 0.90 using the Random Forests algorithm [13]. Harutyunyan *et al.* [14] benchmarked multi-task neural networks algorithms for predicting clinical outcomes on the MIMIC III database [7] and obtained an AUROC of 0.87 for mortality prediction using their multi-task long short-term memory algorithm (LSTM). Purushotham *et al.* [15] benchmarked multiple algorithms for mortality prediction on the MIMIC III database [7] and obtained an AUROC of 0.94 on 24-hour data following ICU admission using the multi-modal deep Boltzmann machines algorithm by treating temporal features and non-temporal features separately. El-Rashidy *et al.* [16] proposed an ensemble stacking model for early mortality prediction with an AUROC of 0.93 for the data obtained during the first 6 hours after ICU admission.

Most studies demonstrated their algorithms performed better than the traditional scoring systems used in the ICU, mainly on the MIMIC III database [7].

2.3. *Techniques for processing critical care data*
Many challenges are associated with the processing of clinical care data such as missing values, periodicity, variable selection, standardization, and class imbalance. Recently, Wang *et al.* [17] proposed the MIMIC-Extract framework for preparing data from the MIMIC-III database [7] for developing machine learning algorithms. They also released clinical taxonomy which could help data aggregation and standardization. Tang *et al.* [18] proposed the FIDDLE framework as a data preprocessing pipeline, evaluating on the eICU [6] and MIMIC-III [7] databases. This could further help the processing and standardization of data across different datasets possible. These frameworks could foster the development of machine learning algorithms on critical care data. Currently, at the time of manuscript preparation, no framework has been specifically designed for preprocessing the MIMIC-IV database [8].

3. Datasets
This study involves three public critical care databases: the eICU Collaborative Research Database [6], the MIMIC-III Critical Care Database [7], and the MIMIC-IV Critical Care Database [8].

3.1. *eICU Collaborative Research Database v2.0*
The eICU Collaborative Research Database [6], released in 2018, is a public critical care database provided by the MIT Laboratory for Computational Physiology in partnership with Philips Healthcare. The eICU contains deidentified data from over 200,000 ICU admissions from 2014 to 2015 monitored by eICU Programs throughout the United States. The database includes vital sign measurements, biochemistry measurements, patient diagnosis, treatment information, care plan documents, severity-of-illness measures, etc.

3.2. *MIMIC-III Critical Care Database v.1.4*
The MIMIC-III Critical Care Database [7], released in 2016, is a public critical care database obtained from patients admitted to ICUs at the Beth Israel Deaconess Medical Center (BIDMC) in Boston, Massachusetts between 2001 and 2012. It contains deidentified data from over 38,597 distinct adult

patients with 49,785 hospital admissions. The database has patient demographics, vital sign measurements, laboratory tests, doctor notes, clinical events, etc.

3.3. MIMIC-IV Critical Care Database v0.4

The MIMIC-IV Critical Care Database [8], released in 2020, a public critical care database covering over 200,000 emergency department (ED) stays and 60,000 ICU stays obtained from hospitalized patients from BIDMC between 2008 and 2019. The database contains the clinical data prior to ICU admission as 65% of patients admitting to the ICU was first seen in the ED. For critical care data, the MIMIC-IV database has similar structure to the MIMIC-III database.

4. Methods

This section describes the steps we employed for cohort selection, variable selection, data cleansing and standardization, data aggregation, data imputation, training machine learning algorithms and evaluating such algorithms. The first part of these steps was designed to transform the data to be in the same format such that the comparison of different algorithms and the stacking of data from different databases were made possible.

4.1. Cohort selection

We applied the same criteria to all the databases for building our study cohorts. We included adult patients who were firstly admitted to the ICU for more than 24 hours into our study. We excluded patients without recorded vital signs or laboratory measurements. For the eICU database, we used the unitdischargestatus variable in the patients table as an indicator whether patients were dead or survived their first ICU stay. For the MIMIC-III and MIMIC-IV databases, the death of patients was indicated in the deathtime and hospital_expire_flag variables in the admission table. Our criteria for cohort selection is in consistent with the studies by [12], [16]. Patients who were dead during their first ICU stay were set as our experimental (or positive) group. Table 1 summarizes the demographics of our study cohorts. Patients who were not dead during their hospital stays were set as our control (or negative) group. We performed random under-sampling for the control group to have its demographics match that of the experiment group during the evaluation of machine learning algorithms.

Table 1. Cohort demographics for the eICU, MIMIC III, and MIMIC IV databases

Demographic	**Dataset**		
	eICU	**MIMIC III**	**MIMIC IV**
Number of patients #	5,519	3,548	4,103
Gender			
Male	3,053	1,881	2,202
Female	2,464	1,667	1,901
Unknown	2	0	0
Ethnicity			
Caucasian	4,300	2,414	2,474
African American	577	195	300
Hispanic	167	70	101
Asian	106	79	124
Others/Unknown	369	790	1,104

Our study cohorts contain adult patients who were admitted to the ICU for more than 24 hours and were dead during their first ICU stay.

4.2. *Variable selection*

Vital sign and laboratory variables selected for the development of algorithms were based on the review of conventional clinical scoring systems [1]–[4] that are relevant to mortality prediction and the consultation with clinical care exports. We chose only the variables presenting in all the databases. Table 2 lists all vital sign variables (7 variables) and laboratory variables (19 variables).

Table 2. Vital sign and laboratory variables obtained from the eICU, MIMIC III, and MIMIC IV databases for benchmarking machine learning algorithms

Variables	Dataset *			Measurement unit
	eICU	MIMIC-III	MIMIC-IV	
Vital signs				
Heart Rate	V	C	C	beats per minute
Respiratory Rate	V	C	C	inspirations per minute
Peripheral O2 saturation	V	C	C	percent
Blood Pressure mean	V	C	C	millimeters of mercury
Blood Pressure systolic	V	C	C	
Blood Pressure diastolic	V	C	C	
Temperature	V	C	C	Celsius
Laboratory results				
Bilirubin	L	L/C	L/C	milligrams per deciliter
Bicarbonate (HCO3)	L	L/C	L/C	milliequivalents per liter
Calcium	L	L/C	L/C	milligrams per deciliter
Chloride	L	L	L	milliequivalents per liter
Creatinine	L	L	L	milligrams per deciliter
Glucose	L	L	L	
Hemoglobin	L	L	L	grams per deciliter
Hematocrit	L	L	L	percent
O2 Saturation	L	L	L	
Inspired O2 Fraction (FiO2)	L	C	C	
Lactate	L	L/C	L/C	millimoles per liter
Partial pressure of CO2	L	L	L	millimeters of mercury
Partial Pressure of O2	L	L	L	
pH	L	L	L	–
Potassium	L	L	L	milliequivalents per liter
Sodium	L	L	L	
Platelet Count	L	L	L	thousand cells per cubic millimeter of blood
Red Blood Cells	L	L	L	
White Blood Cells	L	L	L	

* 'L' denotes the variable(s) extracted from the labevents table
'C' denotes the variable(s) extracted from the chartevents table
'V' denotes the variable(s) extracted from the vitalperiodic table

4.3. *Data standardization*

In the databases, each measurement was associated with a unique type identifier defined by the original EMR system. Glucose, for example, may be recorded under different names such as GLUCOSE, Glucose, BloodGlucose, or Fingerstick Glucose. As provided by [17], we grouped the semantically equivalent measurements together using the clinical taxonomies provided by [18] and [17]. This could reduce the number of missing observations. Since our experiments were performed on different databases, we standardize measurements by performing unit conversion for both intra-database and inter-database. We used the units employed in the MIMIC-IV database (released in 2020) as the standard. In order to handle outliers, we used clinical validity ranges provided by [18], [17], and [19]. Any observation falling outside its validity range was removed.

4.4. Data aggregation
Each measurement was associated with a fine-grained timestamp. Vital sign measurements were mostly frequent, with an average of >1 time per hour in the eICU and MIMIC III databases [18], while laboratory measurements were infrequent. In order to fit the data to machine learning models, we aggregated the measurements hourly for frequently recorded vital signs and every four hours for laboratory measurements. For aggregation, we applied the following summary statistics functions: minimum and maximum. These were similar to [17] and were in consistent with the recommendations made by [18].

4.5. Data imputation
For each variable, missing data were flagged (1 if a data point was missing and 0 otherwise, for each timestep) then first imputed by forward filled from their previous values. If the variable was never observed in this patient, its values were set to -1 for all timesteps.

4.6. Machine learning algorithms
We evaluated six different machine learning algorithms on the eICU, MIMIC-III, and MIMIC-IV databases. Note that we performed standardization, as detailed above, such that the data obtained from each dataset has the same set of variables. Our data were transformed into a binary classification problem in which the event of patients died during their ICU stay or survived was set as a target. For each algorithm, we applied a range of hyperparameter configurations in order to obtain a model with the best hyperparameter setting. The overview and implementation details of each algorithm were as followed:

Logistic regression – Logistic regression [20] was used to statistically model the probability of a certain event (survived/dead) using a binary logistic function. Our data were fitted into a linear logistic regression model. We used L2 regularization and varied a regularization parameter (lambda) between 0 and 1 with an increment of 0.25.

***K*-nearest neighbors** – *K*-nearest neighbors [20] utilizes local spatial information to estimate the outcome of a new sample. For example, the algorithm with 100-nearest neighbors employs the hundred closeted samples, based on a distance metric measured in multidimensional space, to predict the outcome of the new sample. We varied the number of nearest neighbors between 100 and 900 with an increment of 200.

Neural networks – Neural networks [20] utilizes a non-linear computation of input variables to predict an outcome of the event of interest. Inputs were passed into hidden layer(s) in which each neuron in that layer performs a linear combination of all neurons in the previous layer and applies a non-linear transform through an activation function. The resulting values were then passed to the next hidden layer or the output layer. We employed a feed-forward multilayer perceptron model. We varied the number of hidden layers from 1 to 3 and the number of neurons in each layer from 32 to 128 with an increment of 32.

Bagging (Bootstrap Aggregation) – Bagging [20] is an ensemble method for improving the robustness of weak classifiers. It could be implemented with any machine learning algorithms. We employ Bagging with decision tree in this work. Decision tree [20] involves growing and pruning a tree model by determining the best variable to split and the best splitting value from the training data. A single decision tree, however, often has suboptimal predictive performance. Bagging has a collection of many decision trees in which each tree is trained by taking random samples with all features from the full dataset with replacement. Final decision was made by finding a consensus from all decision trees. We varied the number of decision trees from 100 to 500 with an increment of 200.

Random Forests – Random Forests [13] is similar to Bagging with decision trees; however, only a subset of features was selected at random for training each decision tree. We varied the number of decision trees from 100 to 500 with an increment of 200 and a choice of the number of features for each tree (the square root of the number of all features or the common logarithm of the number of all features).

Gradient Boosting – Gradient Boosting [20] involves consecutive training of weak decision trees on the residual of previous models, gradually correcting errors it made. We varied the maximum depth of trees (3, 7, 13, 49 and 98), the number of boosting stages (200, 400, and 600), and the learning rate (0.025, 0.050, 0.100).

All experiments were implemented using the Python programming language version 3.8.5 (Python Software Foundation, https://www.python.org/), the pandas data analysis library [21], and the scikit-learn machine learning library version 0.23.2 [22] on a computer with 64 processing cores and 256 gigabytes of physical memory.

4.7. *Evaluation procedure*

The development of machine learning algorithms requires holding out data for both hyperparameter tuning and performance evaluation. The use of traditional *k*-fold cross validation, in which the dataset is partitioned into *k* distinct parts and one part is iteratively used for validation and the remaining parts are then used for training, may cost information leakage leading to optimistically biased results [23] as it employs the same set of data for both optimizing hyperparameters and evaluating performance.

We employed nested cross validation, in which two rounds of cross validation was performed: one for finding the best hyperparameters and the other for testing the resultant model on a hold-out set. Nested cross validation technique was shown to produce almost unbiased estimation of performance with low variance [24]; hence, it is a recommended technique for algorithm comparisons for small and moderately-sized datasets.

In our experiments, nested cross validation with 10 outer folds and 10 inner folds was used. A dataset was first randomly partitioned into 10 parts for outer cross validation. In each outer round, one part was held out for testing and the remaining parts were used for developing a model. For model development, the remaining parts were then further randomly partitioned into 10 parts for inner cross validation. In each inner round, one part was held out for model validation and the remaining parts were used for training a model. Multiple hyperparameter combinations were evaluated in inner cross validation, then a model with the best hyperparameter was chosen for testing with the hold-out outer test set. Results were then aggregated over different outer rounds.

The receiver operating characteristic (ROC) curves were used to provide insights on the results from different algorithms. The area under the ROC curves (AUROC) was used to provide a summary of the ROC curve.

5. Results

Table 3 shows the comparison of machine learning algorithms on critical care different databases. For each experiment, the AUROC was averaged from all outer cross-validation folds in our nested cross-validation scheme and its standard deviation was also computed. Gradient boosting yielded the highest AUROCs for all three databases, followed by Random Forests. The data obtained during the first 4 hours following ICU admission had acceptable discriminative power for predicting mortality with the AUROCs of 0.89, 0.83 and 0.78 for the gradient boosting models trained on the eICU, MIMIC-III and MIMIC-IV databases, respectively. For each algorithm and each time frame, the models trained and evaluated on the eICU database had the highest AUROCs, compared to those trained on the MIMIC-III or MIMIC-IV databases. K-nearest neighbors yielded the lowest AUROCs in all cases. For all experiments, standard deviations were very low, generally less than 0.02. Figure 1 shows ROC curves comparing different algorithms on different databases for the data obtained from the first four hours.

Table 4 describes the comparison of performances of gradient boosting models for mortality prediction trained on different attributes. The models trained with the top 10 clinical variables, ranking by their feature importance, had similar AUROCs compared to those trained with all clinical variables.

The gradient boosting models trained with only vital sign variables or with only laboratory variables yielded, as expected, lower AUROCs compared to those trained with both groups of clinical variables; nevertheless, these models still have acceptable discriminating power for mortality prediction.

Table 3. Performance of different machine learning models for mortality prediction

Algorithm	Dataset	AUROC (mean ± standard deviation)					
		4 hours	8 hours	12 hours	16 hours	20 hours	24 hours
Logistic regression	eICU	0.83 ± 0.01	0.86 ± 0.01	0.88 ± 0.01	0.89 ± 0.01	0.90 ± 0.01	0.90 ± 0.01
	MIMIC III	0.79 ± 0.01	0.82 ± 0.01	0.83 ± 0.01	0.83 ± 0.01	0.84 ± 0.01	0.85 ± 0.01
	MIMIC IV	0.75 ± 0.02	0.73 ± 0.02	0.74 ± 0.01	0.75 ± 0.02	0.76 ± 0.02	0.79 ± 0.01
K-nearest neighbors	eICU	0.78 ± 0.01	0.80 ± 0.01	0.80 ± 0.01	0.81 ± 0.01	0.81 ± 0.01	0.81 ± 0.01
	MIMIC III	0.76 ± 0.01	0.77 ± 0.02	0.78 ± 0.01	0.78 ± 0.01	0.79 ± 0.01	0.79 ± 0.01
	MIMIC IV	0.68 ± 0.01	0.67 ± 0.01	0.68 ± 0.02	0.68 ± 0.02	0.69 ± 0.02	0.69 ± 0.02
Neural networks	eICU	0.85 ± 0.01	0.86 ± 0.01	0.86 ± 0.01	0.87 ± 0.01	0.89 ± 0.01	0.90 ± 0.01
	MIMIC III	0.79 ± 0.01	0.80 ± 0.01	0.80 ± 0.02	0.82 ± 0.01	0.82 ± 0.02	0.83 ± 0.02
	MIMIC IV	0.70 ± 0.01	0.70 ± 0.01	0.70 ± 0.01	0.72 ± 0.02	0.73 ± 0.01	0.74 ± 0.01
Random Forests	eICU	0.88 ± 0.01	0.89 ± 0.01	0.90 ± 0.01	0.92 ± 0.01	0.92 ± 0.01	0.93 ± 0.01
	MIMIC III	0.82 ± 0.01	0.85 ± 0.01	0.86 ± 0.01	0.87 ± 0.01	0.87 ± 0.01	0.88 ± 0.01
	MIMIC IV	0.77 ± 0.01	0.78 ± 0.02	0.79 ± 0.01	0.80 ± 0.01	0.81 ± 0.01	0.82 ± 0.01
Gradient boosting	eICU	**0.89 ± 0.01**	**0.91 ± 0.01**	**0.92 ± 0.01**	**0.93 ± 0.01**	**0.94 ± 0.01**	**0.94 ± 0.01**
	MIMIC III	**0.83 ± 0.01**	**0.85 ± 0.01**	**0.86 ± 0.01**	**0.87 ± 0.01**	**0.88 ± 0.01**	**0.88 ± 0.01**
	MIMIC IV	**0.78 ± 0.01**	**0.78 ± 0.01**	**0.80 ± 0.01**	**0.82 ± 0.01**	**0.83 ± 0.01**	**0.83 ± 0.01**
Bagging	eICU	0.85 ± 0.01	0.86 ± 0.01	0.88 ± 0.01	0.89 ± 0.01	0.90 ± 0.01	0.90 ± 0.01
	MIMIC III	0.80 ± 0.01	0.82 ± 0.01	0.83 ± 0.01	0.84 ± 0.01	0.85 ± 0.01	0.85 ± 0.01
	MIMIC IV	0.73 ± 0.01	0.74 ± 0.01	0.75 ± 0.01	0.76 ± 0.01	0.77 ± 0.01	0.78 ± 0.01

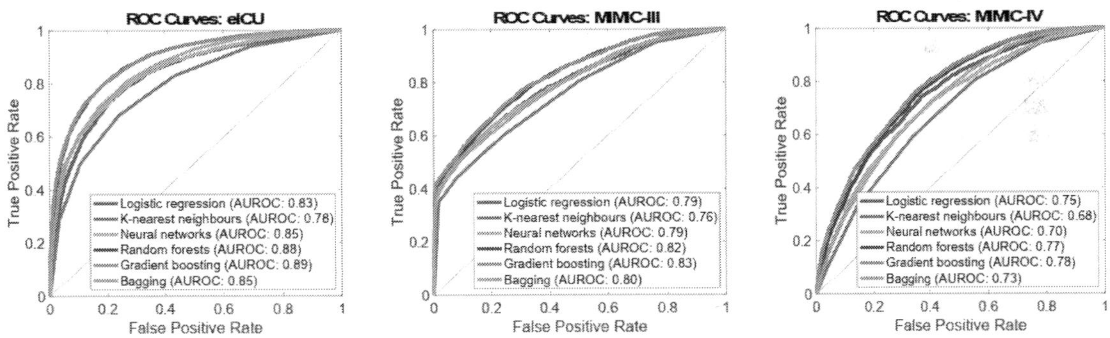

Figure 1. ROC curves comparing different algorithms on different databases for the data obtained from the first four hours of ICU admission

Table 5 details the performance of gradient boosting models for mortality prediction when trained with the data stacked from all the databases combined. Results reported separately for each database. Since the data prepared from all databases were in the same format, we combined all the databases by concatenating them together prior to evaluation. Improvements were seen on the predictions made on the MIMIC-IV database, while similar performances were seen on the predications made on the other databases.

Table 4. Performance of gradient boosting models for mortality prediction on different attributes

Variables	Dataset	AUROC (mean ± standard deviation)					
		4 hours	**8 hours**	**12 hours**	**16 hours**	**20 hours**	**24 hours**
All variables	eICU	0.89 ± 0.01	0.91 ± 0.01	0.92 ± 0.01	0.93 ± 0.01	0.94 ± 0.01	0.94 ± 0.01
	MIMIC III	0.83 ± 0.01	0.85 ± 0.01	0.86 ± 0.01	0.87 ± 0.01	0.88 ± 0.01	0.88 ± 0.01
	MIMIC IV	0.78 ± 0.01	0.78 ± 0.01	0.80 ± 0.01	0.82 ± 0.01	0.83 ± 0.01	0.83 ± 0.01
Top 10 variables	eICU	0.88 ± 0.01	0.90 ± 0.01	0.91 ± 0.01	0.93 ± 0.01	0.93 ± 0.01	0.94 ± 0.01
	MIMIC III	0.82 ± 0.01	0.84 ± 0.01	0.85 ± 0.01	0.86 ± 0.01	0.87 ± 0.01	0.87 ± 0.01
	MIMIC IV	0.77 ± 0.01	0.79 ± 0.01	0.81 ± 0.01	0.82 ± 0.01	0.83 ± 0.01	0.84 ± 0.01
Vital signs only	eICU	0.83 ± 0.01	0.85 ± 0.01	0.86 ± 0.01	0.87 ± 0.01	0.88 ± 0.01	0.89 ± 0.01
	MIMIC III	0.77 ± 0.01	0.79 ± 0.01	0.79 ± 0.01	0.80 ± 0.01	0.80 ± 0.01	0.81 ± 0.01
	MIMIC IV	0.69 ± 0.01	0.71 ± 0.01	0.72 ± 0.01	0.73 ± 0.01	0.74 ± 0.01	0.75 ± 0.01
Laboratory results only	eICU	0.82 ± 0.01	0.86 ± 0.01	0.88 ± 0.01	0.90 ± 0.01	0.91 ± 0.01	0.92 ± 0.01
	MIMIC III	0.78 ± 0.01	0.83 ± 0.01	0.85 ± 0.01	0.86 ± 0.01	0.87 ± 0.01	0.87 ± 0.01
	MIMIC IV	0.77 ± 0.02	0.79 ± 0.01	0.81 ± 0.02	0.82 ± 0.01	0.84 ± 0.01	0.84 ± 0.01

Table 5: Performance of gradient boosting models for mortality prediction when trained with the data stacked from all the databases combined, results reported separately for each database.

Dataset	AUROC (mean ± standard deviation)					
	4 hours	**8 hours**	**12 hours**	**16 hours**	**20 hours**	**24 hours**
eICU	0.89 ± 0.01	0.91 ± 0.01	0.92 ± 0.01	0.94 ± 0.01	0.94 ± 0.01	0.95 ± 0.01
MIMIC III	0.81 ± 0.01	0.84 ± 0.01	0.86 ± 0.01	0.87 ± 0.01	0.87 ± 0.01	0.88 ± 0.01
MIMIC IV	0.80 ± 0.01	0.82 ± 0.01	0.84 ± 0.01	0.85 ± 0.01	0.87 ± 0.01	0.87 ± 0.01

6. Discussion

Here, we discuss our results described in Section 3, provide comparisons of ours with other related work, give insights on the availability of data during the first 24 hours following ICU admission, provide limitations of our study, and give suggestions for further research and development.

6.1. Comparison across different algorithms

Tree-based classifiers (Random Forests, Gradient Boosting and Bagging) had similar trends of AUROCs compared to other algorithms, while Gradient Boosting achieved the highest AUROCs for all time frames. The *k*-nearest neighbors models had the lowest AUROCs for all time frames. This might be because the complexity of ICU data could not be well handled with this algorithm.

Our results are consistent with those of [16] which reported the effectiveness of ensemble tree models for mortality prediction on the first 24-hour time window. On the MIMIC-III database, Our results are slightly higher than those of [17], [18], [25], [26] for most algorithms while having similar trends. These might be because we use different variable sets and different evaluation scheme.

Note that we did not give a direct comparison of our results with traditional scoring systems as further assumptions have to be made in order to produce such scores. According to the APACHE-II, SAPS-I, SAPS-II and SOFA scores computed by [15], [26], our machine learning models attained higher AUROCs than conventional clinical scoring systems for mortality prediction.

6.2. Data availability

We calculated an average number of vital sign and laboratory measurements made following ICU admission for each patient, as illustrated in Figure 2. The volume of vital sign measurements in the eICU database was six to ten folds compared to those made in the MIMIC-III and MIMIC-IV databases. The number of laboratory measurements were similar across all databases. The increment of data has a linear trend.

Table 6: Number of measurements made for each clinical variable, counting for each time period following ICU admission for each critical care database.

Variable	0 - 4 hours			4 - 8 hours			8 - 12 hours			12 - 16 hours			16 - 20 hours			20 - 24 hours		
	eI	M3	M4	eI	M3	M4	eI	M3	M4	eI	M3	M4	eI	M3	M4	eI	M3	M4
Vital signs																		
Heart Rate	43.7	6.3	7.1	58.8	7.1	7.0	59.6	6.7	6.8	59.8	6.4	6.6	59.9	6.4	6.5	59.9	6.2	6.3
Respiratory Rate	40.2	6.1	7.0	53.8	6.9	6.9	54.6	6.5	6.7	54.7	6.3	6.6	54.5	6.2	6.4	54.4	6.1	6.3
Peripheral O2 saturation	39.9	2.2	6.6	54.1	2.5	6.7	55.0	2.4	6.6	54.9	2.3	6.4	54.7	2.3	6.3	54.3	2.2	6.1
Blood Pressure mean	10.4	0.5	1.7	18.7	0.9	2.4	21.5	1.0	2.6	23.0	1.0	2.7	24.1	1.1	2.7	24.9	1.1	2.6
Blood Pressure systolic	10.2	0.5	1.7	18.7	0.9	2.4	21.3	1.0	2.5	22.8	1.0	2.6	23.9	1.1	2.7	24.7	1.1	2.6
Blood Pressure diastolic	10.2	0.5	1.7	18.7	0.9	2.4	21.3	1.0	2.5	22.8	1.0	2.6	23.9	1.1	2.7	24.7	1.1	2.6
Temperature	5.8	<0.1	0.2	9.0	0.1	0.3	9.8	0.1	0.3	10.3	0.1	0.3	10.6	0.1	0.4	10.9	0.1	0.4
Laboratory results																		
Bilirubin	0.4	0.4	0.6	1.2	1.0	1.6	0.9	0.8	1.1	0.9	0.7	1.0	0.8	0.6	0.8	0.7	0.5	0.8
Bicarbonate (HCO3)	1.1	0.8	1.4	1.8	1.4	2.3	1.4	1.2	1.8	1.3	1.1	1.6	1.2	0.9	1.4	1.1	0.8	1.3
Calcium	0.5	0.6	0.7	1.2	1.5	1.8	0.9	1.3	1.4	0.9	1.2	1.3	0.8	1.0	1.1	0.7	0.9	1.0
Chloride	0.5	0.6	0.7	0.9	1.0	1.2	0.8	0.8	0.9	0.7	0.8	0.9	0.7	0.6	0.7	0.6	0.6	0.7
Creatinine	0.5	0.6	0.7	0.9	1.0	1.2	0.7	0.8	0.9	0.7	0.7	0.8	0.6	0.6	0.7	0.6	0.6	0.6
Glucose	0.5	0.8	0.9	1.0	1.4	1.6	0.8	1.2	1.2	0.8	1.0	1.0	0.7	0.9	0.9	0.6	0.8	0.8
Hemoglobin	0.5	0.7	0.9	0.9	1.2	1.4	0.8	0.9	1.0	0.7	0.8	0.8	0.7	0.6	0.7	0.6	0.6	0.6
Hematocrit	0.5	0.6	0.7	0.9	1.1	1.2	0.7	0.9	0.9	0.7	0.8	0.8	0.6	0.7	0.7	0.5	0.6	0.6
O2 Saturation	0.5	0.3	0.2	0.8	0.5	0.4	0.5	0.5	0.4	0.5	0.4	0.3	0.5	0.4	0.3	0.4	0.4	0.3
Inspired O2 Fraction	0.7	0.4	1.3	1.0	0.8	2.4	0.6	0.8	2.1	0.6	0.8	2.1	0.5	0.8	2.1	0.5	0.8	2.0
Lactate	0.4	0.53	0.7	0.7	0.9	1.3	0.5	0.7	1.0	0.4	0.6	0.9	0.4	0.5	0.8	0.4	0.5	0.7
Partial pressure of CO2	0.6	0.8	0.8	1.0	1.5	1.5	0.7	1.3	1.2	0.6	1.2	1.1	0.6	1.1	1.0	0.6	1.0	0.9
Partial Pressure of O2	0.6	0.8	0.8	1.0	1.5	1.5	0.7	1.3	1.2	0.7	1.2	1.1	0.6	1.1	1.0	0.6	1.0	0.9
pH	0.6	0.9	0.9	1.0	1.6	1.5	0.7	1.4	1.2	0.6	1.2	1.1	0.6	1.1	1.0	0.6	1.1	0.9
Potassium	0.6	0.6	0.7	1.1	1.0	1.2	0.9	0.9	0.9	0.9	0.8	0.9	0.8	0.7	0.8	0.7	0.6	0.7
Sodium	0.6	0.6	0.7	1.1	1.0	1.2	0.9	0.9	0.9	0.9	0.8	0.9	0.8	0.6	0.8	0.7	0.6	0.6
Platelet Count	0.4	0.5	0.7	0.7	1.0	1.1	0.6	0.8	0.9	0.6	0.7	0.7	0.5	0.5	0.6	0.4	0.5	0.6
Red Blood Cells	0.4	0.5	0.7	0.7	0.9	1.1	0.6	0.8	0.8	0.6	0.6	0.7	0.5	0.5	0.6	0.4	0.4	0.6
White Blood Cells	0.4	0.5	0.7	0.7	0.9	1.1	0.6	0.8	0.8	0.6	0.7	0.7	0.5	0.5	0.6	0.4	0.4	0.2

eI: eICU; M3: MIMIC-III; M4: MIMIC-III;

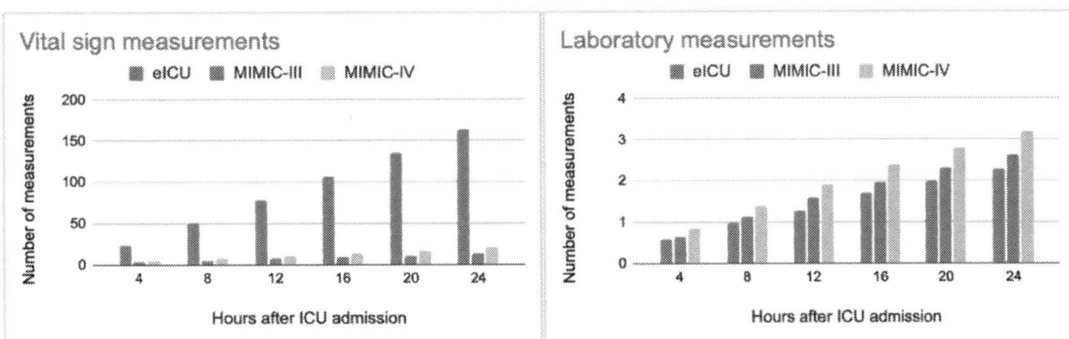

(a) (b)

Figure 2. Average number of measurements made after ICU admission for (a) recorded vital signs and (b) laboratory results for each critical care database.

Table 6 details the number of measurements made for each clinical variable, counting for each time period following ICU admission for each critical care database. The eICU database has more frequent vital sign measurements. Temperature measurements was not frequently recorded in the MIMIC-III and MIMIC-IV databases. The number of laboratory measurements was similar across all databases. During the first four hours of ICU admission, we could expect half of laboratory measurements were made. The measurements were likely to complete during the first eight hour of ICU admission. The different in frequencies of recorded vital signs and laboratory measurements leading to us choosing different time windows (every hour for vital sign measurements and every four hours for laboratory measurements) for data aggregation.

6.3. *Comparison across databases*

The machine learning models evaluated on the eICU database had the highest AUROCs, followed by those evaluated on the MIMIC-III and MIMIC-IV databases, respectively. This might be because the high frequency of data collected in the eICU database providing more discriminative features. Even though the MIMIC-IV database is a successor to the MIMIC-III database, both databases contain similar yet different sets of data: the MIMIC-III database contains data from 2001-2012; and the MIMIC-IV database contains data from 2008 to 2019, both from BIDMC. The differences in their results could be expected. For the models trained on the combination of data from all databases, small improvements in the AUROCs were seen on predictions made on the MIMIC-IV database. The amount of data used for training machine learning algorithms might contribute to these improvements. Note that at the time of manuscript preparation, the MIMIC-IV was still in development.

6.4. *Limitations*

Our study was subject to several limitations. Firstly, we did not attempt to model temporal relationship of continuous variables. All experiments were performed only on traditional machine learning algorithms in which clinical features were aggregated using summary statistics and were treated the same. This was unlike the study by Purushotham *et al.* [15] in which non-temporal features (e.g. laboratory tests) and temporal features (e.g. vital signs) were treated separately using recurrent neural networks. Their study yielded an AUROC of 0.94 on 24-hour data following ICU admission, compared to ours whose AUROC was 0.88 for the same period. A room of improvements could still be achieved.

Secondly, we did not perform any experiments on data imputation apart from using carry-forward imputation and a missing value flag. We did not observe any improvements in our initial experiments in which missing values were imputed as in [17], [26] with global mean, median or normal values. This was also consistent by what reported in [18].

Finally, we performed under-sampling for patients in the control group (survived during hospital stay). This was done as our computation resources were limited and we did not see much improvement over the use of oversampling techniques, reported in [26].

6.5. *Suggestions*

Conventional clinical scoring systems have been developed and employed in the clinical practice for more than two decades. Currently, modern HIS has been a hub of extensive clinical data collections including electronic medical notes, laboratory results, physiological values excerpted from bedside monitoring systems, information regarding fluids delivered to patients via modern infusion pumps, or data derived from mechanical ventilators. These clinical data could be used for the development of clinical support systems. These collections of data are, however, different from hospital to hospital, depending on their context, culture and available resources.

For the automation of clinical scoring processes, one could develop and validate algorithms using the data gathered from the existing HIS together with the data obtained from public clinical databases. However, the utilization of different data requires careful considerations on feature transformations and the algorithms should be less dependent on the features that are hard to obtain or not currently available from the existing HIS.

We found that most tree-based algorithms, such as Random Forests and Gradient Boosting, can work out of the box with slight tweaks on hyperparameter optimization. Carry-forward imputation with a missing value indication could be utilized in the case that some measurements are not yet available. Several clinical outcome prediction algorithms could be combined and formed a multitask algorithm as in [14]. Techniques for modelling temporal features explicitly such as recurrent neural networks could be employed; however, special attentions may need for data transformations as clinical measurements across different systems or databases are usually collected at different frequencies.

7. Conclusion

We performed comprehensive analysis of different machine learning algorithms for mortality prediction on three critical care databases: the eICU, MIMIC-III and MIMIC-IV databases. We achieved similar performances for tree-based classifiers, whilst Gradient Boosting attained the highest AUROC for all the databases. Clinical parameters obtained during the first 4 hours of ICU admission are enough for predicting patient mortality with clinically acceptable performance. The models trained with the top ten clinical features, raking by their importance, yielded similar performances compared to those trained with all clinical variables. The models trained with the data combined from all the databases yielded slight improvements in predictive performance. Clinical features needed for scoring are mostly available after 8 hours of ICU admission. Vital sign measurements in the eICU database are six to ten times more frequent than those in the MIMIC-III and MIMIC-IV databases. These likely to contribute to the higher performance of the algorithms trained on the eICU database, compared to those trained on the other databases. Our study provides guidelines and benchmarks for developing algorithms for the prediction of clinical outcomes in local narratives.

References

[1] W. A. Knaus, E. A. Draper, D. P. Wagner, and J. E. Zimmerman, "APACHE II: a severity of disease classification system.," *Crit. Care Med.*, vol. 13, no. 10, pp. 818–29, Oct. 1985.
[2] W. A. Knaus *et al.*, "The APACHE III prognostic system: Risk prediction of hospital mortality for critically Ill hospitalized adults," *Chest*, 1991.
[3] J. L. Vincent *et al.*, "The SOFA (Sepsis-related Organ Failure Assessment) score to describe organ dysfunction/failure," *Intensive Care Med.*, 1996.
[4] J. R. Gall, S. Lemeshow, and F. Saulnier, "A New Simplified Acute Physiology Score (SAPS II) Based on a European/North American Multicenter Study," *JAMA J. Am. Med. Assoc.*, 1993.
[5] J. Lee, J. A. Dubin, and D. M. Maslove, "Mortality Prediction in the ICU," in *Secondary Analysis of Electronic Health Records*, 2016, pp. 1–427.
[6] T. J. Pollard, A. E. W. Johnson, J. D. Raffa, L. A. Celi, R. G. Mark, and O. Badawi, "The eICU Collaborative Research Database, a freely available multi-center database for critical care research," *Sci. Data*, vol. 5, no. 1, p. 180178, Dec. 2018.
[7] A. E. W. Johnson *et al.*, "MIMIC-III, a freely accessible critical care database," *Sci. Data*, 2016.
[8] A. Johnson, L. Bulgarelli, T. Pollard, S. Horng, L. A. Celi, and R. Mark, "MIMIC-IV (version 0.4)," *PhysioNet*, 2020.
[9] R. M. H. Schein, N. Hazday, M. Pena, B. H. Ruben, and C. L. Sprung, "Clinical Antecedents to In-Hospital Cardiopulmonary Arrest," *Chest*, vol. 98, no. 6, pp. 1388–1392, Dec. 1990.
[10] M. A. F. Pimentel *et al.*, "Towards a robust estimation of respiratory rate from pulse oximeters," *IEEE Trans. Biomed. Eng.*, vol. 64, no. 8, pp. 1914–1923, 2016.
[11] M. Saeed *et al.*, "Multiparameter intelligent monitoring in intensive care II: A public-access intensive care unit database," in *Critical Care Medicine*, 2011.
[12] A. Awad, M. Bader-El-Den, J. McNicholas, and J. Briggs, "Early hospital mortality prediction of intensive care unit patients using an ensemble learning approach," *Int. J. Med. Inform.*, 2017.
[13] L. Breiman, "Random Forests," *Mach. Learn.*, vol. 45, no. 1, pp. 5–32, 2001.
[14] H. Harutyunyan, H. Khachatrian, D. C. Kale, G. Ver Steeg, and A. Galstyan, "Multitask learning and benchmarking with clinical time series data," *Sci. Data*, vol. 6, no. 1, pp. 1–18, 2019.

[15] S. Purushotham, C. Meng, Z. Che, and Y. Liu, "Benchmarking deep learning models on large healthcare datasets," *J. Biomed. Inform.*, vol. 83, no. March, pp. 112–134, 2018.
[16] N. El-Rashidy, S. El-Sappagh, T. Abuhmed, S. Abdelrazek, and H. M. El-Bakry, "Intensive Care Unit Mortality Prediction: An Improved Patient-Specific Stacking Ensemble Model," *IEEE Access*, vol. 8, pp. 133541–133564, 2020.
[17] S. Wang, M. B. A. McDermott, G. Chauhan, M. Ghassemi, M. C. Hughes, and T. Naumann, "MIMIC-Extract," in *Proceedings of the ACM Conference on Health, Inference, and Learning*, 2020, pp. 222–235.
[18] S. Tang, P. Davarmanesh, Y. Song, D. Koutra, M. W. Sjoding, and J. Wiens, "Democratizing EHR analyses with FIDDLE: a flexible data-driven preprocessing pipeline for structured clinical data," *J. Am. Med. Informatics Assoc.*, 2020.
[19] H. Harutyunyan, H. Khachatrian, D. C. Kale, G. Ver Steeg, and A. Galstyan, "Multitask learning and benchmarking with clinical time series data," *Sci. Data*, 2019.
[20] C. M. Bishop, *Pattern recognition and machine learning*, 6th ed. New York, USA: Springer-Verlag, 2006.
[21] W. McKinney, "Data Structures for Statistical Computing in Python," in *Proceedings of the 9th Python in Science Conference*, 2010.
[22] F. Pedregosa *et al.*, "Scikit-learn: Machine learning in Python," *J. Mach. Learn. Res.*, 2011.
[23] S. Varma and R. Simon, "Bias in error estimation when using cross-validation for model selection," *BMC Bioinformatics*, 2006.
[24] A. Vabalas, E. Gowen, E. Poliakoff, and A. J. Casson, "Machine learning algorithm validation with a limited sample size," *PLoS One*, 2019.
[25] S. Purushotham, C. Meng, Z. Che, and Y. Liu, "Benchmarking deep learning models on large healthcare datasets," *J. Biomed. Inform.*, 2018.
[26] A. Awad, M. Bader-El-Den, J. McNicholas, J. Briggs, and Y. El-Sonbaty, "Predicting hospital mortality for intensive care unit patients: Time-series analysis," *Health Informatics J.*, 2020.

Acknowledgement

TNP acknowledges funding for his graduate study from Faculty of Medicine, Prince of Songkla University. He also received funding for research from Graduate School, Prince of Songkla University. This work was supported by the Health Systems Research Institute (HSRI) under grant number HSRI. 63-147. This work was also supported by the Kasikornthai Foundation through the 'PSU Fights for COVID-19' funding. The authors gratefully acknowledge the support of computational resources from the Medical Data Center for Research and Innovation (MDCRI), Faculty of Medicine, Prince of Songkla University.

X-ray and infrared spectroscopy study on contamination of asbestos in Thai commercial cosmetic talc powder product

C Boonruang[1,2], K Won-in[3,*], K Thumanu[4] and P Dararutana[5]

[1] Department of Physics and Materials Science, Faculty of Science, Chiang Mai University, Chiang Mai 50200 Thailand
[2] Center of Excellence in Materials Science and Technology, Chiang Mai University, Chiang Mai 50200 Thailand
[3] Department of Earth Sciences, Faculty of Science, Kasetsart University, Bangkok 10900 Thailand
[4] Synchrotron Light Research Institute, Nakhon Ratchasima 30000 Thailand
[5] Retired Army Officer at the Royal Thai Army, Lopburi 15000 Thailand

E-mail: kritwonin@gmail.com

Abstract. It is well-known that asbestos such as asbestosis and mesothelioma is a fiber causing lung diseases. Talc is used for commercial applications for example paints, plastics, papers, ceramics, construction materials, and cosmetics. It is well-known that the cosmetic talc powder is used for preventing diaper rash as a deodorant. The sampling in this work has been done from various commercial products in Thailand. Their microstructure and elemental composition are analyzed using a scanning electron microscope coupled with an energy dispersive X-ray spectrometer (SEM-EDS). A small-angle X-ray scattering (SAXS) and infrared spectroscopy (IR) based on synchrotron radiation are also carried out to determine their phase composition and functional groups. It has been found that fibers with asbestos-like shapes are present in some samples. The main composition of oxygen, silicon, and magnesium in samples are in the form of the functional groups that associate with absorption band in the range of 3600-3700 cm^{-1} of IR spectra. It is proposed that there are 6 from 17 samples that possibly have asbestos contamination.

1. Introduction

Talc is an industrial raw material that has unique physical and chemical properties, such as hydrophobicity, organophilicity, lamellarity, softness, chemical inertness, high thermal stability, low electrical conductivity, heat resistance, wide particle size distribution, high specific surface area, oil absorption, and polymer absorption capability. However, these features depend on their parent rock types and origin. It is shown that talc which is extracted from various localities shows the difference in mineralogical chemical and physical properties. It is used in many industrial applications including cosmetics, pharmaceuticals, textiles, papers, food, plastics, ceramics, paints, and pesticides [1-5].

Talc is a clay mineral that consists of a hydrated magnesium silicate sheet with the ideal chemical formula of $Mg_2(Si_2O_5)_2(OH)_2$ [1-2, 4]. Moreover, its structure frequently contains small amounts of iron and other trace metals. Talc is widely used as talc powder which is a household product sold globally for use in personal hygiene (baby and feminine powders) and cosmetics [4] as shown in Figure 1.

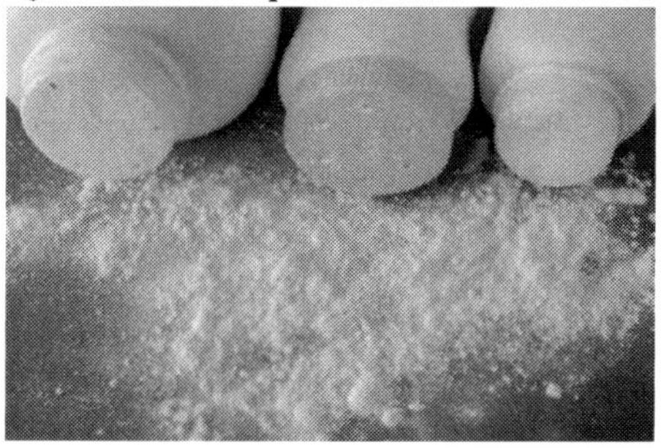

Figure 1. Cosmetic talc powder.

It has been reported that talc powder tends to be contaminated with asbestos fibers [6-10] which contribute to certain types of diseases such as asbestosis and mesothelioma [11]. In Thailand, it has been reported that asbestos contamination has been found in commercial body soothing powders which can result in lung inflammation and talc pneumoconiosis [12].

Asbestos is classified as a carcinogen material of the Catelog1 by the World Health Authorities [13-15]. Several works showed the fiber's toxicity depended on its morphological size, chemical-physical characteristics, surface reactivity, and biopersistence [16-22]. Asbestos is a collective name given to a variety of naturally occurring complex, crystalline, fibrous silicate minerals. Asbestos fibers have a needle-like or sometimes flexible structure that may vary widely in the micron to millimeter ranges. Asbestos minerals are classified into serpentine and amphibole groups. Serpentines have a leafy or layered structure consisted of fine fibers. The leaves consist of alternate layers of silicate tetrahedron (SiO_4), which are held together by hydroxide groups (OH), and magnesium ions. Chrysotile, $Mg_3Si_2O_5(OH)_4$, is only one mineral of this group. It has a very small diameter, tubular, very soft, and bendable. Amphibole has a chain-like crystal structure which stipulates their fibrous nature. They have a larger diameter, straight, firm, and hard, but elastic. This group includes amosite $(Fe^{+2})_2(Fe^{+2},Mg)_5Si_8O_{22}(OH)_2$, crocidolite; $Na_2(Fe^{+2},Mg)_3Fe^{+3}Si_8O_{22}(OH)_2$, anthophyllite; $Mg_7Si_8O_{22}(OH)_2$, tremolite; $Ca_2Mg_7Si_8O_{22}(OH)_2$, and actinolite; $Ca_2(Mg,Fe^{+2})Si_8O_{22}(OH)_2$ [23].

It is well-known that SEM-EDS, XRD, and IR methods are important non-destructive tools for the characterization of materials. It is revealed that morphology, structural, and chemical features of various asbestos minerals have been verified through investigation using a variety of analytical methods, such as electron microscopy [23-32] and infrared spectroscopy [23-26, 33-35]. SAXS is also applied to determine the structure of the element component in minerals [36-40].

This study reports on the determination of contamination of asbestos in talc using synchrotron facilities for the first time. The morphology, microstructure, and elemental composition are characterized using SEM-EDS, SR-SAXS, and SR-FTIR. The safety of using Thai commercial talcum powder has been discussed.

2. Experimental

2.1 Sample

The two samples of talc powders (STD N1 and STD N2) contained with asbestos are used as the reference. Other 17 samples (N3 to N19) are commercial products (including in-country manufacturing and import products) in Thailand.

2.2 Methods

2.2.1 SEM-EDS. Microstructure and elemental composition are characterized using SEM SU350 (Hitachi, Japan) coupled with EDS Ultimax40 (Oxford Instruments, UK) at Coax Group Corporation Ltd. (Headquarter) (Bangkok, Thailand). The system is operated at 5 kV accelerating voltages, in 30-50 Pa vacuum, and using the AZtec Automated program for analysis. Each sample, three X-ray were corrected.

2.2.2 SR-SAXS. Phase composition is determined using a synchrotron radiation-small angle X-ray scattering (SR-SAXS) at the Synchrotron Light Research Institute (SLRI: Nakhon Ratchasima, Thailand). It utilizes from a bending magnet at the photon energy between 6-9 keV, the calculated photon flux density of about 10^{11} photons/sec/0.1% Bending magnet/mrad^2/mm^2, and exposure time of 240 s. The diffraction pattern is obtained from 0° to 40° 2 Theta (2θ).

2.2.3 SR-FTIR. The samples are performed at synchrotron-based FTIR (Fourier Transform Infrared) microspectroscopy. Each sample is prepared using a diamond anvil cell which is a high-pressure cell permitting to operate and adjust under a microscope for transmission studies. The functional groups of the sample composition are evaluated as spectral data and collected at an infrared microspectroscopy beamline (BL4.1 Infrared Spectroscopy and Imaging) at the Synchrotron Light Research Institute (SLRI; Nakhon Ratchasima, Thailand). Spectra were acquired with an SR-FTIR (Hyperion 2000, Bruker Optics, Ettlingen, Germany) coupled with an Infrared microscope (Hyperion 2000, Bruker) using the 36x IR objective lens with an MCT (Mercury cadmium telluride) detector cooled with liquid nitrogen over the measurement range from 3800 to 3000 cm^{-1}. The measurements are performed in transmission mode, using an aperture size of 20×20 μm^2 with a spectral resolution of 4 cm^{-1}, with 64 scans co-added. Spectral acquisition and instrument control are performed using OPUS 7.5 (Bruker Optics Ltd, Ettlingen, Germany) software.

3. Results and discussion

The element composition of samples obtained by using SEM-EDS is reported in Table 1. It is shown that O (26.8 to 38.1), Si (8.1 to 19.1), and Mg (5.5 to 11.9 wt%) are the main composition of samples Small amount of Mg (0.3 wt%) is detected only in sample N19. Ca is found only in samples N1 (4.6) and N2 (6.8 wt%). A large amount of Al is found in samples N17 and N19. K (N17 and N19), Ti (N8, N17, N18, and N19), Fe (N1, N17, N18, and N19), and Zn (N6 and N11) are detected in some samples. Na and P also present only in samples N19 and N17, respectively. However, a large amount of C in all samples is suggested to be coming from surrounding contamination.

Due to the main compositions of asbestos are O, Si, Mg, Fe, and Ca, It is implied that N1 and N2 contain actinolite and tremolite, respectively. The results imply that many kinds of asbestos are possibly contained in samples: chrysotile and/or anthophyllite (N1-N11 and N16); amosite (N17 and N18); and crocidolite (N19).

Table 1. Elemental composition of samples.

Sample	Elemental composition (wt%)											
	C	O	Si	Al	K	Ti	Na	Mg	Fe	Ca	Zn	P
N1	33.7	35.3	15.8	1.3	-	-	-	9.3	2.9	4.6	-	-
N2	36.6	32.4	16.3	-	-	-	-	8.0	-	6.8	-	-
N3	35.0	36.9	16.4	0.4	-	-	-	11.5	-	-	-	-
N4	35.8	37.2	16.1	0.5	-	-	-	10.6	-	-	-	-
N5	31.4	37.8	19.0	-	-	-	-	11.9	-	-	-	-
N6	33.7	37.7	17.0	0.2	-	-	-	11.0	-	-	0.4	-
N7	37.7	36.2	15.9	-	-	-	-	10.2	-	-	-	-
N8	34.8	37.1	16.0	0.3	-	1.2	-	10.7	-	-	-	-
N9	33.8	38.1	17.1	-	-	-	-	11.1	-	-	-	-
N10	38.7	35.6	15.6	0.3	-	-	-	10.0	-	-	-	-
N11	50.2	26.8	8.1	-	-	-	-	5.5	-	-	9.8	-
N12	35.4	35.1	18.0	0.2	-	-	-	11.4	-	-	-	-
N13	36.3	35.3	17.1	0.2	-	-	-	11.3	-	-	-	-
N14	35.4	35.4	17.2	0.5	-	-	-	11.5	-	-	-	-
N15	40.5	33.4	15.8	0.4	-	-	-	9.9	-	-	-	-
N16	36.8	32.7	19.1	-	-	-	-	11.5	-	-	-	-
N17	38.9	32.9	11.5	8.0	2.6	2.0	-	5.7	2.3	-	-	0.3
N18	37.7	31.6	14.4	0.6	-	1.2	-	8.4	6.0	-	-	-
N19	35.4	35.7	12.4	8.6	3.4	2.3	0.3	0.3	3.1	-	-	-

Figure 2 shows morphologies of straight and firm fibers of amphibole-group asbestos in the reference samples (N1 and N2). A small amount of this type of particle is found in N4 and N17 which is proposed to be anthophyllite and amosite as characterized by EDS. SEM can reveal contamination of the small fibrous with a length of 1-2 μm in samples. However, the fibrils of chrysotile may be beyond the resolution limit of SEM. The characterization based on only chemical composition and morphology may give wrong information due to the contribution of other phases including talc, anthophyllite, etc.

N1 N2

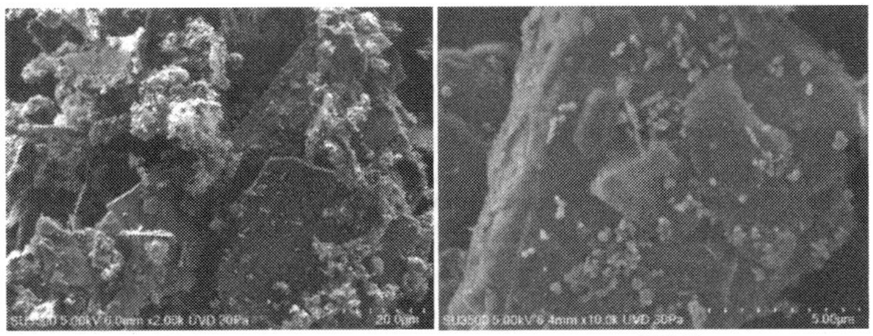
N4 N17

Figure 2. SEM images of some samples that contained asbestos fibers.

SAXS is a primary characterization tool that can provide structural information describing how large molecules are shaped and organize in any sample. Nevertheless, it is a new method to identify the presence of asbestos in talc powder. It is shown fully characterizing mineral assemblage, provides information about the bulk purity, gives information about the origin of the talc deposit and the associated risk, and indicates if problematic levels of any phase are present. Nonetheless, it is cannot distinguish asbestos from non-asbestos forms of the same mineral. The limit of detection may be too high and may give false-negative results if it used as a screening method.

The phase of reference and talc powder samples are shown in Figure 3. It is shown that SAXS is an analytical technique that measures the intensities of the X-ray scattering by a sample as a function of the scattering angle, the same as the X-ray diffraction technique. In previous research which characterized using the XRD technique based on synchrotron radiation, it is reported that the main pattern of chrysotile is present at 2θ of 12.2 and 24.4 °, while amphibole is at ~10.6° [41-44].

FTIR spectroscopy is a powerful tool to qualify and quantify organic substances in the solid, liquid, or gaseous state. However, many sampling techniques require pre-treatment of samples in order to obtain high-quality spectra. It allows solid or liquid samples to be measured with only minimal sample preparation. Because other minerals, including nonasbestos amphiboles, also absorb within these wavelength bands, IR is not always definite for identifying asbestos. The use of an FTIR spectrophotometer improves dramatically the sensitivity and selectivity. With synchrotron light, IR microspectroscopy allows measuring samples at high spatial resolution. Computer-assisted analysis of spectra offers the possibility to reduce matrix interferences and to compare different spectra. It is typically available for testing. However, it is cannot distinguish asbestos from non-asbestos forms of the same mineral, this method is subject to interferences with other minerals, and the detection limit is unknown, up to the samples.

The IR spectra of the reference and talc powder samples are shown in Figure 4. It is shown that absorption bands in the infrared spectra associated with asbestos fibers are all present in the range of 3600-3700 cm^{-1} which correspond to specific hydroxyl bands: the O-H stretching vibration band of the hydroxyl group surrounded by $3Mg(Mg_3\text{-}OH)$ [23-26, 33-35].

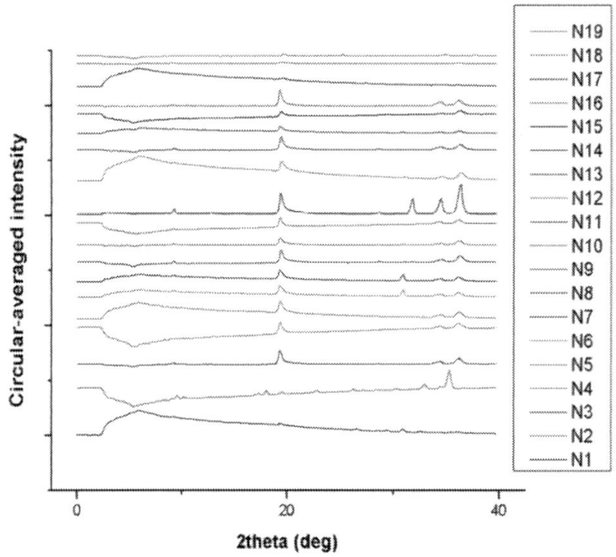

Figure 3. Phase composition of samples using SR-SAXS.

Figure 4. IR spectra of the samples.

The results show that some samples possibly contain asbestos, especially N5, N12, and N17 are the same characteristic as N1, while N4, N10, and N15 correspond to N2. By comparing the information of asbestos morphology, elemental composition, molecular structure, and functional group, it is proposed that there are 6 from 17 samples that possibly have asbestos contamination. The contamination can result from the mining process and production. Asbestos is present in some Thai commercial talc-powder products but not all. It is highly recommended that they should be used with caution.

Further works using other techniques such as Near-IR spectroscopy, TEM (Transmission electron microscopy), XRD, particle size analysis, thermal degradation, and Raman spectroscopy would be done in order to find out more details.

4. Conclusion

Asbestos contamination in Thai commercial talc-powder products has been analyzed using SEM-EDS, SR-SAXS, and SR-FTIR. The main composition of oxygen, silicon, and magnesium in samples are in the form of the functional groups that associate with absorption band in the range of 3600-3700 cm^{-1} of IR spectra. According to information on morphology, elemental composition, molecular structure, and functional group obtained from these techniques is proposed that there are 6 (such as N4, N5, N10, N12, N15, and N17) from 17 samples that possibly have asbestos contamination.

References

[1] Rohl A N and Langer A M 1974 *Environmental Health Perspectives* **9** 95.
[2] Newman A W 1994 *Talc* (New York: Bristol-Myers Squibb Pharmaceutical Research Institute).
[3] Zazenski R, Ashton W H, Briggs D, Chudkowski M, Kelse J W, MacLachem L, McCarthy E F, Nordhauser M A, Roddy M T, and Teetsel N M 1995 *Regulatory Toxicology and Pharmacology* **21** 218.
[4] Ersoy B, Dikmen S, Yidiz A, Goren G and Elitok O 2013 *Turkish Journal of Earth Sciences* **22** 632.
[5] Gokulakumar B, Nedunchezhian G, Vijayaraj C and Balamurugan A 2016 *International Journal of Recent Scientific Research* **7** *14505.*
[6] Rohl A N, Langer A M, Selikoff I J, Tordini A, Klimentidis R, Bowes D R and Skinner D L 1976 *J Toxicol Environ Health* **2** 255.
[7] Vignaroli G, Rossetti F, Belardi G and Billi A 2011 *Nat Hazards Earth Syst Sci* **11** 1267.
[8] Pierce J S, Riordam A S, Miller E W, Gaffney S H and Hollins D M 2017 *Inhalation Toxicology* **29** 443.
[9] Ilgren E B, Sartorio C and Hoskins J A 2017 *Environment and Pollution* **6** 6.
[10] Fitzerald S, Harty E, Joshi T K and Frank A L 2019 *Am J Ind Med* **62** 385.
[11] Yao S, Ventura G D and Petibios C 2010 *Analytical and Bioanalytical Chemistry* **397** 2079.
[12] Prijyamonda B, Marantr N, Pacharee P and Pamorn S 2010 *Siriraj Hosp. Gaz.* **29** 771.
[13] IARC 1987 *Overall evaluations of carcinogenicity: An updating of IARC monographs Volumes 1-42* (Lyon : IARC)
[14] IARC 2002 *IARC monographs on the evaluation of carcinogenic risk in humans : Man-made vitreous fibres Volume 81* (Lyon : IARC).
[15] IARC 2012 *IARC monographs on the evaluation of carcinogenic risk in humans : Arsenic, metals, fibres and dust Volume 100C* (Lyon : IARC).
[16] Pooley F D and Rowlands N 1975 *Inhaled Particles* **4** 639.
[17] Rohl A N, Langer A M, Selikoff I J, Tordini A, Klimentidis R, Bower D R, Skinner D L 1976 *Journal of Toxicology and Environmental Health* **2** 255.
[18] Wehner A P 1980 *IARC Scientific Publications* **30** 373.
[19] Pang T W S, Schonfeld F A and Nazar M A 1987 *Annals of Occupational Hygiene* **31** 219.
[20] Paoletti L, Caiazza S, Donelli G and Pocchiari F 1984 *Regulatory Toxicology and Pharmacology* **4** 222.
[21] Ishida T 2012 *J. Fluresc.* **22** 357.
[22] Dichico M C, Paternoster M, Rizzo G and Sinisi R 2019 *Fibers* **7** 13pp.
[23] Vita R L 2002 *Asbestos: Geology, Mineralogy, Mining, and Uses* (Virginia : U.S. Geological Servey).

[24] Hassan M F, El Wahab M A and Nada A 2007 *Proceeding of the 6th Conference on Nuclear and Particle Physics (Luxor)* (Vienna: International Atomic Energy Agency) 6 pp.
[25] Anbalagan G, Sivakumar G, Prabakaran A R and Gunasekaran S 2010 *Vibrational Spectroscopy* **52** 12 2.
[26] Ristic M, Nagy I C, Music S and Vertes A 2011 *Journal of Molecular Structure* **993** 120.
[27] Gualtieri A F, S Pollastri, N B Gandolfi, Ronochetti F, Albonico C, Cavallo A, Zanetti G, Marini P and Sala O 2014 *American Mineralogist* **99** 1233.
[28] Giacoble C, Wright J P, Giuseppe D D, Zoboli A, Zapparoli M and Gualtieri A 2018 *Minerals* **8** 311 11 pp.
[29] Cossio R, Albonico C, Zanella A, Garofalo S F, Avataneo C, Compagnoni R and Turci F 2018 *Talanta* **190** 158.
[30] McDonald S A, Fan Y, Rogers R A and Godleski J J 2019 *Ultrastructural Pathology* **43** 248.
[31] Johnson K E, Popratiloff A, Fan Y, McDonald S and Godleski J J 2020 *Gynecologic Oncology* **159** 527.
[32] Yang K, Yoo K C and Jung J 2020 *Minerals* **10** 568.
[33] Balducci D and Valerio F 1986 *International Journal of Environmental Analytical Chemistry* **27** 315.
[34] Baietto O and Marini P 2018 *Resources Policy* **59** 44.
[35] Neupane B B, Sharma A, Giri B and Joshi M K 2020 *Heliyon* **6** e03791 11pp.
[36] Allen A J, Thomas J J and Jennings 2007 *Nat Mater* **4** 311.
[37] Yeqing S, Min D and Anqun L 2011 *Materials and Structures* **44** 1735.
[38] Chiang W S, Fratini E, Ridi F, Lim S W, Yeh Y Q, Baglioni P, Choi S M, Jeng U S and Chen S H 2013 *J Colloid Interface Sci* **398** 67.
[39] Das A, Mazumder S, Sen D, Yalmali V, Shah J G, Ghosh A, Sahu A K and Wattal P K 2014 Journal of Applied Crystallography **47** 421.
[40] Aranda M A G 2016 *Crystallography Reviews* **22** 150.
[41] Kogure T, Kameda J, Matsui T and Miyawaki R 2006 *American Mineralogy* **91** 1363.
[42] Millette J R 2015 *The Microscopy* **63** 11.
[43] Kong M, Vogt T and Lee Y 2018 *Current Applied Physics* **18** 1218.
[44] Puig T P, Torices D Y A and Sole J 2020 *Minerals* **10** 388.

Acknowledgments
This work is partially supported by Chiang Mai University (Center of Excellence in Materials Science and Technology, and Department of Physics and Materials Science at the Faculty of Science), and the Department of Earth Sciences at the Faculty of Science of Kasetsart University. The Synchrotron Light Research Institute and Coax Group Corporation Ltd. (Headquarter) are thanked for the analysis instruments. Ms. Witavakorn Sirimontri Ms. Sunisa Sainete, and the BL4.1 team are also thanked for the analysis.

Quantification of climate change mitigation potential at Novi Sad landfill

D Milovanovic[1], M Zivancev[1], D Ubavin[1], V Bezanovic[1], M Petrovic[1]

[1]University of Novi Sad, Faculty of Technical Sciences, Department of Environmental Engineering and Occupational Safety and Health, Trg Dositeja Obradovića 6, 21000 Novi Sad, Serbia

E-mail: dusanmilovanovic@uns.ac.rs

Abstract. Unsanitary landfills and inadequate waste management system represent the 3rd largest global anthropogenic source of methane emissions in the environment. Waste management system in Novi Sad, the second largest city in Serbia, significantly contributes to the national carbon footprint. Unsanitary Municipal solid waste landfill in Novi Sad receives and deposits mixed waste without any treatment, thus causing the organic waste to biodegrade within the landfill body, and produce and emit the landfill gas (mostly CH_4 and CO_2) in the environment. Other potential sources of greenhouse gas emissions from waste management system include vehicles used for waste collection, machinery used at the landfill and landfill fires. The purpose of this study is to determine total baseline emissions of relevant greenhouse gases from the waste management system in Novi Sad, and to estimate and quantify the emission reduction potential for defined organic waste treatment scenario. The potential scenarios for organic waste diversion and treatment from the landfill include composting, anaerobic digestion, mechanical biological treatment, waste to energy, etc. Quantification of baseline emissions and climate change mitigation potential are estimated using the Solid Waste Emissions Estimation Tool, the model developed by Climate and Clean Air Coalition.

1. Introduction

Waste management system represents the third most important anthropogenic source of emissions of methane, which is considered as primary driver of climate change. This sector produces emissions of methane, carbon dioxide and black carbon, as the main source of particulate matter in the air. Novi Sad is the second largest city in Serbia, with a population of approximately 340,000 as of 2014. Novi Sad and the municipalities of Backa Palanka, Backi Petrovac, Beocin, Zabalj, Srbobran, Temerin, Vrbas, and Becej form the South Backa Waste Management Region (SBWMR). The SBWMR has a combined population of approximately 600,000 and encompasses approximately 3,000 square kilometers (km^2). General information about SBWMR is presented in Table 1.

Table 1. General information about SBWMR

	Total population	kg/ capita/ day	MSW (t)	Garden waste (t/year)	Food waste (t/year)	Total organic waste (t/year)	Rural population (%)
Serbia	7,164,132						
Autonomous province (AP) of Vojvodina	1,912,095						
Novi Sad	346,163	1.07	135,194	18,075	40,788	58,863	15
Becej	36,663	0.94	12,579	2,533	4,139	6,672	37
Beocin	15,551	0.94	5,336	713	1,610	2,323	52
Backa Palanka	54,631	0.94	18,744	4,658	6,127	10,785	50
Backi Petrovac	13,222	0.86	4,150	838	1,144	1,982	52
Temerin	28,244	0.94	9,691	1,952	3,188	5,140	10
Zabalj	25,873	0.86	8,122	1,641	2,239	3,880	64
Srbobran	16,073	0.94	5,515	1,114	1,520	2,634	26
Vrbas	41,378	0.94	14,197	2,859	4,671	7,530	42
SBWMR	577,798	1.01	213,526	34,384	65,426	99,810	26

The nine municipalities in the SBWMR produce approximately 214,000 tons per annum (t/year) of solid waste. Nearly half of the solid waste generated in the region is organic. Overall, the region produces approximately 65,500 t/year of food waste and 34,500 t/year of garden waste.

The Novi Sad commercial sector (hotels, restaurants, and cafeterias) produces 805 t/year of organic waste from approximately 730 facilities. In addition, the city's schools produce 283 t/year in food waste. Overall, these large-scale generators produce about 1,100 t/year of organic waste, primarily food waste, representing about 3 percent of the total organic waste produced in Novi Sad.

Landfilling is primary waste management practice in all municipalities. In addition, recycling is present, but only in certain municipalities and in small percentages. Although there were some attempts to introduce partial source separation system, it has not come to life in any of the municipalities. Relatively greater amounts of recyclable materials are separated within the waste separation line in Novi Sad. Since the input material for the separation line is mostly "low-quality" separated dry recyclables waste stream, the percentage of sorted materials is not high, and does not exceed 10%. Also, the capacity of the current plant is insufficient, and only about 10-15% of the total MWS generated in Novi Sad can be processed.

Therefore, all biodegradable waste is disposed of on landfills without any pretreatment and without controlled emissions of Greenhouse gases (GHG), which poses significant environmental risk that can be measured in metric tons CO_2 equivalent. In order to decrease the emissions, it is necessary to diver significant amount of biowaste from the landfill and find the best possible options for treating this waste stream. The most common treatment methodologies, feasible and potentially affordable for the developing countries include composting and anaerobic digestion (AD).

The aim of this paper is to estimate total emissions of methane and black carbon from waste management sector of SBWMR and compare them with emission reductions for different biowaste treatment technologies, in order to facilitate decision making regarding mitigation of climate change effects.

2. Material and methods

As part of the research conducted for the purposes of this project, the calculation of baseline GHG emissions is based on implementation of the SWEET tool developed for the Climate and Clean Air Coalition (CCAC).

SWEET assists stakeholders in estimating emissions and comparing the emissions reduction benefits of different waste management scenarios. The tool can be used to inform MSW management decision-making and priority setting and allows cities to benchmark and project their emissions over time.

The tool assists users in determining first-order city-level estimates of annual emissions of methane, black carbon, and other pollutants (e.g., carbon dioxide) from various sources including:

- Waste collection and transportation
- Waste burning (including open burning and fires at landfills and dumpsites)
- Landfills and dumpsites
- Waste handling equipment (e.g., forklifts, bulldozers)
- Organic waste management facilities (e.g., composting facilities)
- Waste combustion equipment (e.g., waste-to-energy facilities)

Emissions are expressed in metric tons CO_2 equivalent.

SWEET is designed to provide estimates of waste sector Short-lived Climate Pollutants (SLCPs) emissions for cities throughout the world, and to evaluate the effects of alternative waste management strategies on those emissions. Although SWEET uses state-of-the-industry assumptions and calculation methods, the emissions estimates should be considered as approximate and not a substitute for detailed technical analyses and feasibility assessments.

In order to assess the current state of greenhouse gas emissions using the model described in the previous chapter, it is necessary to define the input parameters that include the following information:

- Population covered by the waste collection system – 564,290.
- Climate conditions – Average annual precipitation 578 mm/year and mean annual temperature 11.2°C.
- Waste generation data – 1.01 kg/capita/day with average annual growth rate in quantity of 1%.
- Composition of waste – biowaste 44.6%, recyclables 34.8% and others 20.6%.
- Waste quantities treated by alternative methods (composting, AD, incineration, recycling) – 50,000 tons per year for both scenarios.
- Number and type of waste collection vehicle – 41 heavy-duty trucks and 9 light-duty trucks, all trucks use diesel, 730 working hours per year for each truck.
- Data on uncontrolled incineration of waste (landfill fires and open-air burning) – 3% of waste burned in open by residents and 1% of waste burned at the landfills or dumpsites.
- Number and type of waste handling machinery at the landfill – 5 forklifts, 6 buldozers and 5 tractors.
- Landfill information – site opening year 1980, annual disposal – 107,139 tons per year, average waste depth 25m, without active gas extraction.

3. Results and discussion

The results of emission reductions for composting and AD are shown in the Tables 2 and 3 respectively, in comparison with the emissions from current waste management practice (Business as usual – BAU). The emission results are presented for CH_4, as well as particulate matter ($PM_{2.5}$, PM_{10}) resulting from the black carbon emission.

Table 2. Emission reduction for Scenario with Composting plant (Metric tons)

Year	CH_4	$PM_{2.5}$	PM_{10}	Total emission eq CO_2
2020	0	-25	-10	0
2021	0	-25	-10	0
2022	4,593	-27	-11	3,633
2023	-1,141	-27	-11	-2,110
2024	-6,091	-27	-11	-7,069
2025	-10,376	-27	-12	-11,365
2026	-14,101	-28	-12	-15,099
2027	-17,350	-28	-12	-18,359
2028	-20,197	-28	-12	-21,216
2029	-22,702	-28	-12	-23,731
2030	-24,917	-29	-12	-25,956
2031	-26,885	-29	-12	-27,934
2032	-28,643	-29	-12	-29,703
2033	-30,222	-30	-12	-31,292
2034	-31,649	-30	-13	-32,730
2035	-32,946	-30	-13	-34,037
2036	-34,131	-30	-13	-35,234
2037	-35,222	-31	-13	-36,336
2038	-36,231	-31	-13	-37,356
2039	-37,171	-31	-13	-38,307
2040	-38,051	-32	-13	-39,198
2041	-38,879	-32	-14	-40,039
2042	-39,665	-32	-14	-40,835
2043	-40,412	-33	-14	-41,594
2044	-41,127	-33	-14	-42,322
2045	-41,815	-33	-14	-43,021
2046	-42,480	-34	-14	-43,698
2047	-43,124	-34	-14	-44,355
2048	-43,751	-34	-15	-44,994
2049	-44,364	-35	-15	-45,619
2050	-44,965	-35	-15	-46,233

The construction of a composting plant with a total capacity of 50,000 t/year implies a diversion of that amount of waste from landfills, and therefore a reduction in emissions in comparison to BAU. The composting plant is proposed to start operating in 2022. Therefore, the trend of a constant increase in total emissions will be discontinued in that year, and total emissions will continue to decrease until 2033, followed by a further increase.

Total emissions under this Scenario are projected to increase by about 6% between 2019 and 2050, but compared to BAU, total emissions are down approximately 17.7% over the same time period.

Table 3. Reduction of emissions for Scenario with AD plant

Year	CH_4	$PM_{2.5}$	PM_{10}	Total emission eq CO_2
2020	0	-25	-10	1
2021	0	-25	-10	1
2022	1,288	-27	-11	213
2023	-5,171	-27	-11	-6,256
2024	-10,753	-27	-12	-11,850
2025	-15,595	-27	-12	-16,703
2026	-19,811	-28	-12	-20,929
2027	-23,495	-28	-12	-24,625
2028	-26,730	-28	-12	-27,870
2029	-29,582	-29	-12	-30,735
2030	-32,111	-29	-12	-33,274
2031	-34,363	-29	-12	-35,539
2032	-36,382	-29	-12	-37,569
2033	-38,200	-30	-13	-39,399
2034	-39,848	-30	-13	-41,059
2035	-41,350	-30	-13	-42,573
2036	-42,728	-31	-13	-43,964
2037	-44,000	-31	-13	-45,248
2038	-45,181	-31	-13	-46,442
2039	-46,285	-32	-13	-47,558
2040	-47,322	-32	-14	-48,607
2041	-48,301	-32	-14	-49,600
2042	-49,232	-32	-14	-50,544
2043	-50,122	-33	-14	-51,446
2044	-50,975	-33	-14	-52,313
2045	-51,798	-33	-14	-53,149
2046	-52,595	-34	-14	-53,959
2047	-53,369	-34	-15	-54,748
2048	-54,125	-34	-15	-55,518
2049	-54,866	-35	-15	-56,272
2050	-55,592	-35	-15	-57,012

The construction of an AD plant with a total capacity of 50,000 t/year implies a diversion of that amount of waste from landfills, and therefore a reduction in emissions in comparison to BAU. The plant is proposed to start operating in 2022, and the trend of a constant increase in total emissions will be discontinued that year. Total emissions will decline until 2042, followed by a slight increase again in 2050.

Total emissions under this Scenario are projected to decrease by about 1% between 2019 and 2050, but compared to BAU, total emissions are down approximately 22.4% over the same time period.

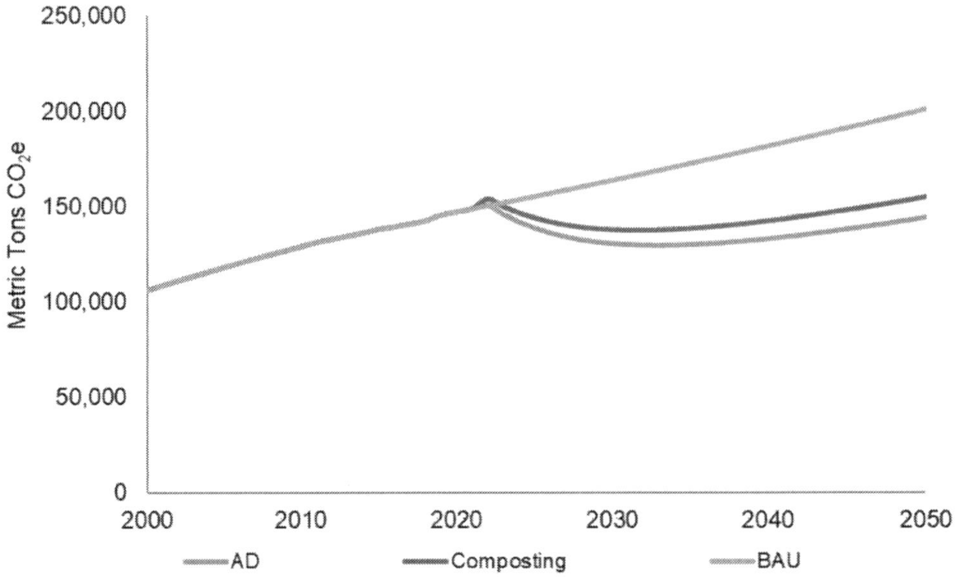

Figure 1. Overview of total emission reduction for both Scenarios in comparison to BAU

A Graph that combines total emissions from both scenarios, as well as total emissions for the BAU is presented in Figure 1. The graph clearly ranks options based on total emissions from the waste management system. Continuation of existing practices is definitely leading to the highest emissions and a constant upward trend is observed. Significant emission reductions are observed for both scenarios, in which equal amounts of organic waste (50,000 t/year) are diverted, but the difference in total emissions is affected by the type of treatment. Therefore, the solution that contributes to the highest emission reductions is the diversion of 50,000 t/year of organic waste from the landfill and treatment within the AD plant, whose emissions in 2050 (after 28 years of use) would reach the level from the beginning of the plant's operation.

4. Conclusions

The research was conducted to consider the impact of waste management system and implementation of different biowaste treatment scenarios in SBWMR on greenhouse gas emissions. Without any changes to waste management practice, emissions will constantly increase and will reach over 200 000 tons of CO_2 equivalent in 2050.

Considering the requirements of the Landfill Directive regarding the reduction of biodegradable waste landfill disposal that should successively reach 65% reductions from the reference year, it turned out that the current system could not result in meeting this objective.

Both proposed scenarios contribute to achieving the objectives of diversion of biodegradable waste from landfills under the aforementioned Directive, which would also result in a more drastic reduction of GHG emissions, around 25% for the Composting Scenario and about 30% for the AD Scenario.

Therefore, in order to reach the biodegradable waste diversion goals, the decision makers within SWMBR will need to start implementing solutions for treatment of biodegradable waste, which include source separation and various treatment options for separately collected organic waste (food waste and green waste).

References

[1] Statistical Office of the Republic of Serbia, Opštine i regioni u Srbiji, 2018
[2] Official Gazette of RS, The national waste management strategy for the period 2010-2019
[3] Faculty of technical scienses, Utvrđivanje količine i morfološkog sastava iz komercijalnih objekata u Novom Sadu u cilju definisanja odgovarajućeg sistema za upravljanje komercijalnim tokom na lokalnom nivou, 2018.
[4] Živančev M, Challenges in development of waste collection systems in emerging countries, 2016 *ISWA-SWIS Winter School Proceedings*
[5] Živančev M, Batinić B, Vujović S, Stanisavljević N and Ribić Č, Possibilities for improvement of municipal waste collection system in transition countries, 2016, CYPRUS2016 4th International Conference on Sustainable Solid Waste Management, Limassol
[6] Milovanović D, Živančev M, Hossain S, Hossain I, Ubavin D and Djogo M, Comparison of landfill operations in developed and developing countries, 2017, The 8th PSU-UNS International Conference on Engineering and Technology (ICET-2017), Novi Sad, Serbia
[7] Faculty of Technical Sciences University of Novi Sad, Report on quantities and morphological composition of waste in 9 representative municipalities in Serbia, 2015.
[8] Solid Waste Emissions Estimation Tool (SWEET) version 3.0: https://ccacoalition.org/en/resources/solid-waste-emissions-estimation-tool-sweet-version-30 / 10.11.2019.

IOP Conf. Series: Materials Science and Engineering 1163 (2021) 012030 doi:10.1088/1757-899X/1163/1/012030

Screening analysis of pesticides in the protected area of Zobnatica Lake

I Mihajlović[1], N Živančev[1], M Sremački[1], D Milovanović[1], I Špánik[2], M Novaković[1] and M Petrović[1]

[1]University of Novi Sad, Faculty of Technical Sciences, Department of Environmental Engineering and Occupational Safety and Health, Trg Dositeja Obradovića 6, 21000 Novi Sad, Serbia

[2]Institute of Analytical Chemistry, Faculty of Chemical and Food Technology, STU, Radlinského 9, 812 37 Bratislava, Slovakia

E-mail: dusanmilovanovic@uns.ac.rs

Abstract. The aim of study was to characterize pollution of water in protected area of Zobnatica Lake in the vicinity of Backa Topola, Serbia. The sampling campaigns were conducted in spring and summer 2019 in order to identify organic pollutants in water samples. Samples were prepared by liquid-liquid extraction and analyzed by GC-MS in SCAN and SIM mode. Screening analyses indicated the presence of pesticides: terbuthylazine, alpha-cypermethrin, aminomethyl phosphonic acid (AMPA) and urea. Terbuthylazine was detected in both sampling campaigns. Results specified the need for comprehensive water monitoring program in order to determine specific pollutants in protected area of Lake Zobnatica caused water pollution by agricultural activities.

1. Introduction

The water consumption in agricultural activities worldwide is around 70%. Pollution caused by agriculture can contaminate not only water, but also the whole environment. The basic task of modern agricultural production is to achieve high and quality yields of agricultural crops. All crops in any form of production are accompanied by characteristic harmful organisms; hence it is necessary to implement appropriate protection measures. The use of herbicides is one of the chemical measures to control weeds. Herbicides may possess selective action, i.e. to destroy only certain types of weeds or total action, to control all kinds of weeds. Herbicides used in agriculture could be transported by rain and melting snow to surface water and from soil to groundwater by leaching processes [1]. Impaired and degraded ecosystems have reduced tolerance and adaptability to environmental condition variations.

Screening analyses and identification of specific pesticides in protected areas caused water pollution by agricultural activities are of great importance in defining the comprehensive monitoring programs as the first step in recognition of water pollution and mitigation of negative impacts of agricultural activities. According to the literature research, only few studies are based on pesticides identification in water samples of protected areas [2, 3].

Main sources of water pollution in Vojvodina, Serbia, are agricultural activities. The poor water quality is not only causing the loss of biodiversity in wetlands, but often makes the surface waters inadequate for irrigation [4-6].

The objective of study was to obtain information about pollution of protected area of Zobnatica Lake by pesticides used in agricultural activities in the vicinity of Lake. Evaluation of water pollution from agricultural sources will result in better management of environmental protection in region. Two sampling campaigns were conducted in spring and summer in 2019 for the purpose of identification of pesticides in run-off water samples in protected area of Zobnatica Lake.

2. Material and methods

2.1. Sampling site

Run-off water samples were collected in the vicinity of Lake Zobnatica, Serbia, in two sampling campaigns, in spring and summer 2019. Water from collectors, with labels K-a to K-f (K-a - 45°53'8.72"N, 19°36'54.73"E; K-b - 45°53'8.78"N, 19°36'55.24"E; K-c - 45°51'21.61"N, 19°36'43.98"E; K-d - 45°51'24.87"N, 19°36'56.60"E; K-e - 45°51'23.45"N, 19°36'56.00"E; K-f - 45°51'21.95"N, 19°36'53.58"E) were used for analyses of run-off water quality. Lake Zobnatica was formed in 1976 in the valley of the river Krivaja. Zobnatica is located 5 km north of Bačka Topola.

Water samples were collected in amber glass 2 L bottles and represent 2-hour composite sample, which is consisted of 8 random samples collected in 2 hours, in 15 minutes intervals. Samples were collected and analyzed in duplicate.

Figure 1. Sampling site – Lake Zobnatica, Serbia

2.2. Sample analysis

Water samples were prepared with liquid-liquid extraction (LLE) using dichloromethane as solvent. The extract was evaporated to 1.5 mL in Kuderna-Danish apparatus. Screening analysis was performed in SCAN and SIM mode on a gas chromatogram (GC) coupled with a mass selective detector (MSD) (Shimadzu QP2010Ultra). GC injector was set to splitless operating mode. The capillary column HP-5MS (30m•0,25mm•0,25μm df) was used for the analysis. Oven temperature gradient program was set to hold time of 10 minutes on 40 °C, then increase of 2 °C per minute was adjusted to 300 °C. Helium was used as carrier gas. The NIST Mass Spectral Library was used to identify chemical compounds.

3. Results and discussion

Detected pesticides within the screening analysis of run-off water with their chemical structure, similarity index with NIST database and names of commercially used products in agriculture are presented in Table 1.

Table 1. Detected pesticides in run-off samples in the vicinity of Zobnatica Lake

Detected pesticide	Chemical structure	Similarity index (with NIST database) %	Comercial product
Terbuthylazine	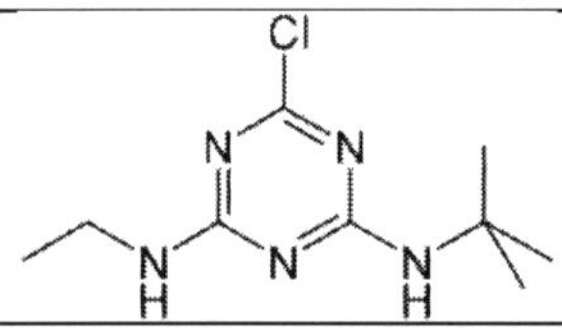	98	Terbis, Rezon, Radazin- herbicide
Alpha-cypermethrin	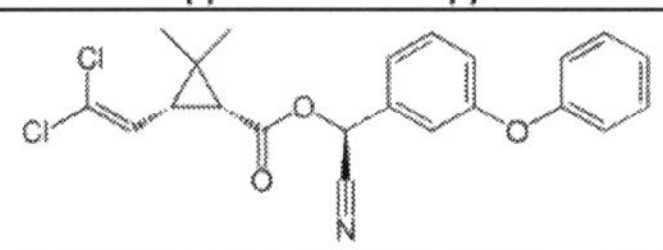	99	FASTAC® 10 EC - insecticide
Urea	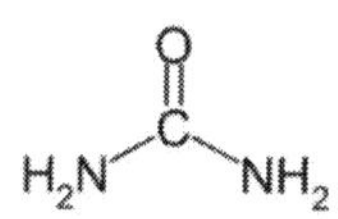	95	UREA N 46 - herbicide
Aminomethyl phosphonic acid (AMPA)	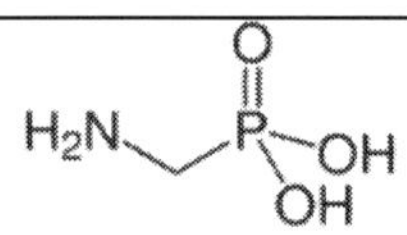	98	Primary degradation product of the herbicide glyphosate (GLIFOHEM, GLIFOL)

The screening analyses indicated the presence of pesticides: terbuthylazine, alpha-cypermethrin, aminomethyl phosphonic acid (AMPA) and urea. Terbuthylazine was detected in both sampling campaigns, in spring and summer 2019.

Terbutylazine belongs to the Triazine group. The herbicide Terbutylazine essentially acts as an inhibitor of photosynthesis. It forms a herbicidal film that prevents weed germination, but is also adopted through weed leaves. This pesticide is a selective for potatoes, maize, fruit trees and other plant species.

Cypermethrin is synthetic compound and insecticide with contact and digestive action. It has a pronounced effect on adult insects and larvae, but it also has a significant effect on eggs with a thin membrane. In higher concentrations Cypermethrin has a repellent effect on pests. Cypermethrin is lipophilic, makes direct contact with the waxy substances of the plant cuticle and enables regulatory function to plant growth. Cypermethrin is toxic to aquatic insects and moderately toxic to human. Toxic effects are expressed by dermal exposure.

Urea herbicides have broad spectrum of action on a large number of weeds, but do not affect perennial weeds. The maximum concentration of the Urea herbicide is maintained at the surface layer of the soil and plant absorbs it by the root. Urea herbicides act primarily through the roots. They are translocated via water to the leaves where they bind to the chloroplast and interfere with the photosynthesis process resulting in the death of sensitive weeds. Urea herbicides bind tightly to soil particles, they are poorly rinsed and persistent. They decompose within 3-6 months and express low acute toxicity [7].

AMPA is the primary degradation product of herbicide glyphosate. Glyphosate is organophosphorus compound and non-selective highly systemic pesticide used for the control of a large number of annual grass and broadleaf weeds. It is most commonly used in the form of the isopropyl amino salt and applied to weed leaves. Glyphosate binds strongly to soil colloids to such an extent that it is no longer accessible to plants. In the soil, decomposition takes place mostly with the help of microorganisms, and to a lesser extent, chemically by hydrolysis, oxidation and photodegradation. Detected herbicides and insecticide could cause surface and ground water pollution with negative impact on biota and human health.

4. Conclusions

Screening analyses indicated water pollution with pesticides and need for detailed analyses. Results specified the need for comprehensive water monitoring program in order to determine the natural and artificial impacts and to maintain the required water quality, as the first step in mitigation of negative impacts of agricultural activities and biodiversity protection. Further research should be aimed at establishing detailed monitoring programs to assess the quality of water media at protected water areas with desirable inclusion of detected organic micropollutants, such as pesticides.

Acknowledgement

This research has been supported by the Ministry of Education, Science and Technological Development through the project no. 451-03-68/2020-14/200156: “Innovative scientific and artistic research from the FTS (activity) domain”, the Bilateral project funded by the Ministry of Education, Science and Technological Development (contract no. 337-00-107/2019-09/16) and Slovak Research and Development Agency under contract SK-SRB-18-0020.

References

[1] Food and Agriculture Organization of the United Nations, FAO 1996 *Control of water pollution from agriculture* 2
[2] Kaiser K 2011 *Bulletin of environmental contamination and toxicology* **86** 56
[3] Ruiz-Toledo J, Castro R, Rivero-Pérez N, Bello-Mendoza R and Sánchez D 2014 *Bulletin of environmental contamination and toxicology* **93** 289
[4] Obrovski B, Mihajlović I, Vojinović-Miloradov M, Petrović M, Sremački M, Brkić M and Radić J 2018 *International Conference the Holistic Approach to Environment*, Sisak, Croatia, 550.
[5] Obrovski B, Mihajlović I, Petrović M, Sremački M, Radić J and Vojinović Miloradov M 2019 *International Conference "Water for all"*, Osijek, Croatia, 100.
[6] Sremački M, Obrovski B, Mihajlović I, Petrović M, Vojinović Miloradov M, Kicošev V, Dragićević P and Radić J 2018 *Water and sanitation technique* **3-4** 550
[7] Watt B, Proudfoot A, Bradberry S and Vale A 2005 *Toxicological Reviews* **24** 161

Hydrothermal blackening on AISI 4140 steel with and without hydrazine hydrate

P Wiman, K Doungkaew and T Siripongsakul*

High Temperature Corrosion Research Centre, Department of Materials and Production Technology Engineering, Faculty of Engineering, King Mongkut's University of Technology North Bangkok, 1518, Pracharat 1 Road, Wongsawang, Bangsue, Bangkok, 10800, Thailand

E-mail: thamrongsin.s@eng.kmutnb.ac.th

Abstract. The blackening coating is a conversion coating with the black colour of crystallised magnetite (Fe_3O_4) on the surface of steel. The coating is to be soaked with oil or wax to prevent the corrosive environment. Recently, the technique to synthesise the black coating, hydrothermal method is introduced. In the synthesis process, hydrazine hydrate ($N_2H_4 \cdot H_2O$) is often recommended as a catalyst but it is also reported toxicity. In this work, AISI 4140 was hydrothermally immersed in the $FeSO_4 \cdot 7H_2O$ added NaOH solution with and without addition of hydrazine hydrate. During the process, the temperature and pressure of the chamber were controlled at 140°C and 2 bar, respectively. To confirm the formation of Fe_3O_4, the samples were characterised by SEM and XRD. These results showed that the coatings were composed of octahedron morphology in all conditions and the thicknesses of the coating with hydrazine hydrate concentration of 0.0, 0.1, and 0.5M were 7.05, 9.25, and 8.75 µm, respectively. With hydrazine hydrate, the Fe_3O_4 layer can be easily formed. However, the results of the oxalic acid spot test and potentiodynamic polarisation indicated the equivalence in quality of the coating, despite the absence of hydrazine hydrate.

1. Introduction

Blackening is a process to convert the surface of steel to be the crystallised magnetite (Fe_3O_4) of which the aesthetic black colour is shown. The oxide surface is expected to be passive and prevents the steel substrate from the corrosive environment. In general usage, oil or wax is further applied to the coating to enhance the corrosion resistance [1]. To dates blackening practically exhibits effectiveness in various applications with low cost and long life time of service.

The conventional blackening process is done in hot alkaline consisted sodium hydroxide (NaOH) with temperature range of 140-150°C. Some oxidizing agents, such as sodium nitrite ($NaNO_2$) and sodium nitrate ($NaNO_3$) are added in the solution to accelerate the chemical conversion and lower the process temperature [1]. Recently, the mixing solution which consisted ferrous sulphate hydrate ($FeSO_4 \cdot 7H_2O$), NaOH and hydrazine hydrate ($N_2H_4 \cdot H_2O$) has been proposed for blackening, and the higher pressure than the atmospheric pressure is applied to lower the concentration of the chemicals in the solution [2, 3]. This blackening is called the hydrothermal method.

Nevertheless, according to the material safety data sheets, most of the oxidizing agents such as $NaNO_2$, $NaNO_3$, and $N_2H_4{\cdot}H_2O$ are harmful, especially in vapour phase they tend to be toxic. Therefore, the solutions which comprised ferrous sulphate hydrate ($FeSO_4{\cdot}7H_2O$), NaOH with or without hydrazine hydrate were studied for the hydrothermal blackening in order to reduce the usage of hydrazine hydrate. The formation of magnetite on a steel substrate with and without $N_2H_4{\cdot}H_2O$ was analysed by microstructural characterisation. Lastly, to assure the quality of the black coating, the corrosion behaviours in terms of the oxalic acid spot test and potentiodynamic polarisation were investigated.

2. Material and experimental producer

For study of the magnetite formation, AISI 4140 steel of which the chemical composition was Fe with 0.43C-0.22Mn-0.82Si-0.20 Mo-0.006 P in wt.% was used as a substrate for blackening. The sample size was 3.0x1.2 cm^2 prepared from a 2-mm thick steel sheet by laser cutting. Each sample was perforated a hole with a diameter of 3 mm on the center top for suspension as immersed into the process solution. Before blackening, surface finishing of all samples was done by grinding with SiC sandpapers up to 1,000-grit number. Then the samples were cleaned in acetone with an ultrasonic bath and dried in air.

For blackening, the hydrothermal blackening solution was based on 0.1M ferrous sulphate ($FeSO_4{\cdot}7H_2O$), 1.0M sodium hydroxide (NaOH) and hydrazine hydrate ($N_2H_4{\cdot}H_2O$). The effect of hydrazine hydrate was investigated by varying the concentration of hydrazine hydrate to be 0.0, 0.1, and 0.5M. Those three conditions of reagents were dissolved in deionised water with magnetic stirring. After completing the solution, the steel samples were suspended and immersed in the blackening bath and they were subjected in an autoclave chamber. The temperatures and pressure of the autoclave were controlled at 140°C and 2 bar, respectively. The coating time is 30 min for all samples.

After blackening, the samples were investigated on characterisation and corrosion resistance. For the characterisation, the surface and cross-sectional morphology were analysed using a scanning electron microscopy (SEM), JEOL JSM-7800F. An X-ray diffractometer (XRD), D8 Discovery, Bruker, with Cu Kα (α =1.5406 Å) radiation was performed to identify the oxide phases formed on the steel surface. The diffraction angles measured from the studied samples was matched with the standard XRD patterns summarised by the International Centre for Diffraction Data (ICDD).

The standard MIL-DTL-13924D test or the oxalic acid spot test was performed after blackening to guage the surface quality of blackening steel in acid media. The samples with the size of 4.0x3.5cm^2 were introduced. This test was done by 3 drops (0.2 ml.) of a 5% oxalic acid solution dropped on surface of blackening steel. The reacted surface with oxalic acid can be observed after 5 min then it will be rinsed and compared. Then the potentiodynamic polarisation testing was performed to confirm the corrosion resistance at room temperature in 3.5% NaCl solution. The apparatus was the Potentiostat/Galvanostat (PGSTAT 302 N). A three-electrode cell configuration was used and consisted a platinum counter electrode, a Ag/AgCl reference electrode and a sample with a surface area of 4.56 cm^2 as the working electrode. All electrodes were placed parallel. Prior to the measurement, the samples were immersed in the 3.5% NaCl solution for 30 min to measure the open circuit potential (E_{corr}). The polarisation curves of samples were measured in a scan rate of 1 mVs^{-1} with the potential range from −0.5 V to 0.5 V with respecting to E_{corr}.

3. Results and discussion

3.1 Characterisation of blackening steel

Figure 1 depicts the XRD patterns of the blackening coating of the steel as prepared by hydrothermal process with different hydrazine hydrate concentrations of 0.0, 0.1 and 0.5M. Only the XRD peaks of iron as AISI 4140 steel and magnetite (Fe_3O_4) were dominantly found. All conditions of blackening steel, the peaks can be readily indexed to cubic Fe with the lattice constant: a=2.87Å, which agrees with the standard pattern ICDD 87-0721. These peaks of Fe correspond to plans (110), (200) and (211). Beside the peaks of Fe, the peaks of magnetite can be defined to cubic Fe_3O_4 with the lattice constant: a=8.38Å as referring to ICDD 88-0866. The major peak of Fe_3O_4 in this work is at (311) plan which is similar to the previously studies on hydrothermal synthesis of magnetite powder and coatings on steel substrates [2,4,5]. Therefore, the blackening coating of Fe_3O_4 can be formed in various conditions.

To observe morphology of blackening steel, SEM images of surfaces and cross-section of the magnetite coating (the left and right columns) samples with 0.0 M (first row), 0.1 M (second row) and 0.5 M (third rows) of hydrazine hydrate were shown in figure 2. The formed Fe_3O_4 crystallite on steel was able to be detected as octahedron morphology in all concentrations of hydrazine hydrate. The thicknesses of the coatings with 0.0, 0.1, and 0.5M hydrazine hydrate were 7.05±0.28, 9.25±0.62, and 8.75±0.47 μm, respectively, in figure 2 (the right column). The Fe_3O_4 sub-layers composed of smaller and bigger crystallite sizes at the inner layer and the outer layer (at the surface) as shown in figure 2 (SEM images at high magnitude in the left column). Zhu et al. [6] and later Chandra-ambhorn et al. [3] reported that the inner submicron grain and the outer large grain of Fe_3O_4 were formed by the reaction between Fe(II) and Fe(III) with hydrazine hydrate in hydrothermal basic solution. Despite the absence of hydrazine hydrate, Fe_3O_4 layer with the appropriate thickness and grain size can be formed in the present study.

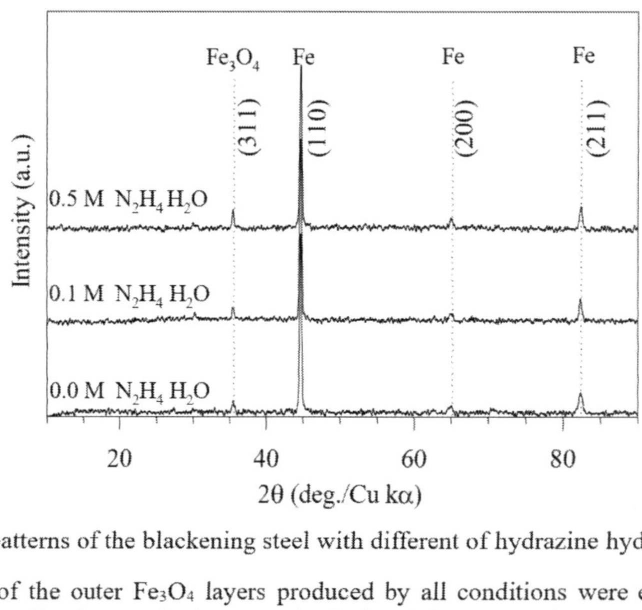

Figure 1. XRD patterns of the blackening steel with different of hydrazine hydrate concentration

The grain sizes of the outer Fe_3O_4 layers produced by all conditions were evaluated using SEM images and the Image J software; furthermore the Debye-Scherer equation was used to compare the results. According to the Debye-Scherer equation, the crystallite grain size (D) of the Fe_3O_4 can be estimated as:

$$D = \frac{k\lambda}{\beta \cos \theta} \quad (1)$$

Where, k is a constant that has a value of 0.94, λ is the wavelength of the X-ray radiation (1.5406 Å), β is the full-width of half-maximum intensity of the peak, and 2θ is the diffraction angle. From the preferred (311) diffraction peaks and the measured outer large grains of Fe_3O_4 layer, the grain size of crystallite Fe_3O_4 prepared by various hydrazine hydrate concentrations were shown in figure 3. The results indicated that the crystallite grain size of the Fe_3O_4 decreased with an increase 0.1 M to 0.5M in the hydrazine hydrates. This result was related with Yan et al. for synthesis Fe_3O_4 nanoparticles which reported that the excess hydroxide ions generated from $N_2H_4 \cdot H_2O$ were absorbed on the surface and underwent the nucleation. The repulsive force between hydroxide ions can inhibit the growth of crystal grains [7].

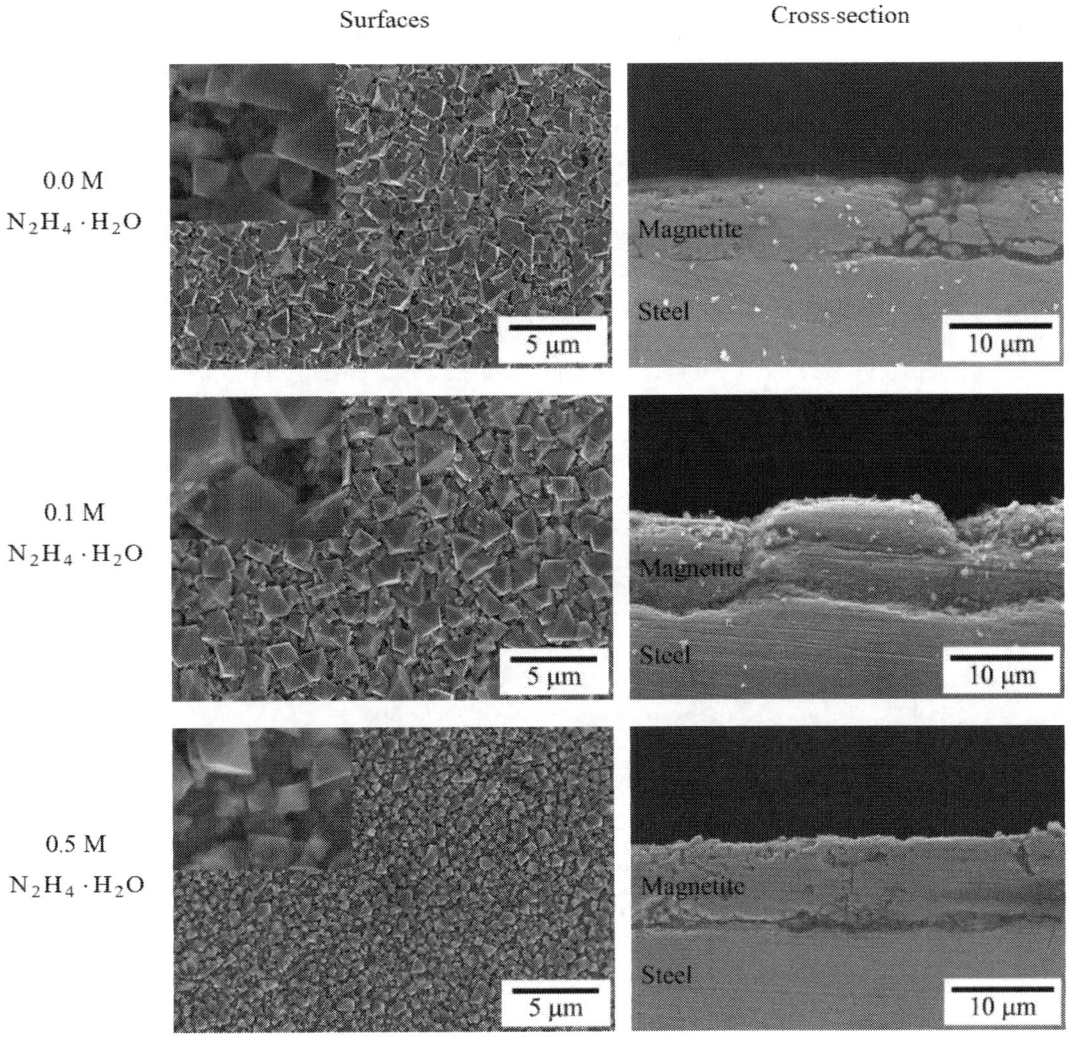

Figure 2. Surfaces and cross-section of the magnetite coating (left and right column) samples with 0.0 M (first row), 0.1 M (second row) and 0.5 M (third rows) of hydrazine hydrate

The possible mechanism of Fe_3O_4 formation was proposed in this experiment [8]. A hydrazine hydrate ($N_2H_4 \cdot H_2O$) was a reducing agent.

$$Fe^{2+} + 2(OH)^- \rightarrow Fe(OH)_2 \quad (2)$$

$$Fe(OH)_2 + N_2H_4 + 4H_2O \rightarrow 2Fe(OH)_3 + 2NH_4OH \quad (3)$$

$$2Fe(OH)_3 + Fe(OH)_2 \rightarrow Fe_3O_4 + 4H_2O \quad (4)$$

According to reactions (2), (3) and (4), Fe $(OH)_2$ contributed Fe (II) which was an iron source (ferrous sulphate) in the alkalescent aqueous media, and it was further oxidised into Fe $(OH)_3$ in the presence of hydrazine hydrate. Under hydrothermal treatment at 140°C and 2 bar, Fe $(OH)_2$ and Fe $(OH)_3$ were oxidised to form the Fe_3O_4 phase. Therefore, Fe_3O_4 crystal nucleus was increasingly formed on the surface of the AISI 4140 steel with submicron size. Then, the crystal nucleus of magnetite grew well at the outer layer. Finally, at the outer surface of the crystallite began to grow and the large size was obtained [6]. Although one condition in the present experiment, the coating was formed without hydrazine hydrate, the Fe_3O_4 layer can be formed. Following reaction (3), the formation of Fe $(OH)_3$ cannot occur. Also, reaction (4) was reduced to the Schikorr reaction (reaction 5) [9].

$$Fe(OH)_2 \rightarrow Fe_3O_4 + 2H_2O + H_2 \quad (5)$$

As mentioned in above the results, from increased thickness of magnetite in present work, it can implied that hydrazine hydrate can increase the nucleation rate as it is a stronger reducing agent than sodium hydroxide. When the concentration of hydrazine hydrate was raised to 0.5M, the nucleation rate is dramatically contributed to Fe_3O_4 crystallite forming on surface of steel with a low growth rate, thus the inner later became submicron crystallite. Compared with the condition without hydrazine hydrate, the crystalline grain size of the 0.1M condition were larger than the others, because possibly the growth of the submicron crystallite was more dominant than nucleation.

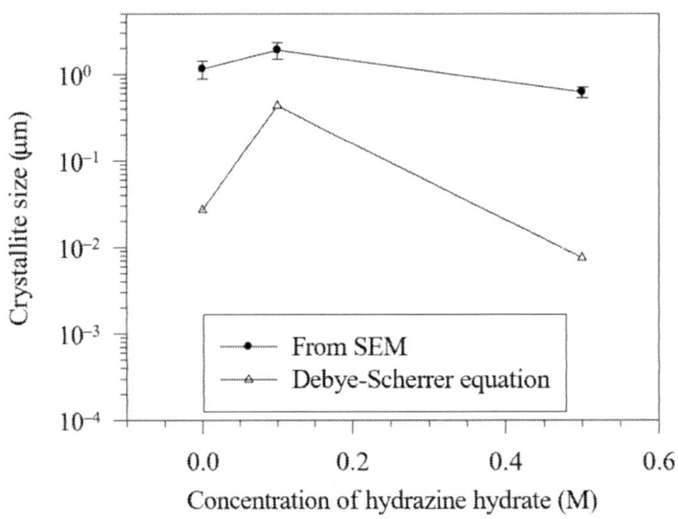

Figure 3. The Crystallite size as a function of hydrazine hydrate concentration on blackening steel

3.2 The oxalic acid spot test

To examine the quality of blackening coating in acid media, figure 4 displays the surface of the magnetite coatings on steel after the oxalic acid spot testing. On the report of MIL-DTL-13924D standard, a remained black colour at the centre of the samples with some or small white stains border indicates a good quality of the coating. In all condition, the blackening coatings were qualified in this test. Some or small white stain border appeared from oxalic acid exposure on the magnetite layer. Therefore, all conditions of blackening process provide good acidic corrosive resistance.

Figure 4. Surface of the magnetite coating on steel after the oxalic acid spot testing

3.3 Potentiodynamic polarisation test

The polarization curves of the magnetite coatings in 3.5% NaCl solution are showed in figure 5. The Tafel extrapolation method was used to determine the corrosion potentials (E_{corr}) and corrosion current densities (i_{corr}). The corrosion potentials of the coatings produced by hydrazine hydrate concentrations of 0.0, 0.1, and 0.5 M were -0.478, -0.507 and -0.483 V vs. Ag/AgCl, respectively. These values were closed to other reports which were -0.505 V [2]. Additionally, the values of the corrosion current densities were 1.48×10^{-5}, 1.58×10^{-5} and 1.74×10^{-5} A cm^{-2} in 0.0, 0.1, and 0.5 M of hydrazine hydrate, sequentially. After the potentiodynamic polarisation test, the blackening samples were remained black colour without the deterioration due to NaCl solution. From this point, it ensures that the polarisation plot was applied the Fe_3O_4 layer, not the steel substrate. The insignificant difference of corrosion current densities indicated that changing the concentration of hydrazine hydrate does not affect to the corrosion rate. Although, the corrosion potentials were slight different due to the steady-state potential of Fe_3O_4 layer with high to low magnetite thicknesses respecting 0.1, 0.5 and 0.0 M of hydrazine hydrate concentration. According to Ohm's law in equation (6), steady-state potential of Fe_3O_4 layer was provided as:

$$E_{Fe_3O_4,steady-state} = |E_{corr}| = i_{corr}\rho\zeta \quad (6)$$

Where, ρ and ζ are the electrical resistivity and the thickness of magnetite, sequentially. Therefore, as the thickness of magnetite increases, the steady-state potential of the Fe_3O_4 layer will shift to a more negative potential or higher absolute potential.

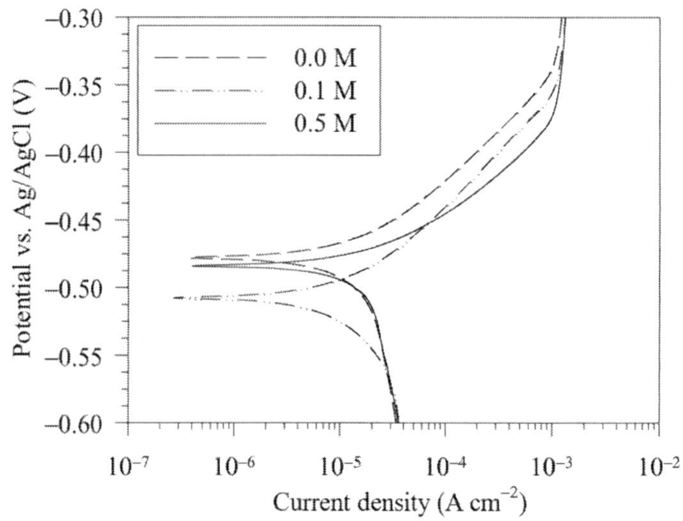

Figure 5. Polarization curves of the magnetite coating at different hydrazine hydrate

4. Conclusion

The AISI 4140 steel was blackened by the hydrothermal process using NaOH solution with varying 0.0, 0.1, and 0.5 M of the hydrazine hydrate at 140°C and 2 bar for 30 min. The results of XRD and SEM confirmed the coating layers of Fe_3O_4 with octahedron morphology forming on the surface of the steel substrates in all blackening conditions. The grains of Fe_3O_4 were composed of two layers: one was micron ordered crystallite at the outer surface and the other was submicron grain at the inner surface. The blackening coating with hydrazine hydrate was thicker than the condition without hydrazine hydrate due to the acceleration of the nucleation rate affected by hydrazine hydrate. The acceleration of nucleation gives rise to the inner fine grain of Fe_3O_4. Among the blackening conditions, with 0.1M $N_2H_4 \cdot H_2O$ the growth step of the outer Fe_3O_4 crystallites was more dominant than the nucleation step. Therefore, the outer large crystallites of blackening steel were obtained. The oxalic acid spot test and potentiodynamic polarisation provided the results showing good corrosive resistance of the blackening coating steel, despite the absence of hydrazine hydrate.

References

[1] Farrel R W 2002 *Met. Finish.* 100 456–62
[2] Zhu H, Cao F, Zuo D, Zhu L, Jin D and Yao K 2008 *Appl. Surf. Sci.* 254 5905–09
[3] Chandra-ambhorn S, Srihirun S and Siripongsakul T 2018 *Anti-Corros. Methods & Mater.* 65 383–88
[4] Wang J, Sun J, Sun Q and Chen Q 2003 *Mater. Res. Bull.* 38 1113–18
[5] Zheng Y-h, Cheng Y, Bao F and Wang Y-s 2006 *Mater. Res. Bull.* 41 525–29
[6] Zhu H, Yang D and Zhu L 2007 *Surf. Coat. Technol.* 201 5870–74
[7] Yan H, Zhang J, You C, Song Z, Yu B and Shen Y 2009 *Mater. Chem. Phys.* 113 46–52
[8] Li Y, Liao H and Qian Y 1998 *Mater. Res. Bull.* 33 841–44
[9] Fattah-alhosseini A, Yazdani Khan H and Heidarpour A 2016 *J. Alloys Compd.* 676 474–80

Acknowledgement
The authors acknowledge financial support from King Mongkut's University of Technology North Bangkok. Contract no. KMUTNB-64-DRIVE-23

IOP Conf. Series: Materials Science and Engineering 1163 (2021) 012032 doi:10.1088/1757-899X/1163/1/012032

Investigation of the Cr volatilisation using the ceramic phase collector technique

T Thublaor[1], P Wiman[1], C Nanna[1], W Jaijitmun[1], S Krittiyanitsakun[1], and T Siripongsakul[1,*]

[1]High Temperature Corrosion Research Centre, Department of Materials and Production Technology Engineering, Faculty of Engineering, King Mongkut's University of Technology North Bangkok, 1518, Pracharat 1 Road, Wongsawang, Bangsue, Bangkok, 10800, Thailand

E-mail: *thamrongsin.s@eng.kmutnb.ac.th

Abstract. An effect of chromium contamination in the lanthanum strontium manganite (LSM) cathode which is a component of solid oxide fuel cell (SOFC) leads to degradation of the device. This is caused by volatilisation of Cr-species from the iron-chromium alloy at high temperature. In this work, the ceramic phase collector technique was used to evaluate the Cr volatilisation. To prove the effectiveness of this method, AISI 430 ferritic stainless steel samples with and without cobalt (Co) spinel coatings were used to compare the Cr condensation in the LSM ceramic. The coatings were produced by electroplating followed by a heat-treatment process. The samples were characterised using SEM with EDX and XRD to confirm the microstructure and the formation of Co spinel. It was found that chromium migration was restrained for coated samples, in contrast the Cr-species deposited in the LSM for uncoated samples. The chromium volatilisation evaluated by the contents of Cr species in the LSM collector was compared to the traditional condensation technique which analysed chemical contents by inductively coupled plasma-mass spectrometry (ICP-MS).

1. Introduction

Recent decades, striving for new sources of sustainable energy has been urgent need to overcome the global energy crisis and environmental problems. Many researches are focused on the chemical reaction to generate electricity directly, such as a battery or a fuel cell due to its efficiency in energy transformation compared to the traditional combustion heat engine. Among many types of fuel cells, the solid oxide fuel cell (SOFC) is an important key device expected to be the next generation of sustainable energy source. An SOFC operates at the intermediate high-temperature range of 600–1000 °C and yields electrical energy by means of oxygen reduction to oxygen ion and ion migration from the cathode to the anode through the ceramic electrolyte. In case of utilising hydrogen as a fuel gas, at the anode side pure water is the product of the oxidation reaction from the fuel cell, thus the SOFC is also known as the clean energy generation device which benefits for low greenhouse gas emission [1, 2].

Interconnect is a component used for connection between SOFC stacks and conveying electricity. The ferritic stainless steel (FSS), which is an iron-chromium (Fe-Cr) alloy, has a potential to be applied as the interconnect due to its operation cost, machinability and mechanical properties. The FSS is not only suitable in the physical term such as compatible thermal expansion with cathode ceramic but also stability in chemical properties both in reducing and oxidising atmosphere [3]. At the operating temperature of SOFC, the passive chromium oxide (Cr_2O_3) is generally formed on the surface of the stainless steel. However, the problem occurs when SOFC stacks are operated for a long period of time, the containing Cr is oxidised and becomes vapour phased Cr-species, $CrO_2(OH)_2$ or $CrO(OH)_2$ [4]. The condensation of vaporised Cr-species is burdened at the cathode which is an ionic conductor. Having been attached by Cr-species makes them lose their functionalities; hence the operation of SOFC fails [3, 5-7]. To prevent this phenomenon, coating the spinel conductive ceramic on the interconnect FSS has been reported on hindering the migration of Cr and reducing the volatilisation of Cr-species [8, 9].

With regard to vaporisation study of Cr-species, many methods for evaluation the amount of volatilised Cr have been developed. A well-known method is the condensation method which can detect the small amount of Cr in the level of 1 ppb using with the inductively coupled plasma-mass spectrometry (ICP-MS) [10]. However, this method needs multiple steps and careful efforts to restore condensates in the column [11]. Another example is the ceramic phase collector technique. This technique is relatively simple compared to the former method. It is performed by enclosing the test material within ceramic plates. Han and co-workers [12] quantitatively reported Cr volatilisation of Crofer22APU using silicon wafer collectors, however, this method needs to analysed by Rutherford backscattering spectroscopy (RBS) which is difficult to access for most researchers.

Therefore, in this work lanthanum strontium manganite (LSM), the cathode material used in SOFCs, has been proposed to be a ceramic collector for evaluation Cr volatilisation. The quantity of Cr was analysed using SEM with EDX. The FSS grade AISI430 samples with and without Mn spinel coating were applied to evaluate the Cr volatilisation in comparison with those evaluated by the traditional condensation method.

2. Experimental

2.1. Sample preparation

The AISI 430 stainless steel was selected for this study. The chemical composition (wt.%) measured using an emission spectroscopy (SPECTRO Model SPECTROMAXx) is shown in Table 1. The samples were provided as the square shape of 25×25 mm^2 coupons with the thickness of 1 mm. All surfaces of samples were polished until 800 grade abrasive paper and cleaned by alcohol in an ultrasonic cleaning machine for 5 min then dried in blow air. For the electroplating process, the electrolyte solution was described in detail in the previous article [13]. The electroplating condition was carried out by using a direct circuit (DC) with current density of 200 mA cm^{-2} for 20 min and cleaned with deionized water (DI). After the electroplating process, the coated samples were subjected to heat treatment in oxygen atmosphere for 24 h at 600 °C which is named the pre-oxidation step. The surface morphologies of heat-treated samples were characterised by a standard scanning electron microscope (SEM) with an energy dispersive X-ray spectroscopy (EDX). The phase structures were identified by a standard XRD diffractometer with Cu Kα radiation.

Table 1. Chemical composition of the AISI 430 stainless steel in weight percent.

Cr	Mn	Si	Ni	C	P	Fe
16.92	0.326	0.295	0.155	0.042	0.018	Balance

2.2. Ceramic phase collector

The yttria-stabilized zirconia (YSZ) powder of 100 g was mixed with the ethanol solution 60 cm^3 and polyvinyl butyral (PVB) 1 g which as the electrolyte support. The mixed solution was ball milled for 6 h. The size of powder was selected with the sieve mesh at 75 μm. The pellet of powder was prepared from 2 gram of weight and pressed into a coin shape with a diameter of 20 mm and thickness of 2 mm. The YSZ pellets were sintered in air at 1400 °C for 2 h with 5 °C min^{-1} of a heating rate. The lanthanum strontium manganite (LSM) was used as the cathode. LSM powder was mixed with the glycerol at a weight ratio of 2:1 to obtain a cathode solid solution. The solution was brushed onto one-side of each YSZ pellet. The thickness of cathode was 3 μm controlled by adhesive tape. The cathodes with electrolyte support were subsequently calcined at 1200 °C for 2 h to obtain the ceramic phase collector.

2.3. Oxidation

The isothermal treatment was performed in a horizontal tube furnace with a gas flow which was 20 vol.% water vapour oxygen (O_2-20%H_2O) for 210 h. In order to generate the water vapour, the oxygen gas with a flow rate 100 ml min^{-1} was delivered downward to the humidity generator with the temperature of water set at 59.4 °C. The sample was placed in the heating zone of the horizontal tube. The furnace was heated from room temperature to 800 °C with a heating rate of 0.5 °C s^{-1}. The oxidation was examined under the isothermal temperature at 800 °C for 210 h in O_2-20%H_2O.

3. Result and discussion

3.1. Formation

Figure 1 shows the XRD result of the spinel coated AISI 430 FSS sample after pre-oxidation in oxygen at 600 °C for 24 h. Two types of cobalt oxide are identified the cubic structure with the peak position matched with the database of Co_3O_4 (ICDD 09-0418) and CoO (ICDD 43-1004). A small amount of CuO matched with the database of CuO (ICDD 41-0254) is observed on the surface. The Cu element is a nobler metal in solution when it is compared with Co and Mn elements. The standard electrode potential of Mn/Mn^{2+}, Co/Co^{2+} and Cu/Cu^{2+} are −1.185, −0.28, and 0.3419 V vs. standard hydrogen electrode (SHE) [14], respectively. This indicate that the Cu ion in solution easily deposition on the surface. The database of Fe (ICDD 87-0722) is showing the existence of the bcc Fe or FSS.

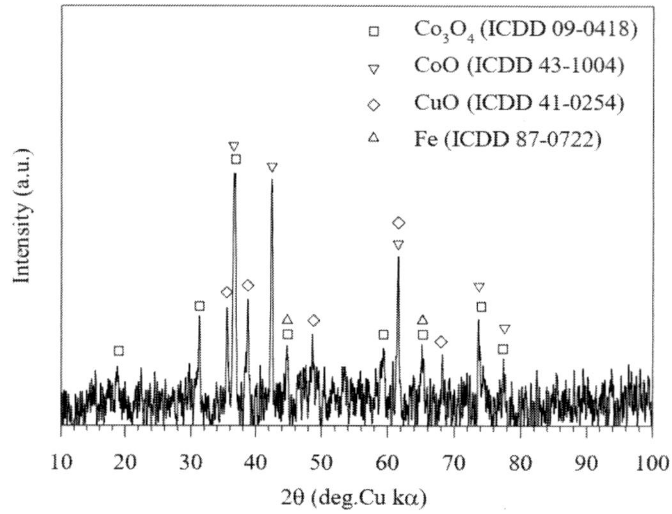

Figure 1. XRD pattern of the spinel coated AISI 430 ferritic stainless steel sample after pre-oxidation in oxygen at 600 °C for 24 h.

3.2. Microstructure analysis

Figure 2 shows the cross-section SEM images using the backscatter electron (BSE) mode of the bare AISI 430 ferritic stainless steel and the spinel coated AISI 430 ferritic stainless steel samples. The images show that the uniform thermal oxide scale has grown on the surface with thickness of 3.3 µm in average and the standard deviation of 1.8 µm for the bare sample. The result of the EDX line scan shows the distribution of Fe, Cr and O across the sample. However, these results are difficult to distinguish between the thermal oxide scale and the steel substrate. Nevertheless, many researchers reported that the Cr_2O_3 and Mn-Cr spinel oxide were generally observed on the stainless steel when they were oxidised in a humidified atmosphere at high temperatures [11, 15-22]. The formation of the thermal oxide scale should be Cr_2O_3. On the other hand, for the spinel coated sample, as shown in figure 2 (b), the average thickness thermal oxide scales is 14.5 ± 6.6 µm. The cross-section SEM image using backscatter electron (BSE) mode and the EDX line scan across interface between the spinel coating and the thermal oxide scale and interface between the thermal oxide scale and substrate. The results obtained from the EDX line scan show that the two different layers can be distinguished in the thermal oxide scale for the spinel coated sample. The relatively high intensity of Fe, Co and O were observed at the interface between the spinel coating and the thermal oxide scale. These results agreed the previous work done by Wu et al. [23] who reported that the top layer of a scale layer is formed by $(Mn, Co, Fe)_3O_4$ spinel for the cobalt-coated and Mn-Co coated samples. For the inner layer of the thermal oxide, the Mn, Cr and O were detected. It indicated that formation of the thermal oxide scale was corresponding to the Mn-Cr spinel oxide. The thermal oxide scale containing the Mn, Cr and O at inner layer should be $MnCr_2O_4$. The assumption was confirmed by Yang et al. [8, 24]. They reported that the coated spinel sample showed the presence of Cr_2O_3 and $MnCr_2O_4$ on the oxidised sample after oxidation in air at 800 °C up to 100 h, with regard to the results of XRD and SEM.

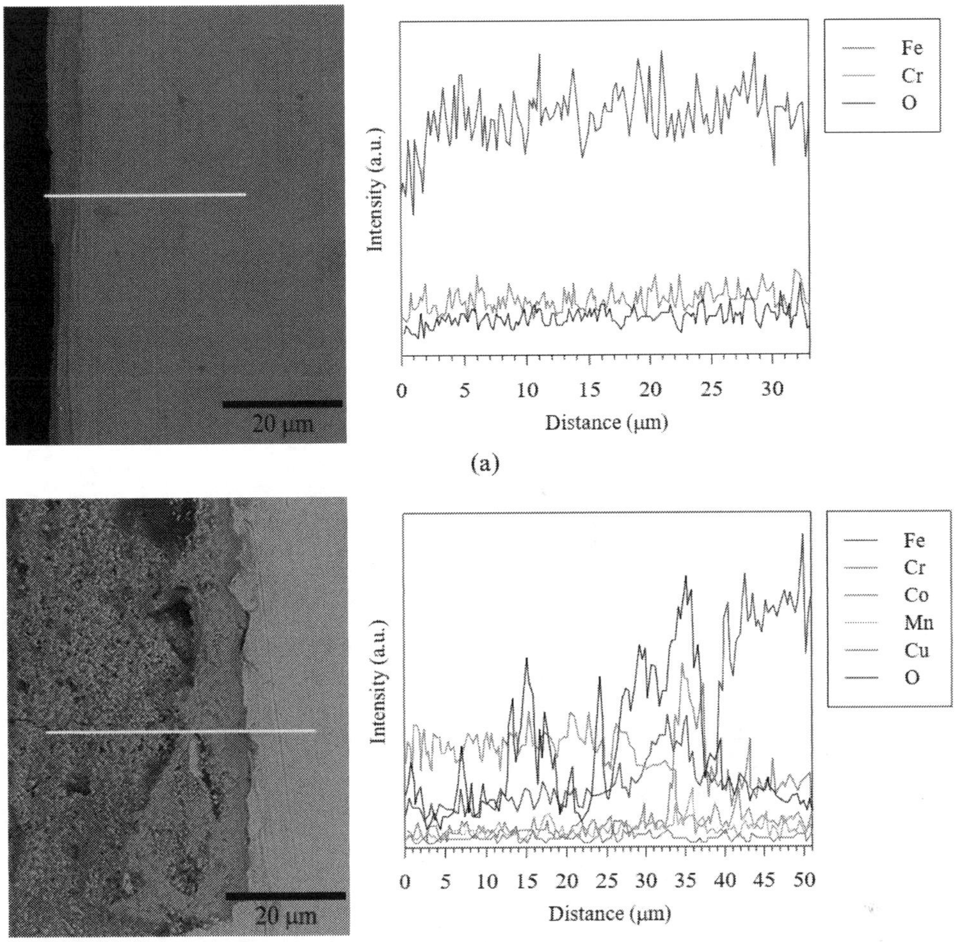

(b)

Figure 2. SEM-BSE image and EDX line scan across interface of sample (a) between the thermal oxide scale and AISI 430 stainless steel and (b) between the spinel coating and AISI 430 stainless steel after oxidation in oxygen with 20%water at 800 °C for 210 h.

3.3. Ceramic phase collector

The SEM images obtained from the fractured sample without epoxy mounting were taken using the secondary electron mode at the interface between electrolyte layers and the LSM cathode collectors as shown in figure 3. The average thickness of LSM cathode is 16.6 µm with 1.6 µm in the standard deviation. In order to measure the Cr-species volatilisation rate, the Cr composition is considered at only local the interface between electrolyte layer and the LSM cathode collector. For the AISI 430 ferritic stainless steel, the formation of the thermal oxide scale is Cr_2O_3 as mentioned in figure 2 (a). The vaporisation of Cr_2O_3 can generate the gaseous of Cr-species as $CrO_2(OH)_2$ according to Reaction (1) [11, 25, 26].

$$Cr_2O_{3(s)} + \frac{3}{2}O_{2(g)} + 2H_2O_{(g)} \rightarrow 2CrO_2(OH)_{2(g)} \quad (1)$$

Volatilised Cr can diffuse through the porous cathode ceramic and accept electron as an oxidising agent. The electrochemical reduction of $CrO_2(OH)_2$ will react at three phase boundaries (TPB) site lead to the deposition of Cr_2O_3 product as explained in Reaction (2) [27, 28].

$$2CrO_2(OH)_{2(g)} + 6e^- \rightarrow Cr_2O_{3(s)} + 2H_2O_{(g)} + 3O^{2-} \quad (2)$$

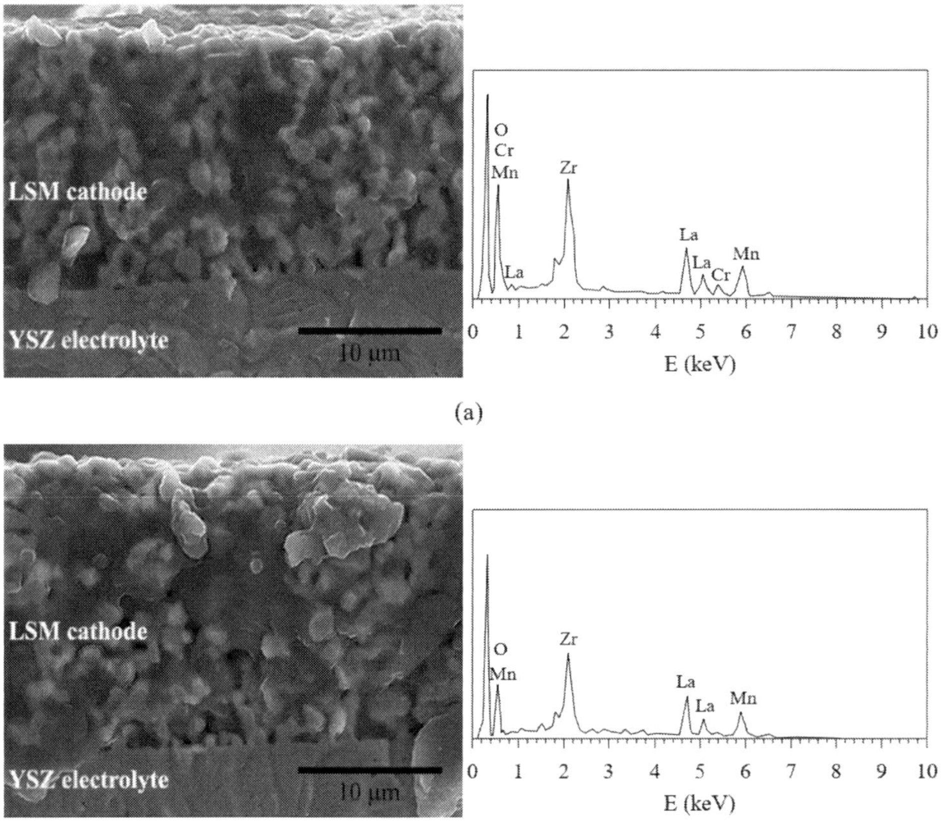

(a)

(b)

Figure 3. The SEM images at the interface between electrolyte layers and the LSM cathode collectors after being affixed with (a) bare AISI 430 stainless steel and (b) the spinel coated AISI 430 ferritic stainless steel after oxidation in oxygen with 20%water at 800 °C for 210 h.

The element concentrations at the interface between electrolyte layers and LSM cathode collectors were shown in table 2. For the AISI 430 ferritic stainless steel, the presence of Cr (1.61 wt.%Cr) is detected at the interface between the electrolyte layer and the LSM cathode collector in figure 3 (a). The result has shown that the accumulation of chromium is mainly detected from the cathode surface to the electrolyte/cathode interface. For the spinel coated sample, the EDX cannot detect the amount of chromium deposition at the electrolyte/cathode interface as shown in figure 3 (b). The results are believed that the spinel coated sample has sufficient efficiency in inhibition outward chromium diffusion.

Table 2. Element composition (wt.%) by using EDX point analysis at the interface between electrolyte layers and LSM cathode collectors after oxidation in oxygen with 20%water at 800 °C for 210 h.

Sample	Weight percent (wt.%)				
	La	Zr	Mn	Cr	O
AISI 430 stainless steel	35.54	24.97	15.79	1.61	21.79
Spinel coated AISI 430 stainless steel	42.61	24.28	17.73	-	15.38

3.4. Microstructure analysis

The chromium deposition at electrolyte/cathode interface was measured from semi-qualitative of EDS analysis. Two assumptions about the Cr deposition were done as follows. Firstly, the Cr accumulation was not only present at the TPB site but also could be observed at the electrolyte/cathode interface [27]. Secondly, the volatilised Cr-species was critically considered to be unable to diffuse through the denser electrolyte support. Thus, it was assumed that the transport of Cr from the substrate was restricted at the electrolyte/cathode interface zone. The equation can be expressed as following:

$$Cr\,collected\,rate = \frac{(wt.\%Cr)\rho_{cat}L_{cat}}{100t} \quad (3)$$

Here $\rho_{cat} \approx \rho_{LSM}$ is the LSM density of 6.5 g cm^{-3} [29, 30]. L_{cat} is the cathode thickness in cm and t is the exposed oxidation time in s. The *wt.%Cr* is weight percent of chromium which is obtained from EDX results.

Figure 4 showed the chromium volatilisation rate for stainless steel measured by different methods. The circle symbols are the rates of chromium volatilisation measured by a condensation method. The square symbols are the rates as measured by the ceramic phase collector method. The white and grey colours refer to the oxidised atmosphere in air and oxygen, respectively. The two results of the chromium volatilisation rates were obtained by the ceramic phase collector. Jin and Lu [31] studied the tri-layer sample YSZ/LSM/AISI 441 with different LSM stoichiometry. They investigated the Cr-species distribution by XPS elemental analysis. Among the porous LSM layer with 30 um thickness, the LSM 105 showed the lowest Cr concentration of 2.4 at.% and 11.5, 17.1, 69.0 at% of Sr, Mn and La, respectively, at the YSZ/LSM interface after heat treatment in air at 800 °C for 500 h. The other three results were obtained using a condensation method with ICP-MS. For humid air with 10%water, Young and Pint [32] compared the different of the calculated and measured Cr loss by evaporation from Fe-20Cr-25Ni-Nb at 800 °C up to 5000 h. They reported that the rate of Cr loss which was predicted by mass transfer calculation. The measured Cr loss rate was 31% lower than that of the calculation. Gannon et al. [33] reported the Cr transport rate was approximately similar to the result of Kurokawa et al. [34]. The AISI 430 stainless steels were exposed in humid air at 800 °C with the Cr transport rate of 1.0–1.1 × 10^{-10} g cm^{-2} s^{-1}. In the figure, it can be seen that, the chromium volatilisation rates obtained by ICP-MS are near the tendency line (the dash line) of those measured by the ceramic phase collector method. This means that the ceramic phase collector method can be used to evaluate the chromium volatilisation.

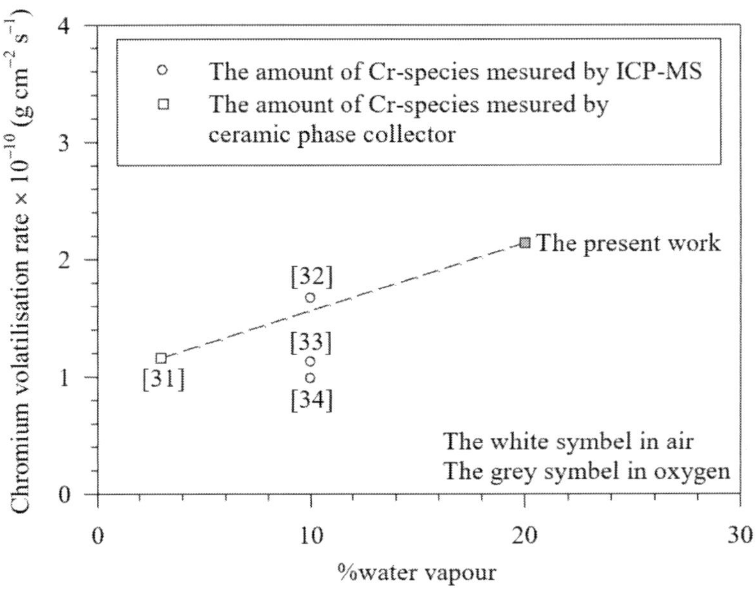

Figure 4. Chromium volatilisation rate with the different chromium collector method as a function of %water vapour compared with literature [31-34].

4. Conclusions

AISI 430 FSS samples with and with coated spinel coatings were investigated to measure the Cr-species volatilisation. The isothermal heat treatment was performed at 800 °C for 210 h in O_2-5%H_2O. In order to estimate the amount of chromium, the ceramic phase collector method was utilised and compared with the traditional condensation method with ICP-MS.

1. The formation of spinel oxide was detected at after pre-oxidation in oxygen at 600 °C for 24 h. The spinel coated AISI 430 FSS was identified to the cubic structure of Co_3O_4 and CoO and a small amount of CuO. After the isothermal oxidation process, the formation of thermally grown oxide was observed on the surface of the uncoated sample and spinel/substrate interface of the spinel coated sample.

2. The vaporisation of chromium from the spinel coated AISI 430 FSS was inhibited by the spinel coating. At electrolyte/cathode interface was unable to detect the amount of chromium deposition. Meanwhile 1.61 wt.% of chromium concentration was detected for the uncoated AISI 430 sample.

3. The estimation of chromium volatilisation was done using the ceramic phase collector method. The Cr measurement was simplified by the conversion the semi-qualitative of the EDS result to the quantitative data. For the ferritic stainless steel, the chromium volatilisation rates were in the same order of 10^{-10} g cm^{-2} s^{-1} compared with the traditional condensation methods.

References

[1] Singhal S C 2002 *Solid State Ionics* **152** 405-410
[2] Singh P,Minh N Q 2004 *Int. J. Appl. Ceram. Technol.* **1** 5-15
[3] Wu J,Liu X 2010 *J. Mater. Sci. Technol.* **26** 293-305
[4] Gindorf C, Singheiser L,Hilpert K 2005 *J. Phys. Chem. Solids* **66** 384-387
[5] Hilpert K, Das D, Miller M, Peck D H,Weib R 1996 *J. Electrochem. Soc.* **143** 3642-3647
[6] Fergus J W 2005 *Mater. Sci. Eng., A* **397** 271-283

[7] Fergus J W 2007 *Int. J. Hydrogen Energy* **32** 3664-3671
[8] Yang Z, Xia G-G, Li X-H,Stevenson J W 2007 *Int. J. Hydrogen Energy* **32** 3648-3654
[9] Hua B, Zhang W, Wu J, Pu J, Chi B,Jian L 2010 *J. Power Sources* **195** 7375-7379
[10] Key C, Eziashi J, Froitzheim J, Amendola R, Smith R,Gannon P 2014 *J. Electrochem. Soc.* **161** C373
[11] Wongpromrat W, Thaikan H, Chandra-ambhorn W,Chandra-ambhorn S 2013 *Oxid. Met.* **79** 529-540
[12] Hong S, Madakashira P, Kim D-I, Cho Y W, Han S H,Han H N 2009 *ECS Trans.* **25** 1437
[13] Thublaor T,Chandra-ambhorn S 2020 *Corros. Sci.* **174** 108802
[14] Haynes W M, Lide D R,Bruno T J 2017 *CRC Handbook of Chemistry and Physics* (New York: CRC press)
[15] Mougin J, Dupeux M, Antoni L,Galerie A 2003 *Mater. Sci. Eng., A* **359** 44-51
[16] Kurokawa H, Kawamura K,Maruyama T 2004 *Solid State Ionics* **168** 13-21
[17] Garcia Vargas M, Lelait L, Kolarik V, Fietzek H,Juez-Lorenzo M 2004 *Mater. Sci. Forum* **461** 823-830
[18] Jian P, Jian L, Bing H,Xie G 2006 *J. Power Sources* **158** 354-360
[19] Chandra-ambhorn S, Wouters Y, Antoni L, Toscan F,Galerie A 2007 *J. Power Sources* **171** 688-695
[20] Falk-Windisch H, Svensson J E,Froitzheim J 2015 *J. Power Sources* **287** 25-35
[21] Wongpromrat W, Parry V, Charlot F, Crisci A, Latu-Romain L, Chandra-ambhorn W, Chandra-ambhorn S, Galerie A,Wouters Y 2015 *Mater. High Temp.* **32** 22-27
[22] Chandra-ambhorn S, Thublaor T,Wiman P 2020 *Corros. Sci.* 108489
[23] Wu J, Johnson C D, Gemmen R S,Liu X 2009 *J. Power Sources* **189** 1106-1113
[24] Yang Z, Xia G, Simner S P,Stevenson J W 2005 *J. Electrochem. Soc.* **152** A1896-A1901
[25] Saunders S R J, Monteiro M,Rizzo F 2008 *Prog. Mater Sci.* **53** 775-837
[26] Opila E J, Myers D L, Jacobson N S, Nielsen I M, Johnson D F, Olminsky J K,Allendorf M D 2007 *J. Phys. Chem. A* **111** 1971-1980
[27] Jiang S P,Chen X 2014 *Int. J. Hydrogen Energy* **39** 505-531
[28] Mahato N, Banerjee A, Gupta A, Omar S,Balani K 2015 *Prog. Mater Sci.* **72** 141-337
[29] Armstrong T J,Virkar A V 2002 *J. Electrochem. Soc.* **149** A1565
[30] Rabelo A A, Macedo M C d, Melo D M d A, Paskocimas C A, Martinelli A E,Nascimento R M d 2011 *Mat. Res.* **14** 91-96
[31] Jin T,Lu K 2011 *J. Power Sources* **196** 8331-8339
[32] Young D J,Pint B A 2006 *Oxid. Met.* **66** 137-153
[33] Gannon P, Gorokhovsky V I, Deibert M, Smith R J, Kayani A, White P, Sofie S, Yang Z, McCready D,Visco S 2007 *Int. J. Hydrogen Energy* **32** 3672-3681
[34] Kurokawa H, Jacobson C P, DeJonghe L C,Visco S J 2007 *Solid State Ionics* **178** 287-296

Acknowledgements
This research was funded by King Mongkut's University of Technology North Bangkok (Contract no. KMUTNB-64-DRIVE-23).

Circular economy as possible solution for asbestos burden

B Zoraja[1], M Živančev[1], D Ubavin[1] and B Nakomčić-Smaragdakis[1]

[1]University of Novi Sad, Faculty of Technical Sciences, Department of Environmental Engineering and Occupational Safety and Health, Trg Dositeja Obradovića 6, 21000 Novi Sad, Serbia

E-mail: bojanazoraja@uns.ac.rs

Abstract. Asbestos has been used worldwide in a large number of products throughout the 1970s and 1980s. After the proven influence of asbestos materials on diseases that could have a fatal outcome was confirmed, bans on the use and production of these materials began. However, there are still significant quantities of asbestos waste and materials in the environment and there is a need for safe and permanent solution. As a support to sustainable development by minimizing waste production and harmful emissions as well as by reducing the consumption of natural resources, the circular economy (CE) is recognized in the EU as a leading solution. The main meaning of CE in waste management field is to turning waste into a resource. Considering above-mentioned, this paper presents how to resolve a problem of asbestos waste that is currently and will continue to be present. Instead of occupying the space, which is also a valuable resource, with landfills that also raise cost, asbestos can be used in construction material industry. Paper presents estimated amounts of asbestos waste, proposal for appropriate waste collection, treatment and market placing of a new product in Serbia. The conclusion of the paper is that the presented model of circular economy for asbestos in Serbia is feasible, in this way it would be possible to achieve some incomes from material selling and, also, releasing some space. In addition, for the implementation of this model of asbestos management, it is necessary to have Government involvement through investment in infrastructure and improvement of legislation.

1. Introduction

The phrase asbestos indicates a range of silicate minerals, which are naturally found in form of fibers and were commercially very usable. Asbestos minerals separate into two main groups based on crystalline structures: serpentine and amphibole, figure 1. [1] The first group refers to chrysotile, hydrated magnesium silicate materialized in the serpentine stones. Has the small-scale asbestos fiber frequently used in asbestos-textile industry (braids, fleece, etc.). Amphiboles are second group and refers to: actinolite, anthophyllite, tremolite, and commercially very used crocidolite and amosite. [2].

IOP Conf. Series: Materials Science and Engineering 1163 (2021) 012033 doi:10.1088/1757-899X/1163/1/012033

Figure 1. Asbestos minerals [3]

Diverse asbestos has various source and configuration, accordingly different physical attributes. Every asbestos has vast specific surface, so contamination can appear all along of the production and preparation processes of the fibers. Raw asbestos may also hold certain content of oil, and traceable metals such as chromium, cobalt, nickel, manganese, well known for their biological activity. [4].

Asbestos usage began over 4500 years ago, but the modern asbestos industry originated from the exploitation of large sites in Russia, Canada, New Zealand, Italy, Australia, Africa, etc. The initiative came from the cotton industry with the goal of producing durable, non-flammable fabric. The main use was for the preparation of cement, plastic fillers, materials for insulation, textile, and brake products. In the golden period of asbestos consumption, thousands types of asbestos-containing products were used in schools, clinics, movie theaters and industrial facilities, etc. [5]. Some asbestos usages are noted in table 1.

Table 1. Usage of asbestos fibers [6]

Category of asbestos products	Asbestos bonded with other materials		Asbestos as textile fiber	Asbestos used as loose fiber mixtures
Percent of total asbestos usage	>98%		<1%	<0.1%
Preparation method	with inorganic materials	with organic materials	roving's, yarn	mixtures with inorganic materials
Main products	Portland cement, hydrous calcium silicate, basic magnesium carbonate	oils, tars, elastomers, plastics, resins	woven, plaited	cement, gypsum, hydrous calcium silicate, basic magnesium carbonate, diatomaceous earth
Main uses	pipelines, flat sheets, corrugated roof sheeting and fencing, insulation boards	roofing products, caulking, joining, packings, gaskets, floor tiles, reinforced plastic sheets, friction materials, thermoplastics, thermosets	cloth, webbing, tubing, jointing's	heat, electrical or sound insulation products

Frequent and widespread use of asbestos, without knowing what its exact biological effect on humans is, without applying any protection measures, has led to the appearance of the illnes and significant spread around the world. Despite all its positive characteristics, asbestos has been recognized as dangerous to human health and life. Today, well-known diseases are associated with risks of exposure to asbestos fibers, such as asbestosis of the lungs, malignant respiratory and abdominal diseases, and others. After it was determined that asbestos is extremely dangerous, countries have begun to prohibit the use and production of asbestos-containing materials. There are currently more than 50 countries, including the European Union, that have banned asbestos completely [7]. The danger posed by asbestos-containing products is related to the inhalation of asbestos from the air. [8-12] It has been estimated that about $125*10^6$ people are occupationally exposed to asbestos, and it has been stated that asbestos-related illnesses cause $1.07*10^6$ deaths per year [5].

1.1. Significance of CE

Circular economy (CE) in comparison to linear economy, that is present, values raw materials differently, it is designed with aim to utilize the maximum value of resources [13]. Materials and products are designed with waste prevention in mind, and they are re-used, recycled, or recovered. New products should have long lifecycle and they can have alternative use also. Effective functionality of the circular economy demands adaptation of market conditions and policy in all sectors. Increasing resource efficiency, reducing cost and environmental effects are goals that demand changing of governments and businesses views. CE generates economic and business opportunities and provides environmental and social benefits. Circular economy in comparison to linear economy is more resilient, innovative, and productive, and finally it would last for longer time.

Three principles are important for circular economy, new product design without pollution and waste, preservation of natural resources and keep material and products in use as much as possible. Strategies like reuse, repair, remanufacture or recycling are key element, in technical cycles, for recovering and restoring materials, components and products. [14].

The aim of this paper is to propose the strategy of CE model in Serbia for asbestos waste and thus eliminate waste by recycling materials within the system and to put a new product on the market.

2. Methods

2.1. Asbestos in Serbia

Data for asbestos quantities, manipulation or waste in Serbia are difficult to access. There is no systematically organized database on the quantities of asbestos produced and used, asbestos inventories or the register of asbestos-related diseases in accordance with European regulations. Calculated quantities of asbestos used in Serbia are shown in figure 3.

In practice, the most common method of storing asbestos waste is in specially prepared landfills. This solution is very debatable, because when asbestos is disposed of, the landfill surface loses any communal value and requires constant investments for its supervision and maintenance. In many countries, such as Poland, storage is the only way to manage asbestos waste that is allowed by law. In the United States, methods are used that consist of coating asbestos elements installed in existing facilities, without removing and damaging the asbestos object. The construction elements are painted, sealed, or covered with a non-asbestos product. Such methods that use coatings increase the mechanical strength and erosion resistance of the asbestos product. [8].

In Serbia possibilities for management of asbestos containing waste are:

- Storage until the moment of export to treatment;
- Treatment, like hazardous waste inertization using Mid - Mix technology with the addition of CaO, or other;
- Landfilling on landfills with permission, municipal landfills, and open dumps.

Treatment product is usually packed in bags and mostly disposed at landfills (not all treatments guarantee there is no more free asbestos fibers).

Problem represents an amount of asbestos that end up in landfills that are not designed for it, such as municipal landfills (not sanitary) and illegal dumping sites. This is particularly characteristic for small settlements, in which minor construction works (repairs, renovation, etc.) are less controlled. Quantities that are believed to be significant, being deferred without control to landfills where immediately represent a danger to the environment.

In accordance with the published data of the Statistical Office of Serbia and the SEPA (Serbian Environmental Protection Agency) for the period 2010-2018, table 2. shows the official quantities of asbestos containing waste generated, deposited, and treated in Serbia [18, 20, 24].

Table 2. Official quantities that are generated, deposited, and treated in Serbia for 2010-2018 year

Year	Generated (t/y)	Landfilled (t/y)	Treatment (t/y)	Export of asbestos (t/y)
2010	516.6	1,034.9	-	-
2011	141	0	310	315
2012	240.24	306	43,025	-
2013	191.83	279	30.47	-
2014	1542.2	1,647	27.54	-
2015	185.49	287	28	-
2016	216	963	-	-
2017	189	495	-	-
2018	-	246	-	-

These are the only available official data, but author's ongoing research on estimated quantities of generated asbestos waste show that these amounts are probably far greater. The highest expected amounts of asbestos waste will be generated in the next ten years, as asbestos installed in buildings during the 1970s and 1980s will be ready for replacement due to decay [17]. These differences indicate the probable existence of larger quantities of waste than officially registered waste, which end up in inappropriate places (unsanitary landfills, stored in households, etc.).

As can be seen in Table 2, landfilling is the most common way of managing asbestos waste in Serbia. The problem is that these landfills were not usually sanitary and were not equipped to receive this specific type of hazardous waste. Of particular concern is that landfilling runs counter to the idea of sustainable land use, recycling and closing the material cycle [6].

2.2. The state of CE in Serbia

The advantages for implementation of CE in Serbia are that part of the legislation is already in line with the acquis communautaire, a process has been started to enable green public procurement to become a standard and innovation has been recognized as a driving force in the introduction of the green economy. One of the strengths is the well-developed IT sector, with exports of services worth almost one billion euros in 2018. Weaknesses have been identified as poor waste management, because of the 12 million tons of waste generated annually, only three percent of municipal waste is recycled, legislation is insufficiently developed, and inspection is ineffective. Energy consumption is also inefficient, 260 grams of waste is generated for each euro of gross domestic product, consumes 4.6 kilowatt hours of electricity, and generates 0.7 kilograms of carbon dioxide (CO_2). Uncontrolled imports of inert and hazardous waste, large imports of short-lived old vehicles, poor penalties and inefficient incentive policies are also aggravating circumstances [21].

Although legislation exists, the CE in Serbia has not been sufficiently investigated. Obstacles to full implementation of Serbia's circular economy are poor waste management, wild landfills, lack of infrastructure and slow procedures.

IOP Conf. Series: Materials Science and Engineering 1163 (2021) 012033 doi:10.1088/1757-899X/1163/1/012033

2.3. *MID-MIX technology*
MID-MIX® TECHNOLOGY is a patented technology that has been applied since the late 1980s and scheme is shown in figure 2. This technology is one of the approved and recommended technologies in Europe BATNEEC (Best Available Technology that does not involve excessive costs). Belongs to the order with the best available technologies that does not involve excessive waste management costs and does not pollute the environment with industrial waste after treatment. Waste treatment plants using this technology are installed in many European countries such as Slovenia, Spain, Austria, Italy, Poland, Croatia, France, Portugal, Serbia, etc. These plants can be mobile and stationary, and are used for inertization of various types of waste from industries such as refineries, petrochemicals, chemistry, pharmacy, food industry, etc. Incurred solidificate and condensed water from evaporation are the final products of the process and do not have a harmful effect on people and their environment. The solidificate created here is a powder of white thick to grey-brown colour with exceedingly hydrophobic attributes. At the end, this solidifier is an inert, non-hazardous material that can be further processed [21, 22].

This powder can be stored in large bags and silos, pressed into briquettes that take up about three times less volume than the initial one, as well as disposed of in landfills for non-hazardous waste. According to EU regulations, solid residue of this process can be further used in the construction industry (concrete blocks), construction waterproofing, road construction (alignment), energy (additive for solid fuels), process industry (additive for industrial fuels), cement industry, asphalt production, etc. [21, 22]

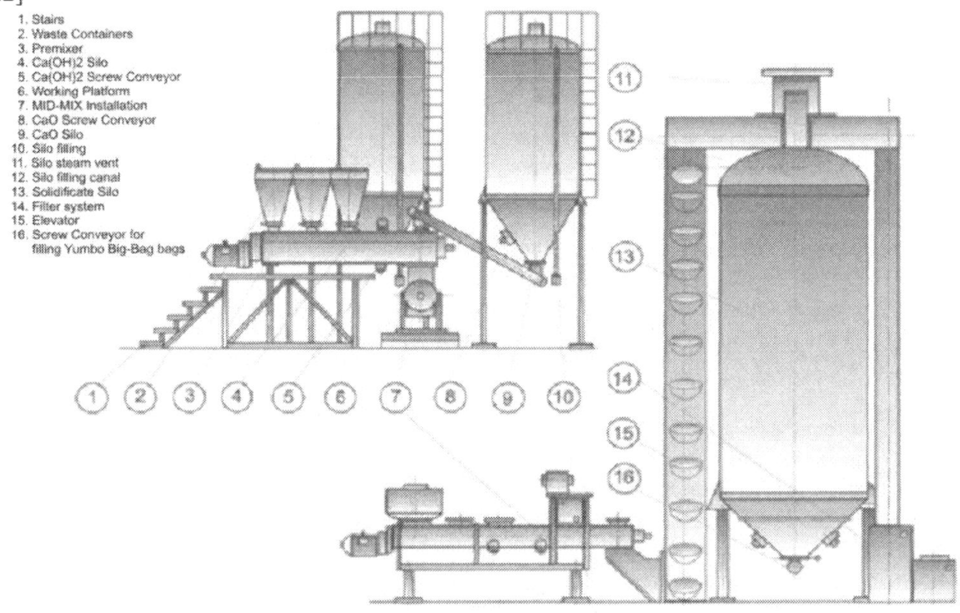

Figure 2. MID-MIX® mobile plant for waste treatment, schematic diagram [21]

2.3.1. Preparation of asbestos mash. I phase - Crushing asbestos waste. Shredding waste for easy insertion into a grinding mill.

II phase - Starting up of the plant. Delivery of prepared asbestos waste, start-up of the plant and release of water.

III phase - Milling of asbestos waste. Dosage of asbestos waste into a mill, grinding, wetting, and obtaining asbestos slurry. During milling, water is wetted, and an asbestos slurry is prepared which is ready for MID-MIX inertization. The wetting prevents the asbestos particles from floating in the air, inhaling them, and creating the necessary conditions for the chemical process to take place.

The process of inerting asbestos waste is carried out by the MID-MIX technological process, which involves complex physical-chemical-thermal processes of dissociation, vacuum encapsulation, and primary solidification of waste.

2.3.2 Processing of asbestos mash. The process begins with the introduction of the waste mixture into the mixer, which homogenizes the waste mixture with process additives (sand, calcium hydroxide and calcium oxide), after which it enters the MID-MIX reactor. The process of primary solidification of the waste mixture takes place in the reactor. The neutral is drained from the reactor, with a bucket elevator, into a silo, in which secondary solidification (ripening) occurs - the completion of a chemical reaction and the permanent transfer of waste into environmentally inert material.

Therefore, after preparation of the asbestos slurry, the slurry is inerted on the MID-MIX plant. The end result is an inert, eco-friendly material - NEUTRAL. The neutral, obtained by the MID-MIX process, is powdered, light in color, odorless and in its chemical composition is a mixture of organocalcium water-soluble salts. Neutral derived from asbestos waste has the following characteristics:

- White to gray powder with zero granulation;
- Lighter than water - swims on water;
- Hydrophobic (does not absorb water);
- Odorless;
- Specific gravity 0.85 kg / dm^3;
- Moisture content below 5% ;
- Excellent hydro and thermal insulation properties.

The neutral is packed in jumbo bags of 1000 liters or directly into the cistern and transported to its final destination (cement plants, landfill, concrete plants)

3. Results

In Serbia use of asbestos strats at the beginning of the twentieth century. Quantities of asbestos fibers used in Serbia are shown in figure 3. [16, 17]

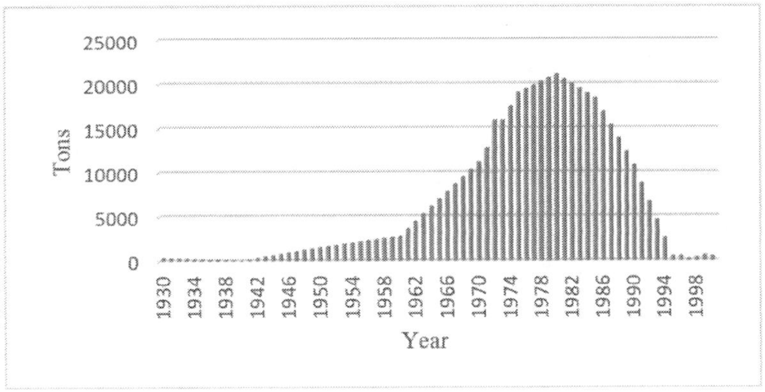

Figure 3. Use of asbestos fibers in Serbia 1930-2000 [18]

Figure 3 shows the trend of asbestos use for the period 1930 to 2000. The total amount consumed for this period was 480,123 tons. [18] As it can be seen on figure 3 period of large asbestos usage in Serbia was during 70s and 80s of the 20th century. In Serbia, asbestos was widely used in the construction of buildings, pipelines, and roof coverings, as well as for the insulation of ships. Asbestos products were produced by the companies: "Jugoazbest" from Belgrade and the company "Fiaz" from Prokuplje. [17] During the 1990s, Serbia was under sanctions and war, which evidently led to a decline in asbestos use.

After that, the use of asbestos never returns to the previous quantities. At same time, the use of asbestos was already banned in some parts of Europe, to be completely banned in 2005. In Serbia asbestos is partially banned in 2011, while in 2015 is complete banned for production and use. [18].

An important reason for significant use of asbestos is the case that Serbia was a big producer and had six asbestos mines, two of them large chrysotile mines Korlace and Stragari. The vicinity of an asbestos production plants has a significant impact on the estimated amount of asbestos use. As can be expected consequential large amount of asbestos waste will arise in next ten to twenty years due to end of life periods for asbestos containing materials that were built in [16].

Table 3 shows the number of companies that have licenses and in which cities they are located. Licenses are for landfilling and treatment of different asbestos wastes according to waste catalogue number. Data on permits are downloaded from the SEPA website (11.02.2020.).

Table 3. Cities with companies with permits for asbestos waste

Waste catalog number for asbestos	Landfilling permission in city	Treatment permission in city
06 13 04*	Belgrade, Kikinda	Belgrade (2), Kikinda
17 06 01 *	Belgrade, Kikinda	Indjija, Belgrade (2)
17 06 05*	Belgrade, Kikinda	Indjija, Belgrade (2)
15 01 11*	-	Sabac, Belgrade (5) Kikinda (2), Nis, Valjevo
16 02 12*	-	Kikinda, Belgrade, Valjevo,
16 01 11*	-	Kikinda, Belgrade (2)
06 07 01*	-	Kikinda, Belgrade
10 13 10	-	Sabac, Kikinda, Ada, Valjevo (2), Ub, Kragujevac, Belgrade (2)
10 13 09*	-	Kikinda, Belgrade

It can be concluded that only two companies in two different cities have permits for landfilling asbestos waste, and that is only for two types of asbestos waste. On the other hand, several companies have permits for any kind of treatment, for each of asbestos waste type. In addition, there is no data on the type of treatment or the exact quantities that these companies treat annually. According to the Decree on the disposal of waste in landfills in Serbia, the disposal of construction waste containing asbestos and other asbestos waste that meets the requirements in accordance with special regulations is possible. For such landfills, it is necessary to keep a plan of locations with precisely indicated places of cassettes in which asbestos is deposited, even after the cessation of operation of these landfills. Storage of waste containing asbestos is possible under strictly controlled conditions for up to a maximum of one year. In practice, this deadline is usually extended. [19]

Contrary to the existing legislation on waste management, there are so-called illegal landfills (dumps) in Serbia where construction material that is generated during the demolition/renovation/maintenance of private residential buildings is quite usual.

Steps for implementation of CE solution for asbestos in Serbia:

- Rising public awareness about asbestos waste. When citizens encounter possible asbestos waste, they should contact companies with specific work permits. In other to do this companies with specific permits should be recognizable and accessible.
- Organized waste collection and transportation to be established. Asbestos waste will be transported to a nearest temporary storage place or directly to treatment facility.

- Treatment of asbestos waste. In this step MID-MIX technology would be applied, transforming asbestos into inert, non-dangerous material. Due to its characteristics, it is used in construction and road construction - production of precast concrete elements, brick products, asphalt, road substrates. Concrete and brick products that come with the addition of neutrals are lighter than classic products. This addition material should decrease consumption of natural resources, that are commonly used.

There are several benefits of commercial applying CE for solving of asbestos problem:

- Asbestos waste is collected and treated properly and there is no need for additional space for waste landfilling.
- Obtained material is used as a resource, for production of a new concrete blocks, that can be used in road construction.
- Eco labelling is a possible solution for putting this new product on market. With the help of government in form of decreasing taxes and promoting this kind of production, problem of asbestos waste would be on a good track to final solution. New product will be long-lasting and at the end of life it could be considered in CE model for construction and demolition waste, that is already supported by EU.
- The economic benefits for the one who applies this principle are twofold, the first is charging for waste disposal, the second is reducing the cost of procuring raw materials.

The negative aspects of commercial application would be cost and exceptional health risk if not treated adequately. Aspects of CE that this paper covers are adequate waste management as well as strengthening the secondary raw materials market. In this way, the following goals are achieved: the recycling rate is increased and, in this way, valuable materials are returned to the economy, and lower consumption of natural resources is also achieved.

4. Conclusions

There are some types of asbestos waste treatment in Serbia, it is questionable how much is treated and processed, and what is done with the remains in the end. Whether a particular treatment is appropriate for the region and the conditions, or for that exactly type and amount of asbestos waste generated can be assessed and determined based on several criteria. There are criteria and limitations related to the price, process efficiency, sustainability, and the material output in the end (is there still activity of fibers).

In paper we have presented one solution for asbestos problem in Serbia in general. Thinking about asbestos waste through the concept of circular economy is feasible with inclusion of all interested parties. The advantages and disadvantages of the commercial application of this asbestos management principle are also presented. For further research, it is planned to conduct sustainability and cost-effectiveness analysis for this model of asbestos waste management.

To implement this model of asbestos waste management, it is necessary for the state of Serbia to invest in infrastructure and to strengthen the implementation of existing regulations. Due to public health effects of asbestos, it is necessary to constantly inform citizens about the state of asbestos waste management.

5. References

[1] Speil S, Leineweber J.P. 1968. Asbestos minerals in modern technology. Environmentar research 2: 166-208.

[2] Dellisanti, F., Rossi, P.L., Valdre, G., 2009. Remediation of asbestos containing materials by Joule heating vitrification performed in a pre-pilot apparatus. Int. J. Miner. Process 91, 61-67. http://dx.doi.org/10.1016/j.minpro.2008.12.001.

[3] https://zgipsman.hr/zdrava-kuca/azbest-neprijatelj-vaseg-doma/ (10.02.2020.)

[4] Virta, R.L., 2011. Asbestos. Kirk-othmer Encyclopedia of Chemical Technology, pp. 1-40. http://dx.doi.org/10.1002/0471238961.0119020510151209.a01.pub3.

[5] Zoraja, B., Beronja. M., Ubavin, D., Hadžistević, M, Mihajlović, V.: Bezbedan rad sa azbestom u Srbiji, 13 Nacionalna konferencija sa međunarodnim učešćem o zaštiti na radu-Unapređenje

sistema zaštite na radu, Tara, 27-30. 10. 2016. , ISBN 978-86-919221-1-5

[6] Spasiano, D., Pirozzi, F., 2017. Treatments of asbestos containing wastes. Journal of Environmental Management 204 82-91

[7] LaDou, J., et al., (2010), The Case for a Global Ban on Asbestos, Environmental Health Perspectives, vol 118 , (7), pp. 897-901, ISSN-L 0091-6765

[8] A. Pawelczyk, F. Božek, K. Grabas, J. Checmanowski, Chemical elimination of the harmful properties of asbestos from military facilities, Waste Management, 2016, http://dx.doi.org/10.1016/j.wasman.2016.11.041

[9] Bernstein, D., Dunnigan, J., Hesterberg, T., Brown, R., Velasco, J.A.L., Barrera, R., Hoskins, J., Gibbs, A., 2013. Health risk of chrysotile revisited. Crit. Rev. Toxicol. 43 (2), 154e183. http://dx.doi.org/10.3109/10408444.2012.756454.

[10] Lee, R.J., Van Orden, D.R., 2008. Airborne asbestos in buildings. Regul. Toxicol. Pharmacol. 50, 218e225. http://dx.doi.org/10.1016/j.yrtph.2007.10.005. Leonelli, C., Veronesi, P., Boccaccini, D.N., Rivasi, M.R., Barbieri, L., Andreola

[11] Whysner, J., Covello, V.T., Kuschner, M., Rifkind, A.B., Rosman, K.K., Trichopoulos, D., Williams, G. M., 1994. Asbestos in the air of public buildings: a public health risk? Prev. Med. 23 (1), 119e125. http://dx.doi.org/10.1006/pmed.1994.1017.

[12] Donaldson, K., Poland, C.A., Murphy, F.A., MacFarlane, M., Chernova, T., Schinwald, A., 2013. Pulmonary toxicity of carbon nanotubes and asbestos similarities and differences. Adv. Drug Deliv. Rev. 65 (15), 2078e2086. http:// dx.doi.org/10.1016/j.addr.2013.07.014.

[13] International Solid Waste Association (ISWA), Circular economy: Resources and opportunities, 2015.

[14] Ferronato, N., Rada, E. C., Portillo, M. A. G., Cioca, L. I., Ragazzi, M., & Torretta, V. (2019). Introduction of the circular economy within developing regions: A comparative analysis of advantages and opportunities for waste valorization. Journal of environmental management, 230, 366-378.

[15] Ansari, F.A., Ahmad, I., Ashquin, M., Yunus, M., Rahman, Q., 2007. Monitoring and identification of airborne asbestos in unorganized sectors, India. Chemosphere 68 (4), 716e723. http://dx.doi.org/10.1016/j.chemosphere.2007.01.040.

[16] Virta, R.L., 2002. Asbestos: Geology, Mineralogy, Mining, and Uses. U.S. Department of the interior - U.S. Geological survey. Report 02e149.

[17] Zoraja, B., Vujovic, S., Ubavin, D., Vujic, G., Stanisavljevic, N.: History of using asbestos – special emphasis on Serbia, Proc. of the ISWA world congress 2016, September 19-21 2016, Novi Sad

[18] Zoraja B, Ubavin D, Vujović S, Tot B, Živančev M: Quantities Of Asbestos Used In Serbia And The World For The Period 1930-2000.16th International conference of occupational health and safety OSH PRIORITY. Ohrid 2019, pp 89-94

[19] Ministry of health of Serbia, National profile of asbestos exposure for republic of Serbia, Belgrade, 2017.

[20] Zoraja B, Ubavin D, Hadzistevic M: Review of state of the art techniques for asbestos waste treatment to support environmental protection projects. 13. M&S International Conference Management and Safety, The European Society of Safety Engineers, pp. 105 - 118, isbn: 978-953-58000-8-8, Ohrid, 2018.

[21] http://euinfo.rs/plac3/wp-content/uploads/2019/11/Prezentacija-ex-ante-analiza-efekata-za-cirkularnu-ekonomiju.pdf (16.02.2020.)

[22] Brkljač N, Šević D, Beker I, Kesić I and Milisavljević S: Procedure for treatment of hazardous waste by MID-MIX procedure in Serbia. International Journal of the Physical Sciences Vol. 7(18), pp. 2639-2646, 2012

[23] www.yunirisk.com (10.02.2020.)

[24] Ministry of Environmental Protection, Environmental Protection Agency, Waste management in the Republic of Serbia from 2011 to 2018, Belgrade, 2018.

IOP Conf. Series: Materials Science and Engineering 1163 (2021) 012034 doi:10.1088/1757-899X/1163/1/012034

Performance Evaluation of Vertical Axis Wind Turbine on Highways

Abdul Rashid Shoib[1], Siti Khadijah Binti Azhar[1], Djamal Hissein Didane[2*], Akmal Nizam Mohammed[2], Mohammed Ogab[1], Lukmon Owolabi Afolabi[2]

[1]Department of Mechanical Engineering, University Tun Hussein Onn Malaysia, 86400, Parit Raja, Batu Pahat, Johor, Malaysia
[2]Center for Energy and Industrial Environment Studies (CEIES), University Tun Hussein Onn Malaysia, 86400, Parit Raja, Batu Pahat, Johor, Malaysia

E-mail: djamal@uthm.edu.my

Abstract. The consumption of electricity at highways is increasing over time, consequently, more fossil fuel is burned to generate more electrical energy. However, this expensive cost of burning fossil fuels can be saved by replacing them with renewable energy sources such as wind energy. Wind energy can be generated by using a wind turbine. Therefore, this study aims to develop a vertical axis wind turbine (VAWT) that can work effectively in highway wind speed conditions while evaluating its performance in terms of torque and power by using a simulation approach. The simulation involves the use of K-omega shear-stress transport (SST) as the turbulence viscosity model. The VAWT undergoes two types of tests during the simulation process. In Test 1, the wind turbine is tested with a various number of blades at constant wind speed to find which model has the best performance in terms of torque and power output. Then, the selected model with the better output proceeds to Test 2 where it is tested with different sizes of blade radius. The results have shown that a Savonius wind turbine with five blades produces more torque and power output compared to the other rotors with two, three and four blades reaching the maximum at 140 RPM. In terms of blade radius, the model with a 400 mm radius has a better performance and capable of producing higher torque and power output than the other blade sizes.

1. Introduction

Energy is the focal theme in the socio-economic development and economic growth of a nation. An estimation has been made that about 10 million MW of wind energy are continuously available on the earth. This is an environmentally friendly option and can be a replacement for fossil fuels as it threatens the long-term sustainability of the global economy [1]. Nowadays, most of the world's countries are relying heavily on natural gas, coal, and oil for their source of energy. However, fossil fuels are non-renewable and concentrated in a limited number of countries. Eventually, it becomes more limited and expensive day by day as they are highly demanded. Fossil fuel also causes environmental pollution. Therefore, various renewable energy sources such as wind and solar energies are continuously replenished on a human timescale. Renewable energies have many advantages, such as low maintenance of their facilities, compared to traditional generators, where their fuels are derived from natural and available resources. Moreover, renewable energy projects also provide economic benefits to many regional areas because most projects are located in large urban centres and suburbs of the cities [2], [3].

Wind energy is becoming cost-effective because the land-based utility-scale wind is one of the lowest-priced energy sources available today, which costs between two and six cents per kilowatt-hour. It depends on the wind resource and the particular project's financing. Besides, wind energy is a clean fuel source, and it does not cause air pollution as the power plants rely on the combustion of fossil fuels [4], [5]. The device that is used to convert wind energy into electrical energy is called a wind turbine. Commonly, the small turbines are used for battery charging for auxiliary power for boats to power traffic warning signs. While large turbines are used for domestic power supply and sending back the unused power to the utility supplier. Wind turbines are divided into two types, which are the vertical axis and the horizontal axis. The vertical axis wind turbine (VAWT) has the main rotor shaft arranged vertically. This arrangement has its advantage where the turbine does not need to be pointed into the wind to be effective. It is very much used for a site that has a high variability of wind direction. Another advantage of this design is the accessibility for maintenance. The gearbox and generator can be placed at the ground due to direct drive from the rotor to the ground-based gearbox [6]. The VAWT itself has two types of designs, which are named Darrieus and Savonius. The Darrieus design has excellent efficiency, but it is poorly reliable due to broad torque ripple generation and cyclical stress on the tower. This design also requires some external power sources to start turning due to the very low starting torque. In contrast, the Savonius is a drag-type device that has the capability of self-starting if it has at least three scoops (blades). The Savonius with two scoops is usually used in anemometers, in some high-reliability and low-efficiency power turbines. They are commonly seen on the roofs of houses, bus or vans. In order to provide smooth torque, this design has been modified with long helical scoops and it is known as twisted Savonius rotor [7]-[9].

Depending on the primary power grid to supply electricity to the highway, the grid power is costly in terms of cost and implementation, and it is also not a long-term sustainable energy source. However, the wind energy restored in the wakes of moving vehicles on the road has its potentials to be a source of energy that can be converted to electricity, but it is yet to be discovered [10]-[12]. Therefore, this study's main objectives are to design an efficient vertical axis wind turbine with a Savonius-type while investigating the performance of VAWT in terms of torque and power on the highway.

2. Materials and methods

This study starts with selecting the Savonius-type blade for VAWT, which is used in this study. The proposed 3D model of the VAWT is designed and developed using the modelling SolidWorks software with specific parameters. The proposed VAWT is then imported into ANSYS to run the CFD simulations. The power and torque outputs are then extracted from the simulation results. Equations 1 to 3 are respectively the theoretical power, actual mechanical power and the maximum power extractable by a wind turbine.

$$P_w = \frac{1}{2}\rho A V^3 \quad (1)$$

$$P_T = T_m \times \omega \quad (2)$$

$$P_{ex} = (0.593) T_m \times \omega \quad (3)$$

Meanwhile, the rotor thrust force equation and the rotor theoretical torque equation of wind turbine are respectively presented in Equation 4 and Equation 5.

$$F = \frac{1}{2}\rho_a A V^2 \quad (4)$$

$$T = \frac{1}{2}\rho_a A V^2 \quad (5)$$

2.1. Rotor design

The S-type design, which is known as the Savonius type, is selected for blades of the VAWT. In the first test, the S-type wind turbine is tested with a various number of blades. The blades are C-shape design with a radius of 300 mm. The original design has three combinations of C-shape, where each of them is cut into a 270 mm radius. Then, the number of C-shape combinations will depend on the number of blades of the wind turbine. In the second test, the chosen wind turbine in Test 1 is tested with different radius sizes of blades, which are 300 mm, 350 mm, and 350 mm. The blades with a radius of 350 mm are cut into 320 mm, and blades with a radius of 400 mm are cut into 370 mm. The height and thickness of the rotors of all models are maintained at 120 mm and 5 mm, respectively in both tests. The geometrical details of the models are shown in Figure 1.

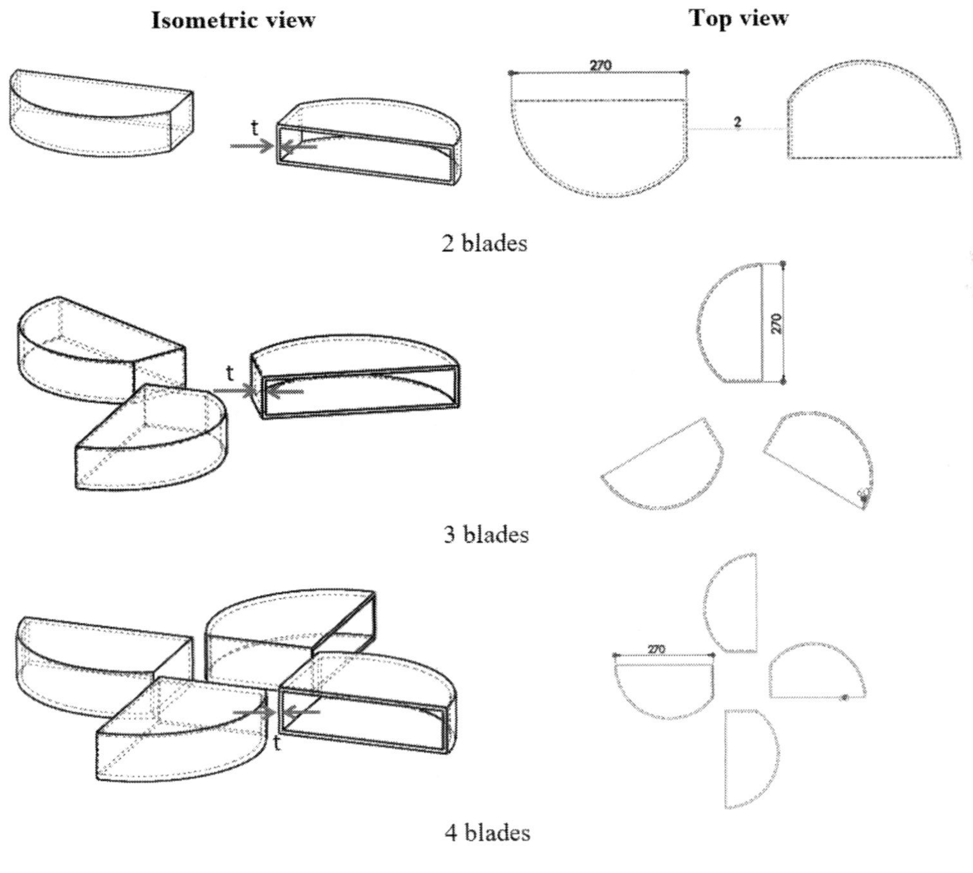

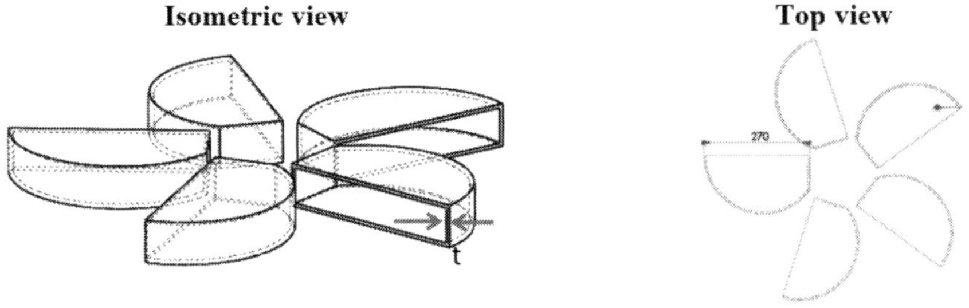

5 blades

Figure 1. Geometrical details of the models.

2.2. Boundary conditions and simulation technique

The CAD model designed is imported into ANSYS, and a proper simulation domain is created. Given that a turbine is a rotating machine, the present model's domain is divided into two regions; rotating and non-rotating regions. The non-rotating region characterizes the simulation's general computation territory, and the rotating region has the rotor.

This study's specific boundary conditions are the velocity inlet boundary condition at the inlet position, pressure outlet at the outlet position, and non-slip wall boundary condition at the rotor surfaces and lateral walls non-conformal interface boundary condition is used between the rotating domain and non-rotating domain. The two-equation K-omega Shear-stress transport (SST) is used as the turbulent viscosity model. For the solution method, the pressure-based coupled algorithm is chosen as the pressure-velocity coupling scheme.

3. Results and discussion

The numerical simulation process starts with a verification process, gird independence test (GIT), and then proceeds with the validation process before presenting the results.

3.1. Grid independence Test and Model Validation

The gird independence test (GIT) was conducted to fix the study's grid sensitivity and determine the mesh's acceptable size so that the number of elements does not influence the solution results. It was observed that all different sizes of mesh have a small amount of relative error. However, the torque output becomes grid-independent when the size of the element is below 0.0125 m. Therefore, the 0.010m size element is selected as it is considered reliable and more accurate to be incorporated into the subsequent simulation processes. Moreover, Figure 2 shows the results of the validation of the present simulation results with the previous experimental results presented in [10]. The graph demonstrates that the current simulation results are close to the previous experimental results. The average percentage of relative error is 12.47%, and it is in the acceptable range. Therefore, the current simulation's reliability is proven to be genuine and valid to carry on the simulation process.

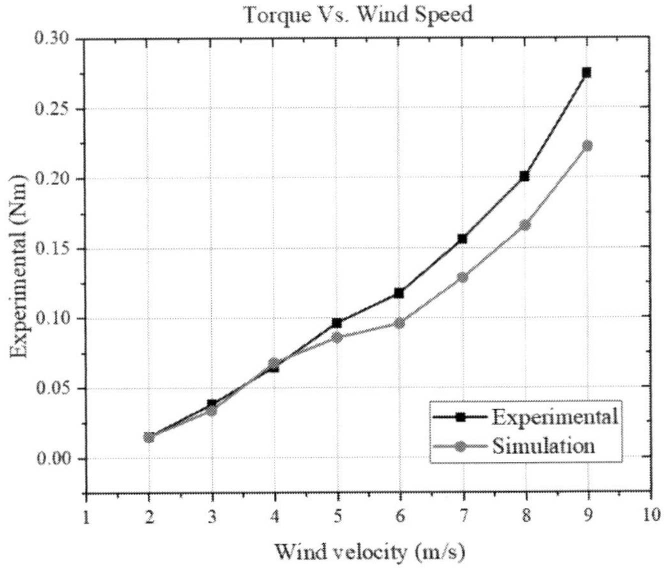

Figure 2. Current simulation and previous experimental torque output [10].

3.2. Evaluation of predicted torque with different number of blades and radius size

In this section, the torque of the various number of blades and radius size of the Savonius VAWT are predicted at highway wind speed at 5 m/s. The results are shown in Figure 3 and Figure 4 with respect to the torque output. Figure 3 shows that the Savonius rotor's torque with five blades increases significantly as the RPM increases compared to other blades and reach a maximum at 140 RPM with 0.17954 Nm. Simultaneously, the rotor with two blades has the smallest torque overall and reaches 0.12093 Nm at 140 RPM. This may be due to the increasing amount of wind force captured by the blades. Thus, when the number of blades increases, thus the torque produced will be higher. However, the three blades have slightly higher torque than the four blades. Moreover, Figure 4 shows that the torque of blades with a 400 mm radius increases significantly when the RPM increases compared to the other radiuses and reaches a maximum at 140 RPM with 0.4143 Nm. While blades with a 300 mm radius have the smallest torque and only reaches a maximum of 0.1795 Nm. This is because the higher the radius, the higher the torque produced.

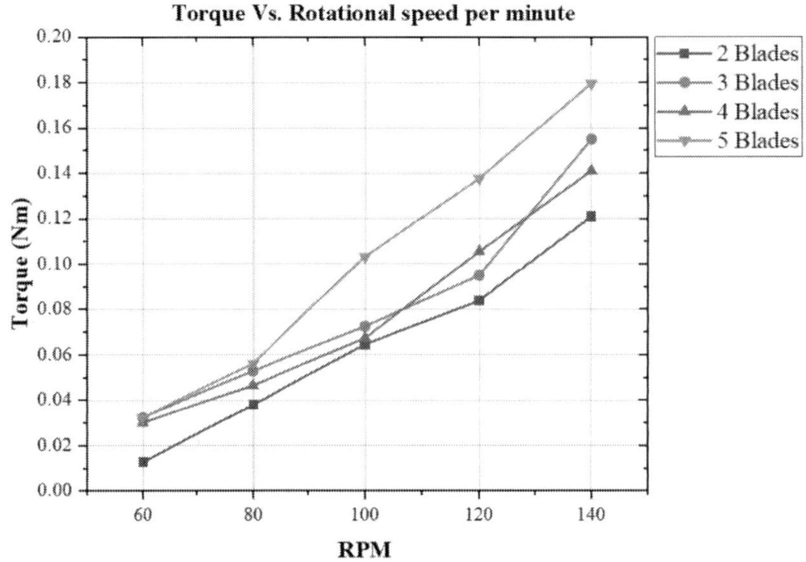

Figure 3. Torque against rotational speed per minute (with different number of blades).

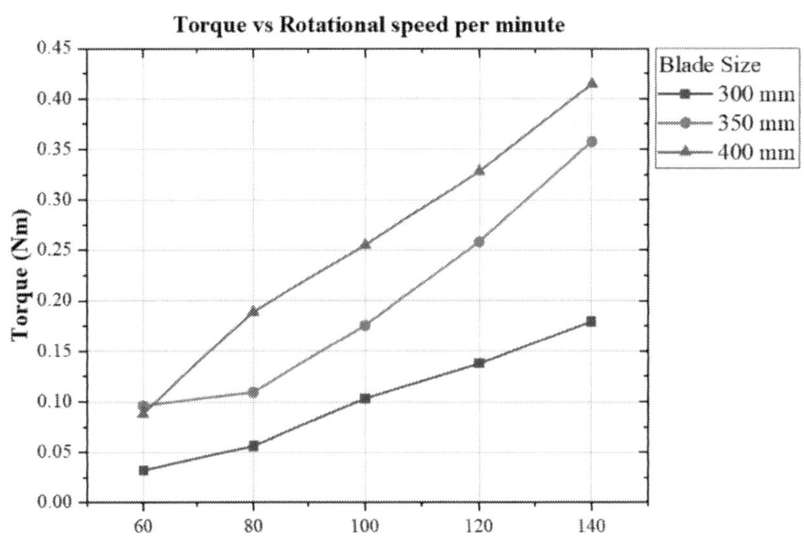

Figure 4. Torque against rotational speed per minute (with different size of radius).

3.3. Evaluation of predicted power with different number of blades and radius size

This section shows the aerodynamic behavior of the Savonius VAWT in relation to power output by using a numerical simulation approach. Figure 5 shows the power output of wind turbines with a various number of blades. From Figure 5, it shows that Savonius with five blades produces the highest power output when the RPM increases, followed by 3, 4, and 2 blades and reach the maximum at 140 RPM with 2.6322 W. While Savonius with two blades has the smallest power output overall and reaches 1.7729 W at 140 RPM. However, the three blades Savonius has slightly higher power output than the four blades. Furthermore, Figure 6 shows that blades with a 400 mm radius produce the highest power output when the RPM increases, followed by 350 mm and 300 mm and reaches the maximum at 140 RPM with 6.074 W. Meanwhile, blades with a 300 mm radius have the smallest power output and only reach a maximum value of 2.6322 W.

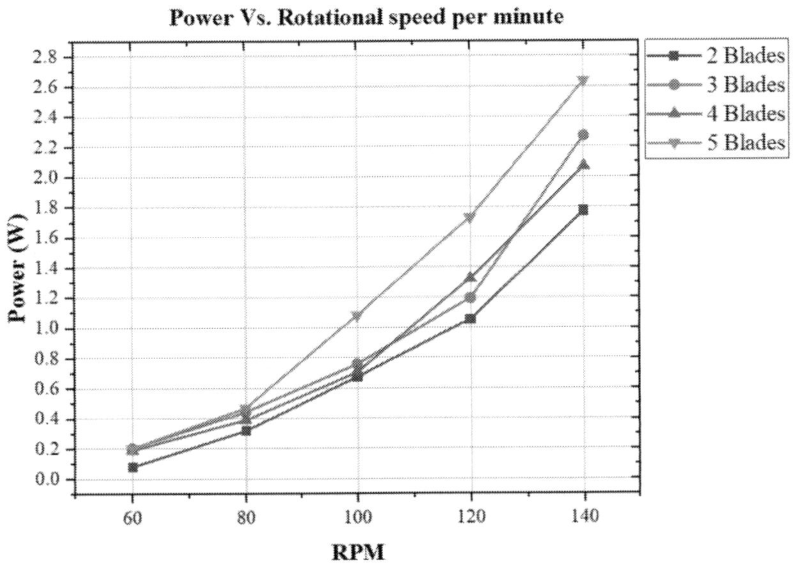

Figure 5. Power against rotational speed per minute (with different number of blades).

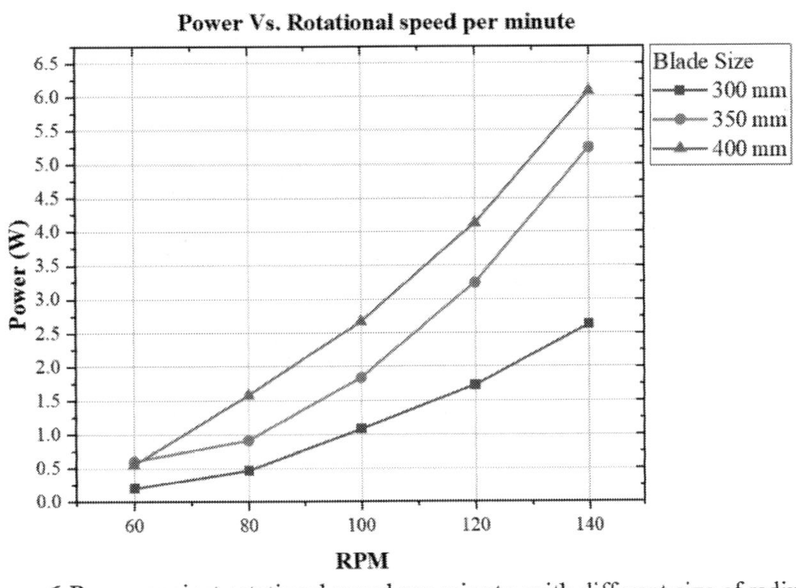

Figure 6. Power against rotational speed per minute (with different size of radius).

4. Conclusions

In this study, a Savonius-type vertical axis wind turbine (VAWT) performance on the induced wind on highways has been studied using the computational fluids dynamic (CFD) simulations approach. The main aim of this study to develop a vertical axis wind turbine (VAWT) that can work effectively in highway wind speed conditions while evaluating its performance in terms of torque and power output. The simulation involves the use of K-omega shear-stress transport (SST) as the turbulence viscosity model. Two types of testing are performed while attaining the torque and power outputs of the turbine. The first test involves the use of a varying number of blades (two blades, three blades, four blades, and five blades), and the second test used the rotor with the better performance in the first test and tested it again using different sizes of blade radius (300 mm, 350 mm and 400 mm).

It is concluded that the five-blade rotor has produced the highest torque and power outputs at various RPM among all the number of blades tested. Hence, the five-blade rotor was tested with various sizes of the blade radius the maximum torque and power values of 0.17954 Nm and 2.6322 W, respectively. Then the five-blade rotor is tested with different sizes of blade radius starting from 300 mm, 350 mm and 400 mm. The 400 mm blade radius has shown better performance in both torque and power outputs with the corresponding maximum torque and power values of 0.4143 Nm and 6.074 W, respectively.

5. References

[1] Herbert GJ, Iniyan S, Sreevalsan E, Rajapandian S. A review of wind energy technologies. *Renewable and sustainable energy Reviews*. 2007 Aug 1;11(6):1117-45.

[2] Didane DH, Wahab AA, Shamsudin SS, Rosly N, Zulkafli MF, Mohd S. Assessment of wind energy potential in the capital city of Chad, N'Djamena. In*AIP Conference Proceedings* 2017 Apr 21 (Vol. 1831, No. 1, p. 020049). AIP Publishing LLC.

[3] Didane DH. Wind and Solar Assessment in the Sahelian Zone of Chad. *International Journal of Integrated Engineering*. 2018 Dec 31;10(8).

[4] TrueWind Solutions LL. Wind energy resource atlas of Southeast Asia.

[5] Al-Aqel AA, Lim BK, Noor EM, Yap TC, Alkaff SA. Potentiality of small wind turbines along highway in Malaysia. In2016 *International Conference on Robotics*, Automation and Sciences (ICORAS) 2016 Nov 5 (pp. 1-6). IEEE.

[6] Alquraishi BA, Asmuin NZ, Mohd S, Abd Al-Wahid WA, Mohammed AN. Review on Diffuser Augmented Wind Turbine (DAWT). *International Journal of Integrated Engineering*. 2019 May 5;11(1).

[7] Ramu S, Abhilash M, Ajay M, Aravind S, Hariprasad M. Low Expense Vertical Axis Wind Turbine Using Permanent Magnets. *International Journal of Mechanical Engineering and Technology* (IJMET). 2016 Mar;7(2).

[8] Kim MG, Dalhoff PH. Yaw Systems for wind turbines–Overview of concepts, current challenges and design methods. In*Journal of Physics: Conference Series* 2014 Jun 16 (Vol. 524, No. 1, p. 012086). IOP Publishing.

[9] Varnon R. Derecktor converting boat into hybrid passenger ferry. *Connecticut Post*. 2012.

[10] Didane DH, Maksud SM, Zulkafli MF, Rosly N, Shamsudin SS, Khalid A. Performance investigation of a small Savonius-Darrius counter-rotating vertical-axis wind turbine. *International Journal of Energy Research*. 2020 Oct 10;44(12):9309-16.

[11] Didane DH, Maksud SM, Zulkafli MF, Rosly N, Shamsudin SS, Khalid A. Experimental Study on the Performance of a Savonius-Darrius Counter-Rotating Vertical Axis Wind Turbine. In*IOP Conference Series: Earth and Environmental Science* 2019 Jun 1 (Vol. 268, No. 1, p. 012060). IOP Publishing.

[12] Didane DH, Mohd S, Subari Z, Rosly N, Ghafir MA, Masrom MM. An aerodynamic performance analysis of a perforated wind turbine blade. *InIOP Conference Series: Materials Science and Engineering* 2016 Nov 1 (Vol. 160, No. 1, p. 012039). IOP Publishing.

Acknowledgments

This research was supported by Ministry of Education (MOE) through Fundamental Research Grant Scheme (FRGS/K296). The authors would like to thank the Ministry of Higher Education Malaysia for supporting this research under Fundamental Research Grant Scheme Vot No. FRGS/1/2016/TK04/UTHM/02/5 and partially sponsored by Universiti Tun Hussein Onn Malaysia

AUTHOR INDEX

Institute of Physics
Dirac House, Temple Back
Bristol BS1 6BE UK

ISSN: 1757-8981
ISBN 978-1-7138-4626-0

9 781713 846260

CURRAN ASSOCIATES INC.
proceedings.com